Valery Popov

Mental Karate

Methodology of scientific and technical creativity and conceptual design

4rd edition

Seoul
<Pushkin House>
2025

Published by Pushkin House
Seoul, Republic of Korea
www.pushkinhouse.co.kr

MENTAL KARATE

Translated into English by © Alexey B. Nelyubov
Book design by © Kniga-Memuar
Illustrations designed by © Kniga-Memuar using internal images
of © Shutterstock, Inc

ISBN: 979-11-7036-167-1 (03170) (2025)

ISBN: 978-5-00117-372-4 (1st Edition, 2018)
ISBN: 978-5-00117-628-2 (2nd Edition, 2018)
ISBN: 978-5-0055-6617-1 (3rd Edition, 2021)

Printed in the Republic of Korea
4th Edition

About the Book

The book is devoted to the methods and means of scientific and technical (and other) creativity, ensuring the rationalization of the processes of identifying and implementing important human needs and is intended to replace the methodology of creativity developed by G. S. Altshuller, called "The Theory of Inventive Problem Solving" (TRIZ), which in various forms has been published many times in many countries of the world (including the USA, France, Switzerland, Japan, Spain, South Korea, the GDR and FRG, Hungary, Bulgaria, Czechoslovakia, Poland, the countries of the former USSR), although over more than 50 years TRIZ has, of course, become morally obsolete (with which almost everyone agrees).

It is recognized that any human activity in all spheres and in everyday life is the formation and implementation of various designs, therefore, the book, having preserved the best and eternal from TRIZ, presents the system methodology of design activities developed by the author, including the stages of conceptual design of technologies, equipment, products, determining the success or failure of the design as a whole.

The author pays particular attention to heuristics, which, like martial arts techniques, act as a kind of "mental karate" that develops creative potential and increases the effectiveness of creative activity of each reader. The effectiveness of heuristics is confirmed by many vivid examples of their implementation given in the book.

The book presents the basics of creative pedagogy for training creative specialists. The bookmark attached to the book is a patented device for self-training of the reader's personal growth.

The book is intended for a wide range of readers: scientists, specialists, teachers, students, schoolchildren and others, and in general it allows each reader to develop creative thinking, form the foundations of a new innovative culture, which aids to become even more successful.

"The greatest mystery of the Universe is life.
The greatest mystery of life is man.
The greatest mystery of man is creativity."

Friedrich Gorenstein,

writer, playwright, screenwriter

to the films "Solaris", "Slave of Love"

This book is dedicated
to the blessed memory of my mother,
Nina Martynovna Popova

Contents

Chapter 6

Methods and means used at the stages of setting, solving the problem and commercialization of the result

Chapter 7
Training of creative professionals

Chapter 8

Appendix

List of notes

Preface by the author

From a young age I was attracted by any creativity, but especially by scientific-technical one. The word "inventor" sounded to me magical, and the desire to become an inventor seemed to me a pipe dream, some kind of mystery. And only at the age of 70 when I was already awarded the honorary title "Honored Inventor of the Russian Federation", I read deep philosophical definitions of the main mysteries of life, formulated by a writer Friedrich Gorenstein (the author of scripts for the famous films "Solaris", "Slave of Love"): "The greatest mystery of the Universe is life. The greatest mystery of life is man. The greatest mystery of man is creativity". I will add on my own that creativity itself is full of mystery.

The book you are reading is about the mystery of creativity, how to understand and use it in your life. For example, two and a half centuries ago Claude Adrian Helvetius, a French philosopher, called the mystery principles. The statement of Helvetius, full of deep philosophical and practical meaning, is known: "Knowledge of some principles easily compensates lack of knowledge of some facts". The statements of Gottfried Leibniz, a German philosopher, mathematician, physicist are close to this: "It is useful to study the discoveries of others in a way that would reveal to us the source of inventions", as well as Nobel Prize winner Lev Landau: "A method is more important than discovery, because the correct method of research leads to new, even more valuable discoveries". It was acquaintance, comprehension and assimilation of the mystery, principles, methods of creativity that allowed me, a person with a basic mining education, to obtain many patents for inventions and utility models not only in the mining industry, but also in the areas where I was certainly not a specialist: energy, automotive, oil and gas production and transportation, mechanical engineering and instrument making, rescuing people from high-rise buildings, medicine, etc. My close colleagues, whom I call "blood brothers", have the same high results of intellectual activity.

Schoolchildren, students, specialists and others, whom my colleagues and I (and not only we) have taught the mystery of creativity, have remarkable creative results, in comparison with those who are unfamiliar with this mystery.

Of course, discovery and comprehension of this mystery, principles, methods is an endless process, each person has its own and very instructive one. That is why, in the preface to the book, I share my own experience of understanding the key, in my opinion, aspects of creative activity. This, I hope, will help readers understand the order of presentation of the book's material, aimed ultimately at developing a creative potential of each reader, forming the basics of a new innovative culture.

Especially I showed an attraction to scientific-technical creativity in 1966–1968, when I worked as a design engineer in the design department of the mine named after 40th anniversary of the Komsomol (All-Union Lenin Communist Youth Union) in the city of Leninogorsk, East Kazakhstan region (now this city is called, as before, Ridder). During that period, I submitted many rationalization proposals on technologies for ore deposits development, which were not only accepted and implemented, but, what was especially pleasant, were very generously (as it seemed to me at the time) rewarded. Almost all solutions were obtained by a method called (as I learned later) the "trial and error" method.

This method is the innate form of learning, described by E. Thorndike in 1898, based on randomly committed motor and mental actions, due to which some significant problem for an animal (including humans) is solved. Although the method of "trial and error" also uses discursive (rational) thinking, in common parlance it is called "the method of (scientific) trial and error", "the method of direct search" and even the "monkey method", because a person, using this method, acts practically like a monkey that, feeling the fire warmth and wanting to get warm, carelessly sticks its paw into the fire (makes an unsuccessful test) and, having burned itself, does not repeat this mistake, but makes other, more successful tests that are not associated with direct contact with the fire.

The method of "trial and error" as an element of the "old" innovative culture still remains the most widespread method of scientific-technical (and not only) creativity of any person: a child, a schoolboy, a housewife, a worker and even a prominent scientist. Of course, the number of trials and errors is different for everyone and depends on the level of competence, creativity and other abilities. But even such a genius of scientific and technical creativity as Thomas Edison, when he was developing an alkaline battery, conducted 50 thousand experiments (tests).

For a person with a developed imagination, an effective technical solution can be suggested by analogies between objects of animate and inanimate nature and technosphere objects (for example, by analogy with honeycombs, honeycomb architectural structures of houses have appeared) and random associations — verbal and others related to human feelings (smell, hearing, touch, sight, taste). For example, the word "air" can associatively call the concepts of "breathe", "pulmonary diseases", "mountain air", "sea air", "smell", "aroma", "resort", "heal".

The latter concept is very productive and can become a clue to an invention, for example, of a "healing air conditioner", which not only purifies air, regulates its temperature and humidity, but also prevents and even treats pulmonary diseases.

In general, the named methods are unproductive and, as a rule, require a lot of time and efforts, and their use most often leads to trivial results and only occasionally ends in an unconscious but very productive mental process — "insight": a sudden enlightenment, in which an original technical solution to any need is almost instantly formed.

That is why the words of the famous Russian scientist V. A. Zvyagintsev, that he said almost half a century ago, were offensive, but obviously fair; and they are still relevant today: "However paradoxical it may seem, the actual situation is that, a modern man, in terms of his intellectual work technique, is at the level not much higher than the level of a Neanderthal." These words can be counted as a compliment, comparing to the currently widespread "monkey" method.

In 1973 or 1974 my colleague gave me a book to read that changed my understanding of scientific-technical creativity. The author of this book was Heinrich Saulovich Altshuller, it was called "The Invention Algorithm". From the book I learned for the first time that scientific technical creativity could not only be a "blind" direct search or an almost unconscious mental process related to analogies and associations, but also consisted of a sequence of meaningful rational steps of creative activity, application of a methodological advice and special systematized information resources that dramatically increased the efficiency of scientific and technical creativity. Of course, after that I found and read all the books of G. S. Altshuller — the author of the theory of inventive problem solving (TRIZ) — and more than once admired his ingenious methodological findings and descriptions of remarkable inventions that confirmed his advice.

However, for me, a man with a higher technical education, it was difficult to understand a lot of incorrect (from the science of science, mechanical engineering, mechanics, physics, design, logic and other fields of knowledge view point) concepts and definitions and the set of conventions that existed in TRIZ: "bi-system", "poly-system", "folding and deployment of technical systems", "x-factor", "x-element", "MATCEM", "supereffect", "anti-function", "max- and mini-tasks", "su-field analysis", "the law of energy through passage", "the law of mono-bi-poly-mono transition", etc. Who knows TRIZ, he will understand what it is about, and who does not know, it is better not to get acquainted with these concepts at all or to get acquainted after getting a good basic technical education in order to be able to "separate the wheat from the chaff" (that is, useful from harmful).

But then (in the 70s of the last century) G. S. Altshuller was revered as a demigod and it was incredibly difficult to doubt his approach, let alone, refute any of his ideas.

And only the meeting with the outstanding methodologists of scientific-technical creativity, Ph. D. Heinrich Yazepovich Bush and Doctor of Technical Sciences, prof. Aleksandr Ivanovich Polovinkin (whom I consider to be my teachers) saved me from the torment on the topic "probably everything is correct, it is something I just don't understand." Meeting with them, their books opened for me a whole world of little-known heuristic principles, strategies, tactics, methods, techniques, called heuristics, which were developed by ancient Greeks, but were hidden like a secret weapon. Only a small part of the information about these heuristics has survived to this day. However, it is known that as early as the 17th century, Rene Descartes developed a rational method for discovering truths, including four rules of search, 22 rules for mind guiding, methods of intuition and creative doubt. And the 18th century began with the principles verification of Isaac Newton and was embellished with the fruitful activities of M. V. Lomonosov, who developed the most successful methodological means of heuristics for his time — the logographic method for finding problems solutions[1].

"Mental karate" — the way how the famous Russian scientist in psychology of creative activity V. A. Molyako[2] figuratively called heuristics. In creativity, heuristics act as techniques (and most importantly — ideology) of karate (or other martial art). Everyone knows that a person who possesses a martial art will always defeat another, even physically stronger opponent who does not possess such an art.

G. Ya. Bush, a Latvian scientist, who was a great polymath and spoke five foreign languages (he specially learned the ancient Greek in order to collect and study heuristics), before the collapse of the USSR, collected and dreamed of publishing descriptions of more than 3000 heuristics. However, he died early, and the result of his invaluable work was lost.

In my practice, the effective use of the heuristics known to me in scientific and technical creativity was impeded by a lack of even the simplified understanding of creativity essence as a design activity that had mandatory stages: setting a task, solving a problem, forming a variety of results of problem solving and their commercialization (or other use). Later, I understood that any human activity was a design of varying complexity (choice or formation and performance of a design). This led me to create a systems design methodology that included eight stages, three strategies, and three basic design tactics (the stages are presented on the next page in a figure without strategies and basic tactics).

The first seven stages refer to conceptual design stages, at which the external and internal properties of a designed technical system (technology, equipment, products, etc.) are formed, and therefore determine the success or failure of a design as a whole.

In addition, the system design methodology has included the use of metaknowledge (megascience) to implement the so-called basal principle of heuristics, which excludes or eliminates harmful consequences of "narrow" specialization of design subjects and provides a sharp increase in efficiency of the mechanism of knowledge convergence (closing-in, transfer).

Design stages in the system design methodology of design activity

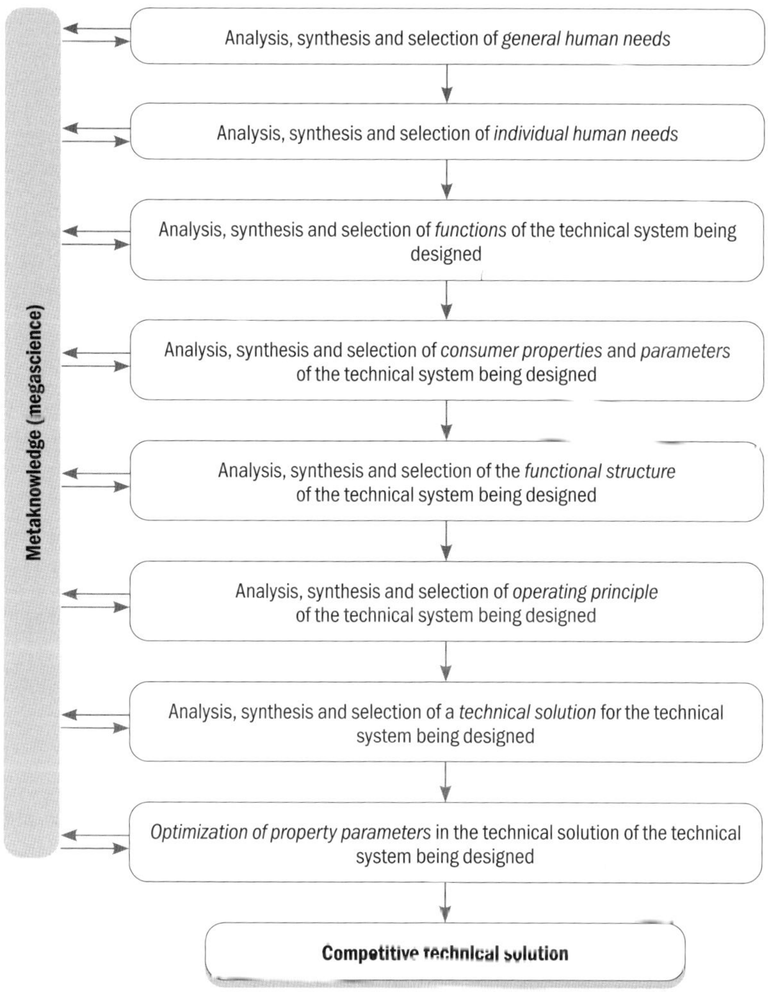

And if earlier for me (and all the specialists I interviewed), design stages were "blurred" in the process of problems solving and a **design result** embodied the results of mental analysis, synthesis and selection (see the diagram "Design stages in the system design methodology of design activity") of technically realizable general and individual human needs, functions, consumer (external) properties, functional structure, operation of a designed technical system principle — and **appeared** in the mind **immediately in the form** of a certain **shape** of a technical decision (then, if necessary, the parameters of properties were optimized), then the division of this single process at the stages of conceptual design allowed me to find (and work out myself) heuristics, including methods, technologies and means, the use of which at each stage of conceptual design dramatically increased the efficiency of a problem (design) solution as a whole.

However, many heuristics developed by ancient scientists are presented in hard-to-find editions in little-used languages. And many of the concepts, methods and techniques described in the works of G. S. Altshuller and his followers, need significant adjustments and indications of limited appliance. So, G. S. Altshuller appreciated highly the potential of empathy. Nevertheless, some physicians believe that the excessive empathy (presenting oneself as a technical object or part of it) in childhood leads to a split personality, that is, to schizophrenia.

That is why the main goal of this book is not just to acquaint the reader with the most effective heuristics, but also to present them as a part of the methodology of scientific and technical creativity and conceptual design, which combines interrelated and scientifically correct information about the incentive force of any human actions — human needs in scientific and technical creativity as a design activity, contradictions in technical systems, methods and means of intensifying their resolution at various stages of scientific and technical creativity.

I believe that the title of the book "Mental Karate: Methodology of Scientific and Technical Creativity and Conceptual Design" like no other imaginatively and exactly reflects its purpose and content.

I am proud that some of the heuristics and other concepts presented in the book have been significantly developed by me (or with colleagues I mentioned). These developments include:

- A corrected classification (variety) of technical systems;
- Development of concepts, components and levels of novelty in scientific and technical creativity;
- Characteristics of nine classes of resources, hidden in technical systems and their environment, allowing identification of new resources in technical systems to reduce a cost of their operation or to develop new competitive technical solutions;
- A methodology for formation and use of criteria for progressive development of technical systems (CPD), which makes it possible to identify the most probable and promising directions for the development of technical systems (in fact, a need for formation and use of CPD was substantiated by the ancient Roman philosopher Lucius Annei Seneca:

"If one does not know to which port one is sailing, no wind is favorable");

— Criteria for unusual properties of materials, allowing to identify new materials with unusual properties, the use of which can serve as an impetus for creation of a new industry of goods and services;
— Correction of development patterns of technical systems and development of generalized patterns descriptions, allowing the most complete presentation of technical systems development stages;
— Development of the heuristic techniques method, adjustment and addition of the intersectoral fund of heuristic methods, allowing the expansion of the possibilities of its use to improve technical systems or create new technical systems generations;
— The "ideal final result" concept development, contributing to technical solutions formation close to ideal;
— "strong" or "controlling" verbs concept transformation into more creative concepts of "good" and "bad" verbs, allowing the use of "good" verbs to form new useful functions, and "bad" verbs — to identify possible harmful functions.

I am even more proud of the design activity systematic methodology worked out by me, as well as heuristics, which include:

— The method of using the most important human needs for new important needs formation in creation and development of technical systems[3];
— The method of "ascent-descent" for formation of necessary new objects variety, phenomena, functions, properties, functional structures, operation principles, technical solutions of technical systems;
— The concept of "good" and "bad" adjectives and the methodology for their use, which makes it possible to form technical solutions that implement new useful properties, exclude or reduce manifestations of harmful properties;
— Creative activity models based on the method of "trial and error" with the use of productive knowledge, contributing to a deeper understanding of the methodological foundations of creative activity[4];
— The concept of "generalizing function", contributing to systematize many types of generalizing functions based on varieties of functions and synonyms of function names that further serve as hints to form of a variety of competitive technical solutions.

The development and creation of new heuristics have led me to work out the basics of creative pedagogy, where the systematic methodology of design activities have been used: the creative pedagogy formula of A. G. Aleinikov (reflecting in a concise form the concept of effective pedagogical technology of creative orientation), productive knowledge, criteria and indicators of creativity in training programs, textbooks and manuals. The creative pedagogy testing has shown that many people receive significant scientific and practical results already in the learning process.

I have always wanted to write a book so that it would contain additional opportunities to read its main thoughts in 15–30 minutes, even if this book is 500 or more pages (an option for lazy and very busy people). I would not mind myself the possibility of such a reading. There are many options for implementing such opportunities (see, for example, the method of a content summary in the posters of the book "The Revolution in Learning")[5]. Therefore, many of the even-numbered pages of my book summarize the essence of the sections and chapters in the form of posters and / or summaries of main points. Appendix 13 provides definitions of terms in italics in the main text. If a reader knows the definitions of these terms, then he can skip them. The famous English physicist Stephen Hawking perhaps jokingly, but it seemed that, rightly said: "Each formula would reduce the potential readership by half". Therefore, there is not a single mathematical formula here.

In this book, I am sure, every reader of any category and age (a schoolchild, a housewife, a worker, a scientist and others), who is determined just by the ability to understand the stated, will find something new and, most importantly, useful.

Even the bookmark in the book I have patented is creative: it has a "magic" window and heuristics-tips, so it can be successfully (personally tested with children and adults) used to search for new ideas, personal growth training instead of a popular, but often life-threatening search for Pokemon. How to look for new ideas using the bookmark is described in Appendix 14.

It is known from the history of scientific and technical creativity that on the way to the knowledge and development of its mystery there may be a lot of difficulties. The great philosopher of the XX century, Martin Heidegger, with his natural paradoxical feature, once exclaimed: "Humanity is heading back into the future!" — having in mind psychological barriers to thinking, the consequence of which was a lack of understanding of heuristics, disbelief in them and in the possibility of developing creative abilities. ("If God has not given such abilities, it is already impossible to do anything else!" — so or about so many people, even highly educated, mistakenly say.)

Extremely negative consequences arise if such barriers exist for decision-makers and people responsible for innovative activities development in corporations, economy and industry sectors of the country, when there are not enough funds for their extensive development, and heuristics, aimed at intensifying innovation activities, are not developed and are not used. The actions of such leaders (and ultimately — of corporations and entire industries) resemble a fable: "A man walks and sees how a peasant, dripping with sweat, is sawing a big tree with a dull saw. "Don't you better file your saw?" — advises man. "I have no time. I have to saw the tree!"

I hope that this book will turn every reader, having such barriers, towards the future, and that will dramatically increase the effectiveness of any of his activities.

Chapter 1
Human needs – the main motive force of human actions

1.1 Basic concepts related to human needs

The main driving force of any human action is human numerous and various *needs* associated with everyday life, family, household, job, study, health, sports, spiritual life and, in general, with multiform interaction of a person and his environment.

There are organic and social / inorganic needs. For a long time, it was believed that organic needs are genetically inherent in a person, and their satisfaction is carried out through instincts (for example, a need for air, water, food, sleep, disposal of waste products) and only social / inorganic needs are formed during development of social forms, social and economic, climatic and other conditions of human life. However, the latest advances in psychology[1] convincingly prove that ALL needs are formed (even organic ones are not instinctive).

In general, a need is the necessity for something that lies outside the body and is its subject. The need that has "found" its object is transformed into a motive, on the basis of which the motivation of a particular force is formed.

Interest is a motive concretization, it serves as a driving force of all social human actions, it is distinguished by manifestation of human cognitive needs and attention to certain objects and phenomena.

To some extent, the concept of *a need* is similar to the concept of *aspiration*, which, depending on the degree of awareness, is expressed in the form of *attraction* or in the form of *desire*. All these concepts are not identical. For example, a desire is less important than a need in the spiritual world of a person. It does not always coincide with a need for stable functioning of human personality, ensuring his life activity, therefore, a desire often belongs to the sphere of *illusory* notions, when a person, creating ideas and concepts, in his consciousness endows material objects with the qualities that are not inherent in them.

The transition from a need to an activity consists in transformation of a need into a *goal* and *interest*.

If a need remains the essence, and does not become the incentive to activity, then a model of asocial behavior of a human consumer is formed. Homeless, beggars, parasites, as well as people suffering from alcohol or drug addiction are typical representatives of this behavior. However, asociality has also positive manifestations based on faith, motivation for solitary existence. Monks, hermits, as well as communities that have rejected the benefits of civilization, etc. refer to these representatives[2].

Of course, not every activity meets social norms if it is based on realization of, tentatively speaking, bad needs (some of them are given above, and others — in sections 1.2.4 and 1.2.5).

Most of human needs can be satisfied with the help of technical systems created for these purposes — artificial material objects — or by human actions with material objects (including natural ones). Such needs are called technically realizable needs. A variety of technical systems (TS) is shown in Fig. one.

Figure 1. Variety of technical systems

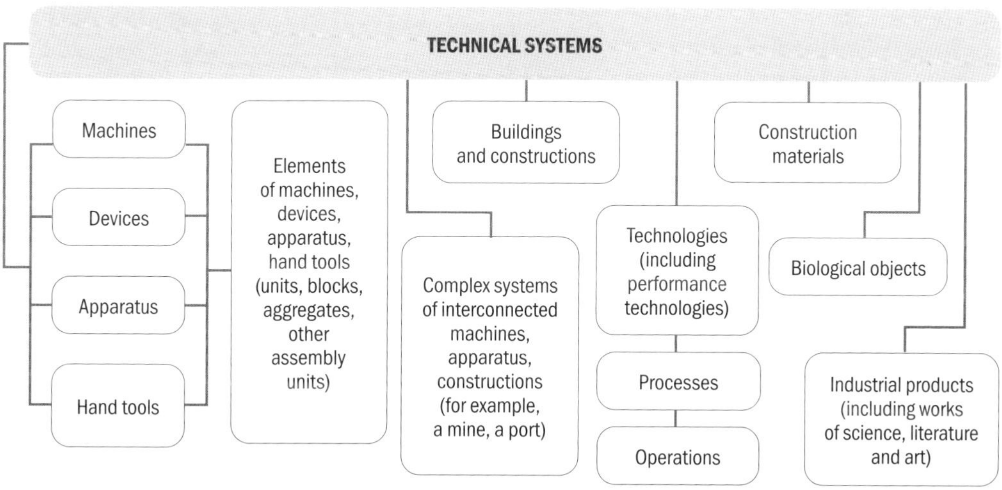

The concept of *technology*, which implies several meanings, is the most difficult in a variety of TS.

Technically realizable needs are described in the following order:

1) The indication (description) of the action (A) leading to realization of a need;

2) The indication (description) of an object (O), a subject of labor;

3) The indication (description) of special conditions (C) and restrictions under which an action (A) is performed.

Here and further posters are presented every 3-10 pages,
briefly reflecting the main thoughts
on the pages following each poster.

The main motivating force of any human actions is a variety of human needs.

"It is well to remember that the entire universe, with one trifling exception, is composed of others."

John Andrew Holms,
American politician

Table 1. Descriptions of technically realizable needs

A	*O*	*C*	*Technical system option*
heat	liquid (water)	to a boil	boiler
illuminate	area (the objects)	by a distance	spotlight
grind	grain	into flour	mill

Examples of descriptions of technically realizable needs are given in Table 1.

Actions (A) in needs are described by more than 15 thousand verbs. Taking into account plenty (millions) of objects names, objects of labor (O) and special conditions and restrictions (C), the number of technically realizable needs is enormous and cannot be calculated.

Further, with rare exceptions, we will consider only technically realizable needs. The exception is due to the fact that not all human needs are technically realizable. For example, the need for love or the need for respect cannot be realized with the help of TS.

Many *analogs* of TS have already been created to satisfy almost every human need, they differ in functions, properties and parameters of properties. For example, there are many products for sitting to satisfy the need to "sit …": a chair, a stool, a sofa, a bench, an armchair, a pouf, a desk (Fig. 2).

Figure 2. Products for a person's sitting: a — a chair, b — an armchair, c — a stool, d — a pouf, e — a bench, f — a sofa, g — a desk

There are various swimming technologies (which are more often called techniques) to satisfy the need to "swim ...", differing in various speed rates and muscle loading: crawl, breaststroke, butterfly (Fig. 3).

Figure 3. Swimming techniques: a — crawl, b — breaststroke, c — butterfly

Therefore, having a lot of analogs, to satisfy a need, it is often enough just to choose the analog that is closest in its functions and properties, which is called a prototype (Fig. 4).

Figure 4. Need satisfaction through the choice of analog

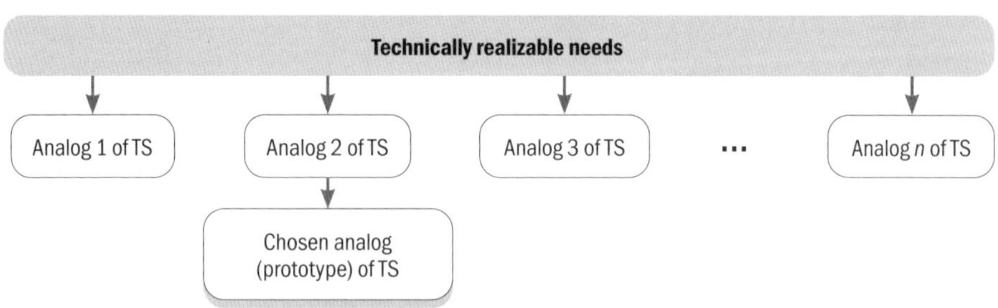

A simple (from decision-making point of view), but very rare case occurs when there is no analogue of TS to satisfy a need. Then it is necessary to develop a pioneer (the first in the history of mankind) TS, and most often this is very difficult (Fig. 5).

Figure 5. Need satisfaction in the absence of a TS analogue

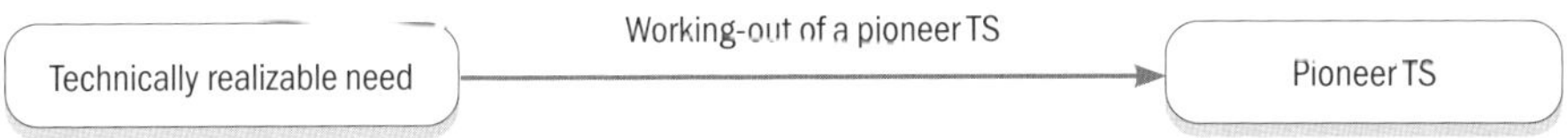

The most common case is when the functions, properties or parameters of properties implemented in well-known TS analogs do not satisfy any human needs. For example, keeping money, jewelry, food, expensive fur coats in ordinary closets led to theft and / or premature damage of items. Or, for example, the aircraft speed with a propulsor based on a piston engine and a propeller (subsonic speed) was insufficient.

Then the idea of improving TS or even creating a new generation TS can most often arise when considering the possibilities of implementing new technically realizable needs in the analogue (prototype) of TS (Fig. 6), formulated as antonyms of human needs causes. For example, a lack of protection against theft and premature damage of objects have causes different types of needs "to protect ...". As a result, a safe, refrigerators for food, expensive fur coats and many other things have been worked out. And a lack of subsonic speed rates (or, in other words, inability to reach supersonic speed rates) of aircraft has led to creation of propulsors that have ensured supersonic flight speed rates. Or, for example, the inability to charge a cell phone far from a stationary energy source has become an impulse for the emergence of a new technically realizable need to "generate ...", or rather, "convert" energy. This need has served as an incentive for many options formation for portable power converters integrated into cell phones, tablets, and other gadgets that provide them with charging in extreme (and not only) situations.

Figure 6. Need satisfaction with a lack of any features, properties or parameters of properties in the analog (prototype) of TS

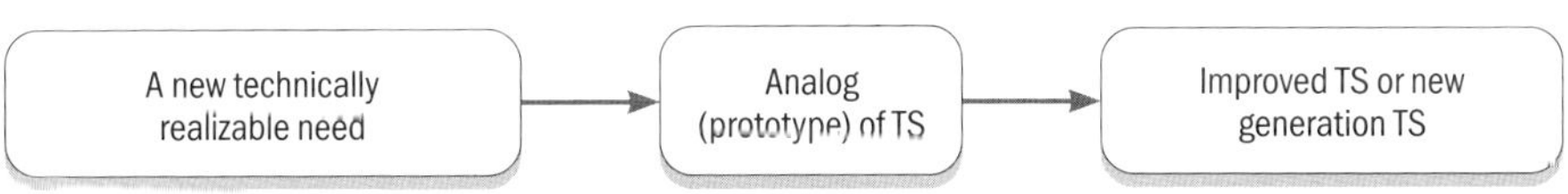

Thus, in all cases of a pioneer TS creation, TS development or creation of new generation TS, the determining idea is one, the emergence impulse of which will always be a new technically realizable need formation.

Peter F. Drucker, a classic of the modern management, said in this regard: "It is the want that is unique, and not the means to satisfy it." In the following sections of the book, the reader will find many examples confirming this brilliant statement.

It should be noted that it is often necessary, in order to fulfill a specific need, to fulfill a lot of other needs sequentially. For example, the need for bread industrial production necessitates implementation of the following new requirements for TS creation (Fig. 7).

Figure 7. The example of new human needs emergence for realization of a need

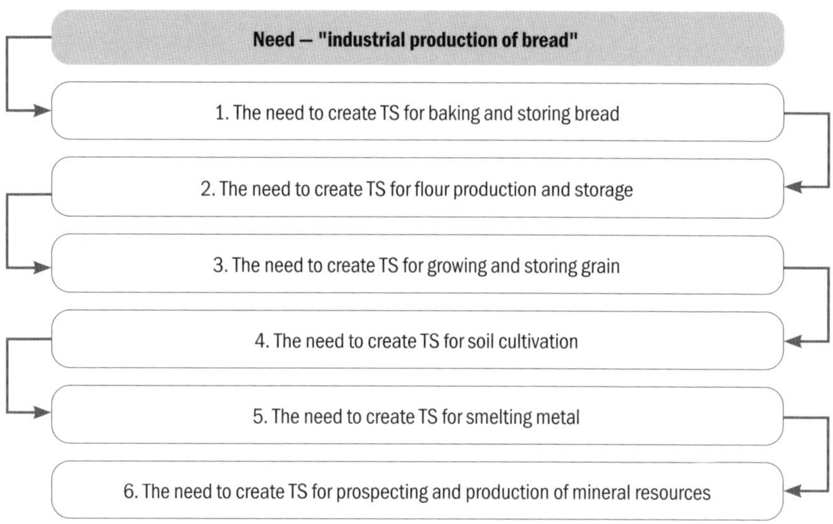

To understand the mechanisms of new human needs formation and their use to create pioneer TS, improve TS or create TS of new generations , it is necessary to consider the most common reasons for the emergence of new human needs, as well as the taxonomy of needs.

1.2 The reasons for the emergence and development of human needs

1.2.1 Environment is the main source of human needs

The human life runs in a complex and continuously changing environment, which can be represented as a simplified model that reflects a set of interacting basic fields (spheres): natural environment, production environment (technosphere), sociosphere (Fig. 8).

Figure 8. Model of the human environment

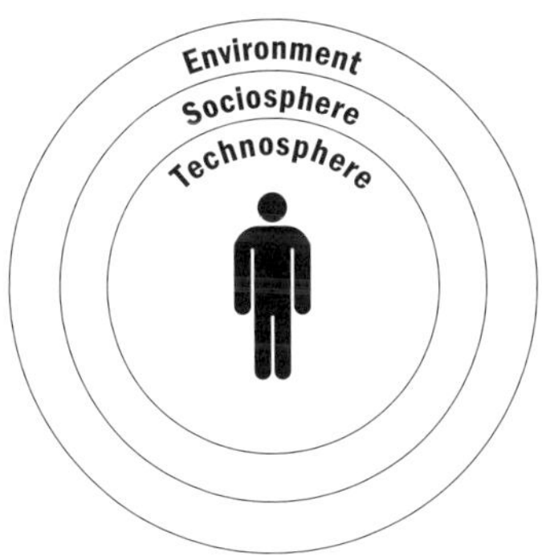

A man constantly enters into relations with all components of his environment. The degree of harmonization of these relations has a tremendous impact on the formation of human needs as the main driving force of any of human actions. In turn, man's activity determines his evolution and associated opportunities in the knowledge of the surrounding world. The consequence of this is creation of pioneer TS, improvement of TS and creation of new generation TS, designed ultimately to meet human needs.

The natural environment — the mega exosphere — consists of four exospheres — the near-surface shells of the Earth (Table 2, Fig 9)[3].

Table 2. Classification of the "natural environment" concept

The natural environment (the mega-exosphere)			
Exospheres (the near-surface shells of the Earth)			
Atmosphere	Lithosphere	Hydrosphere	Biosphere

Under the influence of *exogenous* (including space) and *endogenous* factors in relation to the Earth, on the one hand, and human activities, on the other hand, there is a constant interaction and interpenetration of exospheres.

Figure 9. Exospheres (near-surface shells of the Earth): atmosphere, lithosphere, hydrosphere and biosphere

The atmosphere is the gaseous (air) shell of the Earth, extending over a distance of up to 3000 km.

The atmosphere affects the essence and dynamics of all exogenous processes occurring on Earth: oxidation of organic life remains, physical weathering and chemical rocks decomposition, circulation of ground and underground waters, formation of glaciers and permafrost, occurrence of catastrophic phenomena — floods, tropical cyclones, typhoons, tornadoes, etc.

The Earth's atmosphere is a dynamically changing system, which predetermines the constant need for qualitative improvement of existing TS and / or the creation of new TS generations to meet human needs for using the atmosphere and preventing its negative effects[4].

The lithosphere — the outer solid shell of the Earth, includes the earth's crust and the upper part of the Earth's mantle, consists of sedimentary, igneous and metamorphic rocks.

Terrible processes (earthquakes, volcanic eruptions, mudflows, landslides, landslips, erosion) occur continuously within the lithosphere, which have a huge impact on the formation of environmental situations in certain regions or even lead to global disasters[5].

Satisfaction of the overwhelming majority of human material needs is directly or indirectly associated with the lithosphere: for housing, food, clothing, energy sources, etc.

The growing needs of humanity increasingly reinforces the need for careful attitude to the lithosphere, creation of technology and equipment for comprehensive development of natural resources, deep processing of mineral resources, prevention of global human-made violations of the lithosphere.

The hydrosphere is the aquatic environment of the Earth, represented by a set of oceans, seas, continental bodies of water, snow and ice sheets.

Human needs associated with water cover a wide range of problems related to ensuring adequate water quality, rational use of sources and water supply systems, rational exploitation of biological and mineral resources of the hydrosphere and its *balneological* capabilities, improvement of methods and means of all water resources development, including the World Ocean.

The biosphere. The existence area of living organisms on Earth is called the biosphere — the sphere of life. Living things (plants, animals, microorganisms, etc.) inhabit the atmosphere, hydrosphere and the upper part of the lithosphere of our planet.

When realizing needs associated with the use of the biosphere, one should take into account the patterns of development and functioning of the biosphere as a whole and its individual ecosystems, as well as the relationship between the processes occurring in the biosphere under the influence of the Earth's magnetic field, solar and cosmic radiation. For example, plants productivity, peaks of cardiovascular diseases of people, periods of mass reproduction of many animals are closely related to the activity of the Sun.

The sociosphere (according to E. Nef) is a part of the geographic shell of the Earth, which includes, along with the natural landscape, mankind with its inherent industrial and social relations at the present stage of development, as well as the part of the natural environment mastered by man — the *technosphere*[6].

In economic, sociological, socio-political and in some other areas of knowledge, the technosphere is understood as a component of the sociosphere, in which technical human activity is actively manifested, representing a set of environmental elements within the geographic shell of the Earth, created from natural substances by labor and man's conscious will, having no analogues in the wild[7].

In the technical fields of knowledge, the technosphere is the totality of all functioning and no longer operating technical objects and all products of their activity, as well as changes that have arisen on Earth and in space. Such products

The most acute reason
for the emergence of new needs
is insufficient protection from natural
and man-made disasters, military
and other threats, crimes, diseases.

and changes include: altered compositions of the atmosphere, water and soil; biogeocenotic changes caused by deforestation, plowing or overgrowing of land, drainage of swamps, creation of reservoirs; changes in relief or subsoil associated with minerals extraction, etc.

Having originated in the form of technical systems (TS), designed to compensate and enhance labor and creative capabilities of a man, the technosphere not only forms a kind of living environment, but also becomes a structural system of society, functioning under conditions of continuous exchange of material, energy and information resources. Initial resources are converted into necessary products, articles and services and directed to the sociosphere, distributing them among society members in accordance with the current economic laws and norms, and the society exists as an open non-equilibrium system that irreversibly transforms the non-renewable part of resources and accumulates waste. A resulting industrial and household waste is returned to the natural and man-made environment.

The technosphere has theoretically unlimited possibilities to change both within geospheres and in space. Like any relatively independent system (subsystem), the technosphere obeys the internal laws and patterns of structure, functioning and development, in other words, it is capable of self-development. Insufficiently conscious use of the laws of the technosphere self-development by society, when creation and use of technical systems has been limited just by the resources of the biosphere, has increasingly begun to lead to negative environmental, social and material consequences. All this prompts to develop the technosphere on a scientific basis in the future, turning it into harmonious part of the noosphere[8]. More about the noosphere see in Section 1.2.6.

Let's consider the most common reasons for the emergence of new human needs. These include human needs for various protections to improve the level and quality of life, as well as mania, phobias and false needs.

1.2.2 Insufficient protection of a man from threats

Along with very significant reasons, for example, thirst, hunger, causing organic human needs for water, food, etc., an acute reason for the emergence of new human needs (sometimes even more significant than those mentioned) is insufficient protection from natural and man-made disasters, military and other threats, crimes, diseases, terrorist acts, etc. The strength of motivation depends on the frequency and severity of events consequences. The main common causes of new protection needs are:

1) Insufficient protection of people from natural disasters and an increase in the severity of their consequences;
2) Increase in the number of dangerous man-made accidents and the severity of their consequences;
3) Sharp deterioration in the environmental situation;
4) Military threats;
5) Outbreaks of dangerous diseases;
6) Shortage of food and water;
7) Increase in crime;
8) Increase in the number of terrorist acts and the severity of their consequences;
9) Unwanted climate changes;
10) Negative consequences associated with the geographic isolation of a country, region.

Natural catastrophes, or natural disasters, have always taken place. Natural disasters include: volcanic eruptions, earthquakes, tsunamis, landslides, landslips, mudflows, avalanches, floods, hurricanes, typhoons, tornadoes, hail, lightning, forest fires. In the event of a sudden, sharp attack or at unusually high intensity, heavy rainfall, snowfall, frost, ice can be the nature of disasters. With the development of knowledge other natural phenomena that lead or may lead to natural disasters has become recorded: giant solar prominences (their consequence is magnetic storms), ozone holes, the possibility of collisions of the Earth with large space objects (comets, asteroids).

It is difficult to say with certainty whether the frequency of natural disasters has been increasing throughout human history. However, it can certainly be concluded that severity of natural disasters consequences has been constantly rising. This happens due to increase in the total number of people on Earth, high housing density in megalopolises and regions that are outwardly favorable for life, periodic concentration of a large number of people in various technical systems: ships, trains, airplanes.

The most dangerous natural disasters are earthquakes, cyclones, typhoons, dry and desertification, which often lead to environmental disasters — complete destruction of the ecological balance in natural systems.

Even country's geographic location may be a reason for the emergence of new needs related to the natural environment. For example, many people believe that island isolation of Japan is one of the main reasons for intensive telecommunication technologies development there[9].

It should be noted that the appearance of many potentially dangerous anthropogenic objects in the lithosphere, hydrosphere, atmosphere, and even in space

(Fig. 10) is a result of dynamic human activity in the natural environment. Errors in design, construction or operation of these objects (sometimes natural disasters are added to errors) can lead to dangerous man-made accidents, which on a large scale are called *man-made disasters.*

Dangerous man-made accidents include accidents at industrial facilities, transport; fires, explosions, emissions of harmful gases and spills of harmful substances; uncontrolled release of various types of energy. One of the largest dangerous man-made accidents (actually a disaster) of the 20th century is the accident at the Chernobyl nuclear power plant, the consequences of which have not been completely eliminated until now.

Figure 10. Space debris in low Earth orbit

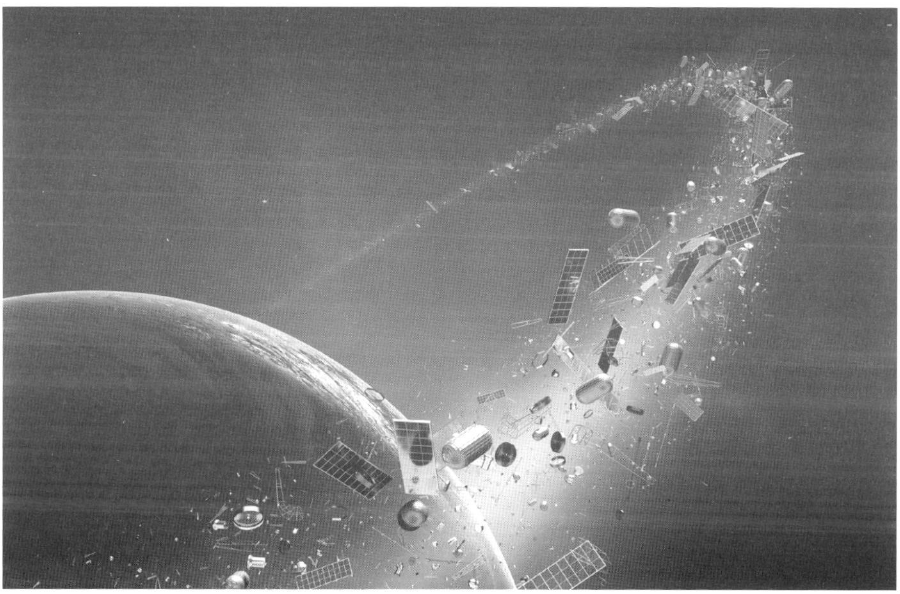

With the changing strength of motivation, the needs are arising, that are related to military threats (sometimes artificially exaggerated), outbreaks of dangerous diseases, shortages of food and water, increase in crime, in number of terrorist attacks and severity of their consequences[10].

1.2.3 Insufficiently high level of population living standards

Many types of needs arise in connection with the insufficiently high level of material consumption (living standard) and *living standards* of an individual and population as a whole. Living standards, in addition to material benefits, in particular, depend on purity of environment, level of personal and national security, political and economic freedoms.

In addition, in recent years, the concept of "living standards" has included a number of other non-material needs, the measurement of which using physical indicators is almost impossible and which are assessed exclusively by the subjective attitude of those covered by a survey. This applies, for example, to the content of work and leisure, satisfaction of needs for self-affirmation, respect from others, creativity, social and political life, intra-family relations and relations with friends, a perceived degree of democracy development and confidence in public and private institutions. Below are the characteristics of some similar and other needs, formulated by G. A. Murray[11]:

1) **Dominance** — the desire to control, influence, direct by word, order, persuade, hinder, limit others;
2) **Aggression** — the desire to dishonor, condemn, curse, mock, humiliate, destroy the opponent by word or action;
3) **Search for friendly ties** — the desire for friendship, love, goodwill, sympathy for other people, suffering in the absence of friendly relations, a desire to bring people closer together, to remove obstacles between people;
4) **Rejection** — the desire to reject attempts of converging, hypercriticality, rudeness, solitude, inaccessibility, shamelessness.

In recent years, the destruction of monopolies has become an increasingly significant reason for the emergence of many new needs, emergence of a large number of small firms and aggravation of competition between them, which in general affects a level and quality of each person's life[12].

It should be noted that almost any type of human activity ultimately comes down to creation of products (goods) and / or services. It is widely believed that in market conditions the effectiveness of this activity (*competitiveness*) is determined mainly by two parameters — "quality" and "price", time of delivery of goods or services is often important as well. In the competition, a winner is the one who has higher quality at a comparable price (subject to time requirements). However, with a very high force of motivation (danger to life, hunger, irresistible desires, etc.), quality and price of goods (in the absence of analogues) may not be decisive, but the time factor rises sharply.

One can distinguish the competitiveness of the following objects:

goods, producers (firms, enterprises), industries, countries. It should be noted that if the measure of product competitiveness on the market is a comparison of its price and quality with the price and quality of similar goods, then the competitiveness of a commodity producer (firm, enterprise), industry, country, except for specified (price and quality) and many other factors (for example, profitability of production, a level of labor productivity, safety of products or services), depends on the number of classes of goods and their varieties presented on the market (including new ones with new consumer properties).

In turn, qualitatively new consumer properties, which generally determine the competitiveness of a product, are being formed during innovations creating — as a rule, high-tech goods — in the process of innovation activity[13].

It has been established that the competitiveness of a commodity producer, industry, country is directly dependent on the intensity of innovation[14].

The effective strategy for increasing competitiveness of a commodity producer, industry, a country can be diversification in its various forms: diversification of production, goods, personnel, exports, loans, insurance, portfolio of securities, liquidity, etc.

For example, diversifying production, firms, one masters new industries and spheres (the most profitable), expands the classes and varieties of goods presented in the markets[15]. At the same time, the important consequence of diversification is the decrease in financial risks and, accordingly, the increase in financial and other stability of activities based on the implementation of the international proverb: "Do not put all your eggs in one basket!" In other sections of the book, the reader can get acquainted with many methods aimed at intensifying generation of various diversification forms.

Another effective strategy for increasing competitiveness of an industry, a country (less often a commodity producer) can be creating competitiveness clusters that ensure stable functioning in the system of related industries. In this aspect, creating clusters, or clustering, is considered as integration of various independent producers, industries, and even countries interacting with each other within a single chain of added value formation for competitive goods.

For example, the efficient economy of Singapore is based actually on two clusters of competitiveness. The first cluster has brought together research and training centers, engineering companies, manufacturers of computers, telecommunication equipment, peripheral devices, components, software products, as well as service firms and market institutions connecting them. The second cluster is based on technologies for deep oil refining and sales of resulting products[16].

The methodological principles of clustering are based on economic calculations, organizational, structural, legal and other transformations and are not yet sufficiently developed.

1.2.4 Manias and phobias[17]

A huge number of human needs arise in connection with manias (pathological irresistible attraction) and phobias (irrational uncontrollable fear). According to many estimates, about 30% of people are amenable to manias and phobias to varying degrees (another 30% are not sure whether they have manias and phobias). Most manias and phobias are caused by mental disorders and diseases of people, and many manias and phobias arise in childhood, accidentally and accompany a person all his life.

The list of manias, phobias, as well as other related manifestations (-lalia, -lexy, -noia, -phagy, -philia, -phrenia) includes more than 700 names and has been continuously updated throughout the history of mankind.

Figure 11. Blinders

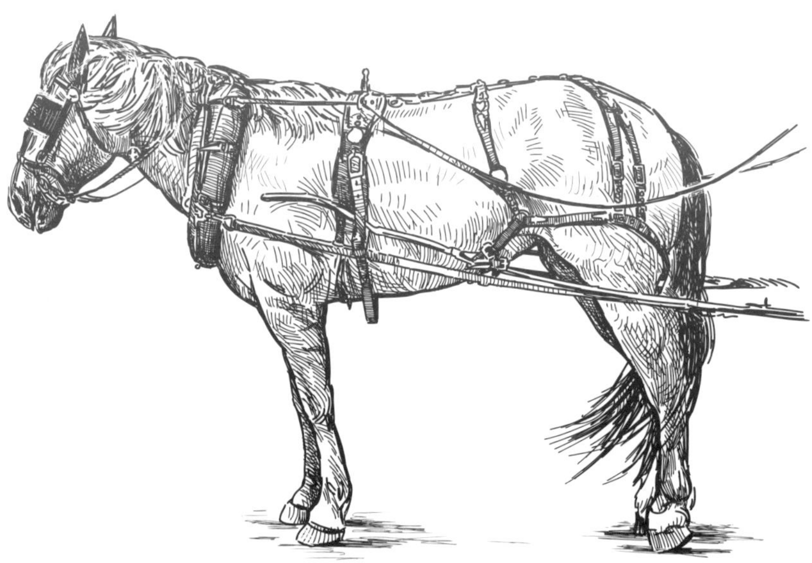

For manias realization (including very harmful and even dangerous) and protection against phobias, they come up with various TS, sometimes science-intensive. However, most people amenable to mania, phobias and other similar manifestations need psychological help and / or treatment. However, a treatment procedure itself requires new effective TS.

It should be noted that animals are also amenable to manias and phobias. To level these manifestations, some domestic animals are trained, and sometimes special TS are created and used (for example, blinders for draught horses, preventing them from looking around and reducing their fearfulness; Fig. 11).

Manias and especially phobias can become the basis for conscious or unconscious (wrong) formation of false needs.

1.2.5 False needs

With the steady decrease in income from goods sale (technologies, equipment, products), rational problem solutions can be qualitative improvement of goods, in which their consumer properties are significantly improved, and / or creation of new goods generations, which are characterized by a sharp (jumping) improvement in consumer properties. At the same time, decisions are based on new human needs implementation, as well as activities diversification. However, new human needs formation is not an easy task, the effective solution of which requires special knowledge, experience and skills that develop a person's creative abilities.

Therefore, the inability (sometimes unwillingness) to form important real needs, as well as religious prejudices, is reasons for conscious or unconscious (wrong) formation of false needs. To do this, they use phobias, spread of rumors, unfounded predictions, exaggeration of consequences of certain problem situations, which are often accompanied by pseudoscientific explanations, basically, scientific myths[18].

The sensational "mistake 2000" became an example of a brilliantly performed action to form false needs. The year 2000 problem related to the fact that software developers recorded a year data using two characters, for example, January 1, 1961 as 01.01.61. With two-digit representation of a year data after 99, 00 was coming (99 + 1 = 00), which was interpreted by many programs as 1900. And this could lead to serious failures in the work of banks, financial organizations, production management systems, etc. The problem was inflated for the sake of additional profits that manufacturers and distributors intended to realize through the sale of updates that fixed the problem. Nobody explained to ordinary users that the problem was not initially critical. On the contrary, they offered to save a computer from unavoidable breakdown and loss of all data.

Many well-paid publications were organized, as well as speeches of even titled scientists and specialists on radio, television, scientific conferences, who

Not all human needs
can be technically realized.

Those needs that can be technically realized are satisfied through selection or stage-by-stage development of a technical system (TS)

imaginatively talked about how trains would collide around the world with the onset of 2000, planes would fall, nuclear power plants would explode, and other mass accidents and man-made disasters would take place, comparing coming troubles with the end of the world. All these activities were so convincing that in many countries, at the behest of governments, so-called national action plans were developed to solve the "error of 2000". And although with the onset of a new century no troubles happened due to the "error 2000", according to the most rough estimates, $ 300 billion was spent on its prevention around the world, of which $ 3 billion fell on Russia (at a difficult time for it)[19].

False needs, often very harmful for a person, are sometimes formed as compensation for some sensual dissatisfaction of needs. For example, deafening music, which causes the resonance of cellular protoplasm and the state of sound intoxication, is a pseudo-aesthetic harmful need that does not actually differ from the harmful need for drugs, alcohol, etc.

Religious prejudices and manifestations of social paranoia, for example, the repeatedly predicted "end of the world", are a significant reason for the emergence of many new false needs.

1.2.6 Noosphere and problems of sustainable development

Technological progress and development of scientific knowledge have significantly changed the way of human life and increased the well-being of a huge number of families. However, with the development of civilization and economies militarization of many states, negative global, primarily environmental problems have arisen.

Already in the late 19th — early 20th centuries, many prominent Russian scientists (D.I.Mendeleev, V.V.Dokuchaev, V.I. Vernadsky, N.A. Umnov) pointed out the inevitability of humanity's transition to a fundamentally new paradigm of its technical and social development driven by knowledge, reason, morality and environmental ethics.

In 1927, E. Leroy, a French philosopher, introduced a concept of the noosphere (from Greek νόος — mind and σφαῖρα — ball, sphere), which were used by V.I. Vernadsky in his research on the biosphere.

He formulated a number of conditions necessary for formation and existence of the noosphere[20]. These conditions, relevant to the present day as well, are as follows:

1) Human settlement of the entire planet;

2) Sharp transformation of communication and information exchange

means between countries;
3) Development of contacts, including political ones, between all countries of the Earth;
4) Noticeable predominance of man's geological role over natural geological processes in the biosphere;
5) Expansion of the biosphere boundaries and human spacewalk;
6) Discovery of new energy sources;
7) People of all races and religions equality;
8) Growing role of the masses in solving domestic and foreign policy issues;
9) Freedom of scientific thought and scientific research from the pressure of religious, philosophical and political constructions; creation of favorable conditions for free scientific thought in a state system;
10) Well-thought-out system of public education and increase in the well-being of workers, creation of a real opportunity to prevent poverty, malnutrition and hunger, extreme reduction in disease risk;
11) Reasonable transformation of the primary nature of the Earth, with the goal of making it capable of satisfying all material, aesthetic and spiritual needs of numerically growing population;
12) Exclusion of war from society life.

It should be noted that on behalf of the Club of Rome — the international group of business leaders, politicians and scientists — the Massachusetts Institute of Technology conducted a research[21], the results of which led to the stunning conclusion that the human use rate of many important resources and the production rate of many types of pollution in the world had already exceeded the permissible limits, that could lead to uncontrolled reduction in the following indicators: food production, energy consumption, industrial production, etc.

For this reason, in the second half of the 20th century, the contours of the new model of Earth civilization existence in the form of the *concept of sustainable development of society* were outlined by the works of many scientists and politicians (N. N. Moiseev[22], V. A. Koptyug, F. Mitterrand, etc.), which was recorded in the declaration adopted at the UN Conference on Environment and Sustainable Development (Rio de Janeiro, 1992). The ideological basis of this concept is that the states of the world undertake to ensure that their needs are met with the preservation of the environment in the interests of the present and future generations.

It is recognized that implementation of the concept of sustainable society development is possible only as a result of uniting all countries and peoples around well-defined scientific and technical ideas, developing a new moral and material values system, increasing the importance of cultural and spiritual components in the sphere of personal and public interests.

The sustainable development of society concept necessitates formation and ranking of a corresponding human needs list, implementation of which should ensure sustainable development of society. In this case, it is important to have a list (or lists) of human needs. Systematization of needs always adds creativity with the use them[23].

1.3 Taxonomy of human needs

The famous American politician John Andrew Holms described a huge number of human needs very aptly: "It is well to remember that the entire universe, with one trifling exception, is composed of others."

K. Obukhovsky, a Polish psychologist, counted over 120 classifications of human needs[24], but none of them became generally accepted.

There is not and cannot be an exhaustive list of human needs, since with the development of scientific, cultural and social levels of man and society, new needs constantly arise and their list should be constantly updated. "Life itself can be called the reason for needs," the Russian academician P.M. Ershov says[25].

The hierarchy of needs developed by A. Maslow[26] is one of the most famous in the world (Fig. 12).

According to A. Maslow, there is a clear priority in the sequence of achieving needs: if the first level needs are not satisfied to a large extent or completely, then the rest of the needs are not of great interest for a person; after satisfying the first level needs, the needs of the second level are the most important for a person, etc. However, many people consider such a hierarchy to be controversial[27].

Professor A. I. Polovinkin offers the following taxonomy of human needs[28]:

1) Providing food;
2) Provision of housing and clothing;
3) Protection from global destruction of humanity;
4) Protection from criminal attacks on individuals and society;
5) Protection against natural disasters;
6) Protection against man-made disasters and accidents;
7) Protection from disease and pain;
8) Creation and preservation of ecologically normal environment;
9) Realization of the need for beauty of environment;
10) New information sourcing, its processing and transmission;
11) Provision of socially useful leisure for people, etc.

This classification, of course, has a number of advantages, but it also makes

Figure 12. Hierarchy of needs according to A. Maslow

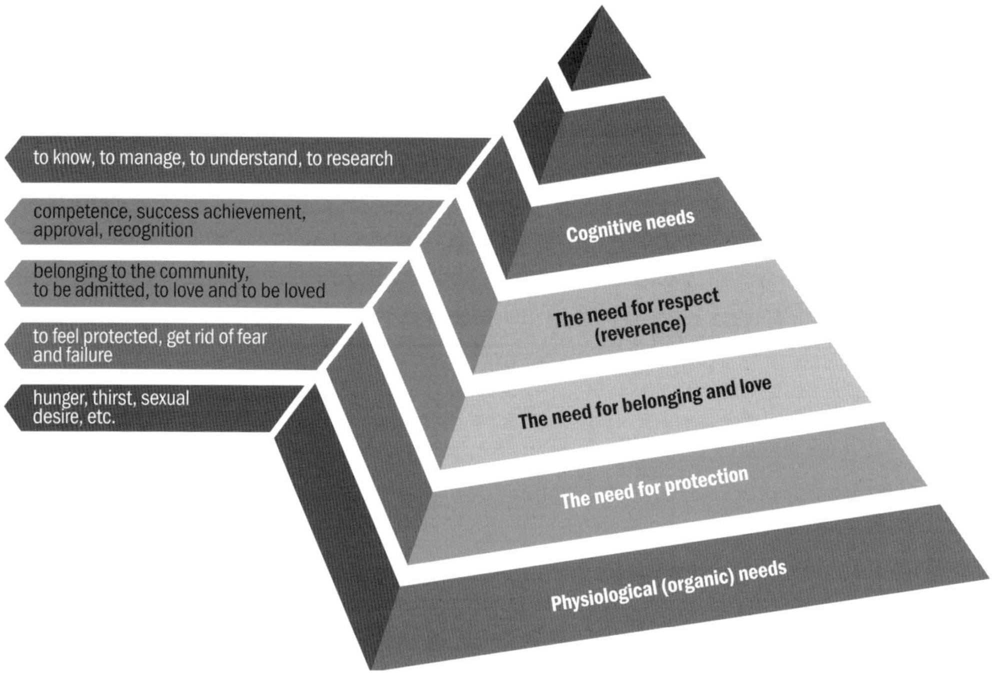

you want to correct it.

That is why it is proposed to set aside attempts to form a universal classification of human needs, and move on to the steps that contribute to their technical implementation, for example, by specifying needs.

One can highlight *general human needs* aimed at satisfying basic natural needs (food, water, sleep, etc.), preventing major threats to life, developing important areas of life, including those that characterize living standards (for example, transport, leisure, etc.), and individual needs of a person, representing various specific options for general needs (for example, general needs for food can be specified in individual needs for bread, meat, vegetables, and the general need for transportation can be represented in individual needs for moving on land, water, air).

The basis for formation of general human needs is *general causes of needs emergence*, each of them is considered as a fact, the given of the events taking place[29] (see, for example, general reasons for the emergence of protection needs — section 1.2.2). Let me remind you that general human needs can be formulated as *antonyms* of general causes of human needs emergence. For example, facts of people's insecurity from natural disasters and increase in severity of their consequences cause the general need (as the opposite of a cause) — for protection (security) from natural disasters and reducing severity of their consequences. In turn, when analyzing general reasons for human needs emergence, one can reveal the *particular reasons* that have led to general reasons (as facts). Thus, the particular reasons for high degree insecurity of people from natural disasters and increase in severity of their consequences are: the unexpectedness of a natural disaster; the destruction of houses and structures caused by natural disasters (tsunami, earthquake, etc.); large human losses associated with impossibility (due to a lack or imperfection of technologies and means) to free people quickly from a rubble, deliver them to a safe place, extinguish fires, etc. At the same time, individual human needs, as well as general ones, are formed as antonyms, but as private reasons for needs emergence. For example, in this case, individual needs are: the need for early warning of natural disaster; the need to prevent destruction of houses and structures from natural disasters; the need to release people from rubble quickly, deliver them to a safe place, extinguish fires, etc.

Technically realizable human needs are formed as the need to create appropriate technologies, equipment and products — TS for elimination or leveling of private causes of arising needs and satisfaction (implementation) of individual human needs originating from private reasons.

For instance, in the example under consideration, the technically realizable needs may be: the creation of effective methods and means of forecasting and early warning of natural disasters; the development of high-strength materials, stable structures; the improvement of communication facilities; the creation of advanced technologies, equipment and products for rapid elimination of natural disasters consequences[30].

The reasons for emergence and the main groups of needs for protection formed in this way are presented in Appendix 1.

What should be done for the successful implementation of the formed technically realizable general and individual human needs? See this in the next sections of the book.

1.4 Phasing of human needs realization

1.4.1 The principle that determines the sequence of stages

The principle that determines the sequence of stages of human needs satisfaction through the creation (and use) of TS is essentially based on the philosophical principle of ascent from the abstract to the concrete, implemented (starting with Hegel) in the method of theoretical thought movement from the "poor" (as philosophers say) to an increasingly complete, comprehensive and holistic reproduction (description) of the object or phenomenon under study.

The sequence of specification of general **(GN)** and individual **(IN)** human needs is that (Fig. 13) **GN** should be represented by a set of individual needs **(IN)** of a person in the created TS; then the **IN** should be realized in a set of functions **(F)** of the TS; at the next stage, each **F** should be specified by external (consumer) properties **(P)** of the TS, which together with the **F** should be realized in a set of functional structures **(FS)**; at the next stage, each **FS** should be realized in a set of operating principles **(OP)** or existence of the TS (further, for simplicity, we will use only the formulation "principle of operation of the TS"); finally, each **OP** at the next stage should be realized in a set of technical decisions **(TD)** of TS. At the same time, starting from the FS stage, at each subsequent stage, external (consumer) properties (P) are specified in the internal properties of the FS, OP, TD. If it is necessary to optimize the quantitative parameters of properties **(OQP) of the TD**, the **OQP** stage is possible (in Fig. 13, possible elements of the **OQP of the TD** and the graph arcs are shown with a dotted line). For more

Figure 13. Phasing of human needs realization

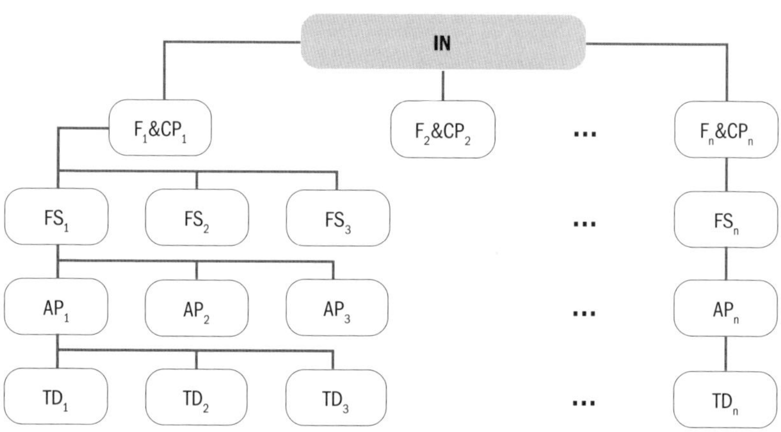

information on optimizing properties in the TD, see 1.4.6.

Of course, an option is possible when a specific TD can be obtained due to only one implementation of F&CP, FS, AP. However, the probability of obtaining a competitive TD increases in proportion to the number of formed F&CP, FS, AP and TD.

Another version of the principle representation: ascent from the abstract

Figure 14. Representation of the principle of ascent from the abstract to the concrete in the implementation of human needs

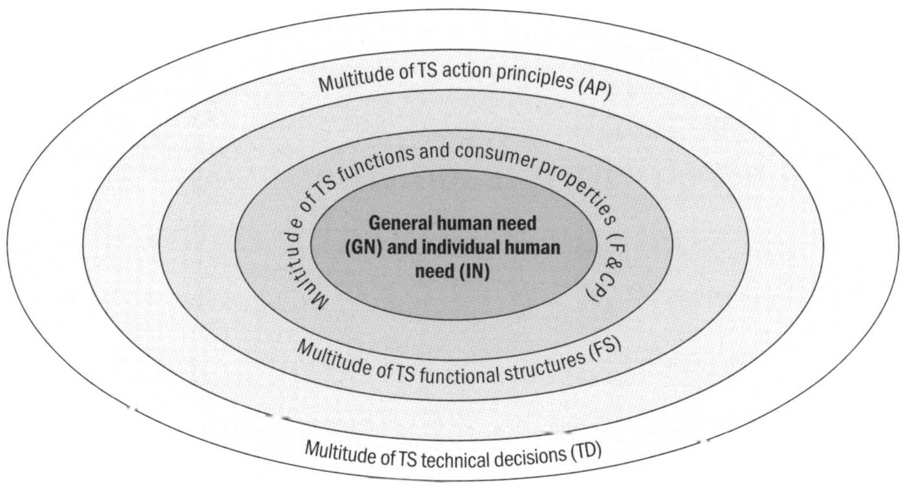

to the concrete when realizing a human need is shown in Fig. 14.

Taking this opportunity, I want to share with the reader a revelation: the presentation of scientific and technical creativity as the activity for consistent transformation of an abstract technically realizable need into a specific technical solution helped me understand long ago the deep meaning of abstract art, which I did not understand and did not accept, considering that people just pretended they liked the art. And the fact that there were many such people made me very sad. At the same time, it was surprising that realist artists were even dismissive of paintings that were practically photographs (which, in my opinion at the time, evidenced about the highest degree of skill), therefore the realizing that the abstract representation, both in scientific-technical and in artistic creation gave each person his own opportunity to conjecture and concretize abstraction, brought me great joy (although the preferences of the realistic direction in art have remained). The examples of the abstract painting are shown in Fig. 15.

A common expression comes to mind: "The devil is in the details." A continuation arises involuntarily: "And the angel is in the abstraction." This is worth thinking about.

The formation stages of F&CP, FS, AP, TD will be considered more detailed in the following sections of the book.

Developing a technical system (TS), an abstract technically realizable human need (TRN) is concretized sequentially in TS functions, TS properties, TS functional structure, TS operation principle and TS technical decision.

To concretize TRN effectively means to form a multitude of functions and properties of TS that increase its utility, and to reduce the number and influence of functions and properties that increase a charge for utility.

Figure 15. The examples of abstract art

1.4.2 Formation of technical system's functions and properties

When developing a pioneer technical system (TS), improving an analogue (prototype) of TS or creating a new generation TS, first it is necessary to concretize a formed technically realizable human need in a variety of functions and properties of TS.

Function — the action (impact or reaction) of a material object or phenomenon aimed at changing or preserving the properties of another material object or phenomenon.

By the nature of manifestation, functions can be *useful* (for example, in a car "move" is a useful function), *harmful* (in the same car, due to the friction of parts, a harmful function is "to convert mechanical energy into heating one", or "to heat"). Functions can be neutral (for example, when winding a metal cable onto a winch drum, the cable heats up due to alternating bending compression, but only slightly, by a degree fraction; therefore, the arising function, when winding the cable, "transforming mechanical energy into heating one", doesn't affect the environment and can be referred to as neutral).

In turn, useful functions are divided into *principal, main and subsidiary.* A principal external function is the one for realization of which a technique, technology, a product are actually created. For example, the principal function of a car (like an excavator, transporter, plane, ship) is to "move". The principal function is provided by several (in complex technical systems — several dozen) *main* functions. For example, the main functions in a car include "converting energy", "changing direction of travel", "restriction of movement" (braking system), etc.

Each main function is supported by numerous subsidiary functions. For example, the main function of "converting energy" using engine operation provides plurality of subsidiary functions that realize operation of ignition system, mechanisms that convert thermal energy into mechanical energy, etc. However, for the developer of a unit or part of TS which serve as the main or subsidiary function in TS, each of these functions can become the main one. Various types of TS functions are shown in Fig. 16.

Figure 16. Conversion of an individual human need into the TS functions of various types

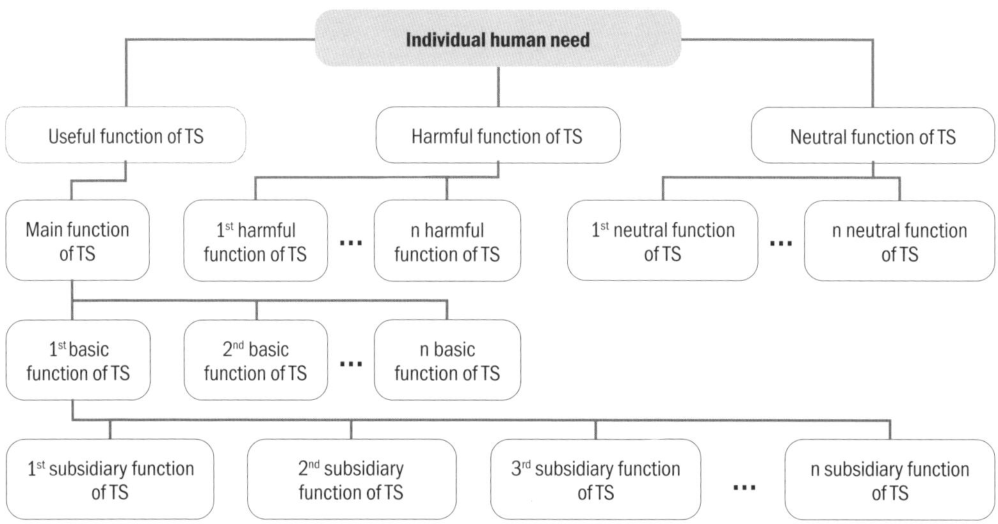

Formation of a list and definitions of technical system functions during its development is, on the one hand, perhaps the most difficult task in a creative process, on the other hand, it is the set (complete or incomplete) of correctly or inaccurately formulated functions that determines success or failure of scientific and technical (and not only) creativity in general. A correctly formulated function can suggest non-trivial technical solutions for its implementation.

TS developing properties are important attributes that concretize a technically realizable human need. In general terms, properties of TS are the form of TS features existence (qualitative certainty), with the help of which one can define similarity or difference with other TS. There are qualitative and quantitative characteristics of properties.

The qualitative characteristic is determined by presence or absence of properties (for example, if there is radiation or not), and a quantitative characteristic (parameter) determines a property's measure (in the example given, the level of radiation is measured in becquerel (one becquerel is equal to one radioactive decay per second)).

Some properties of objects, phenomena, technical systems can arise, change and disappear.

There are plenty of properties classifications, formed on various grounds, and many classifications include the large number of properties types. For example, the classification of physical and technical characteristics (properties) formed under my leadership includes about 1400 units. The fragment of this classification is shown in Fig. 17.

Figure 17. Classification of physical and technical characteristics (properties). Fragment

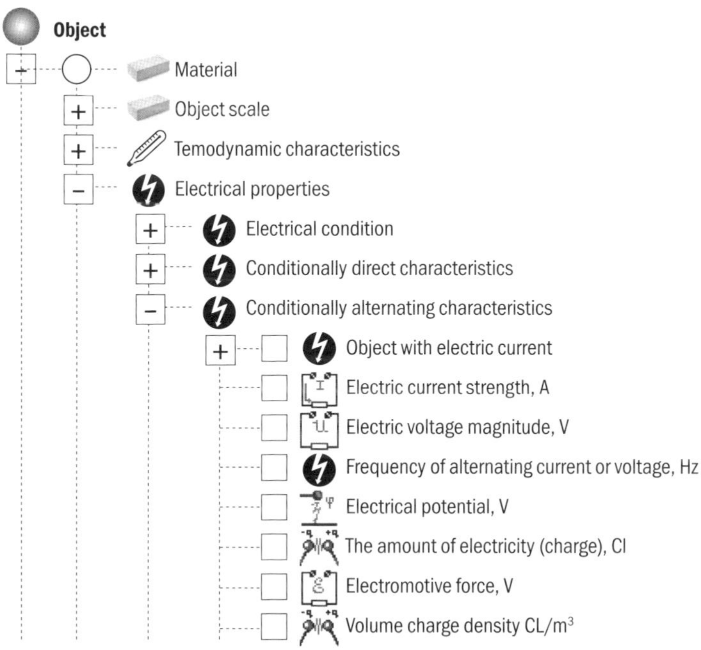

A multitude of properties and parameters form the *consumer value (worth)* and the utility of TS. The assessment of utility depends on particular consumer desires, as well as on his characteristics (nationality, culture, habits, tastes, etc.) and on the circumstances in which a consumer is, for example, the degree of given product rarity[31].

The properties of TS are grouped on two different bases (Fig. 18). By significance (base 1), *general significant, target* and *special significant properties* are distinguished. By target (base 2) there are *functional, technological, environmental and ergonomic, economic properties.*

Figure 18. Properties of a technical system, grouped by different bases

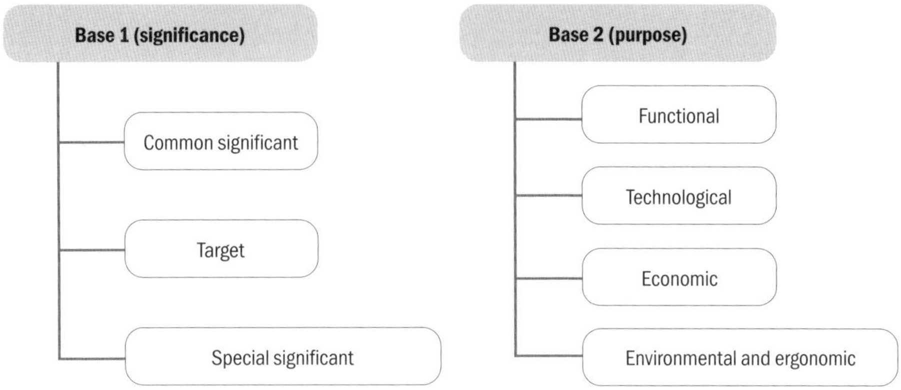

The general significant properties of TS include properties inherent in various classes of technical systems: performance, dimensions, weight, energy consumption, cost, environmental and ergonomic properties, manufacturability, operation and repair, aesthetic properties, reliability and durability, etc.

The target properties of TS include properties inherent in this class of TS and which directly determine the consumer value and the utility of TS. For example, the target properties of electric passenger cars include maximum speed, engine power, and number of passengers. These properties are inherent in any passenger car and directly determine their usefulness — ensuring the possibility of people's rapid transfer.

Strange as it may seem at first glance, but such seemingly important properties for an electric car as energy consumption per 100 km and cargo maximum mass do not directly affect the ability to transfer people quickly and therefore are not referred to a target, but to *special significant properties*, which in general case characterize the functioning efficiency of TS main units, the processes of their operation, transportation, storage and utilization of technical systems. That is why, for example, minimum turning radius, acceleration time to maximum speed, car dimensions and much more can be referred to the special significant properties of an electric car.

In general, TS properties are divided into two large groups:

1) The properties that directly affect the utility (consumer value) of TS;

2) The properties that determine the charge for utility.

The effective concretization of a technically realizable need consists in the formation of a multitude of TS functions and properties that increase its utility, and reduce the number and influence of functions and properties that determine the charge for utility.

Herbert Spencer, an English philosopher and sociologist, one of the founders of evolutionism rightly noted: "Progress is supposed to consist in the produce of a greater quantity and variety of articles".

To achieve this goal, it is necessary to form, develop and use taxonomy of physical, chemical, biological and industry-specific properties. The methods of using functions and properties (including the so-called *"unusual properties"*) for pioneer TS development, improving TS analogs (prototypes), creating new generations of TS are presented in the following sections of the book.

1.4.3 Formation of technical system's functional structures

The formation of TS functional structures variety is the next stage of the concretization of a technically realizable human need after its presentation by a variety of functions and properties of the technical system (TS) being developed.

The functional structure of TS is a set of interconnected functional elements or subsystems that ensure the implementation of TS functions. It presents a directed graph, the vertices of which are the names of elements or subsystems of TS, and the edges are their functions. The common types of functional structures are shown in Fig. 19.

Figure 19. Types of functional structures: a — linear, b — branched, c — closed, d — stellar, e - honeycomb, f — matrix

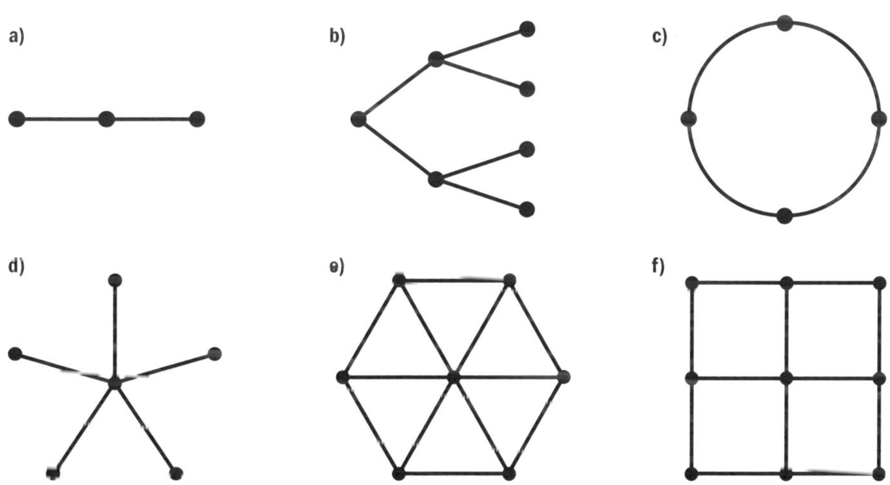

The principle of TS operation reflects the main idea of TS functioning, which determines its potential efficiency and the level of competitiveness in general.

It should be noted that sometimes even a simple elements combination of TS functional structure (for example, *inversion*) leads to the noticeable increase in TS efficiency. Fig. 20 shows a linear functional structure of barges towing (used for many decades), when a tug is pulling a barge along with itself on a rope.

Figure 20. Graphical scheme of the functional structure of barge transportation: a tug is in front (pulling)

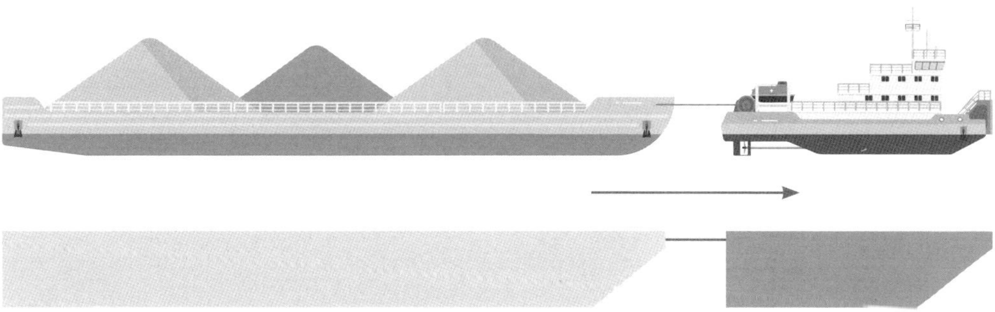

It was enough just to change the barge and the tug in places (the tug is pushing the barge), and due to the reduction of water turbulence caused by the tug in the previous scheme, with a new TS functional structure, the towing speed increased by 10–30% and the fuel consumption decreased by 10–30% (fig. 21). By so doing, they not only avoided spending additional funds, but even saved on the rope.

Figure 21. Graphical scheme of the functional structure of barge transportation: a tug from behind (pushing)

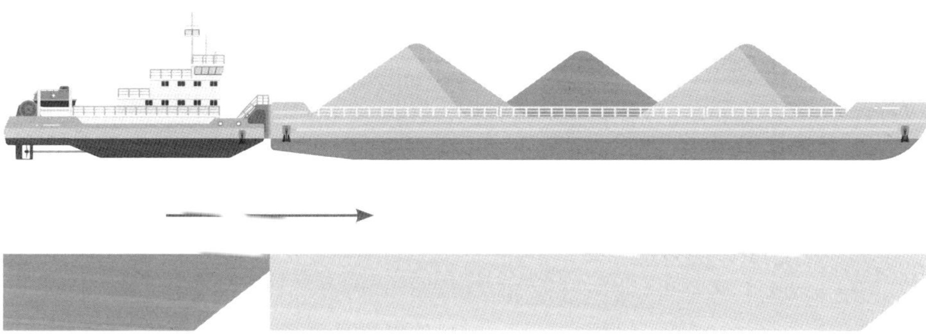

Section 6.3.4 outlines one of the methods to improve the efficiency of formation and use of TS functional structures during new TS development.

1.4.4 Formation of technical system's operation principles

After a technically realizable human need is concretized by the sets of functions, properties, functional structures of TS being developed, a new stage begins — a variety of TS operation principles is formed, each of which should reflect the main idea of TS functioning, which determines the potential efficiency and level of TS competitiveness as a whole. In the hierarchy of TS descriptions, their action principles are particularly important, since, on the one hand, they represent TS at the fairly high level of abstraction, which allows generating a plurality of qualitatively different action principles; on the other hand, each of operating principles has sufficient specificity to assess their feasibility in technical solutions.

The principle of any TS operation can be represented in the form of the oriented graph, the vertices of which are properties of TS elements or subsystems, and the edges are flows of the transformations of energy, substance and information in TS and the environment. The constituent parts of the graph are, in essence, effects and phenomena. Depending on TS, their principles of action may include physical, chemical, biological, geometric, physicochemical, biochemical and other effects or phenomena.

In other words, the principle of TS operation is the cause-and-effect transformation of energy, substance and information in the elements of their flow (effects and phenomena), bound on the basis of the same qualitative and accurate or close (in the range) values of flow elements quantitative properties (parameters).

In the modern natural science[32], a specific, identifiable, measurable, stable and potentially multiply repeated causal relationship is called the effect that qualitatively and / or quantitatively fixes the new property of objects of material-field nature, manifested in their interaction with external environment, or the existing relationships of individual properties in the form of the theoretical or empirical mathematical relationships of quantities.

There are *physical* (mechanical, electrical, magnetic, optical, thermal, sound, atomic) *phenomena*, *chemical phenomena* (chemical reactions), *biological phenomena* that occur with living organisms.

The effects are usually divided into natural scientific (NSE) and scientific-technical, or scientific and technological (STE). The former reflect the laws and patterns that are phenomenological in nature (for example, the heating of a conductor when electric current passes through it), or they fix the relationship of indicators (quantities) that appear in their definitions (for example, the dependence of pressure on force). The STE are considered as indivisible, elementary units. In contrast to them, STE are the chains of compatible NCE that correspond to the typical principles of TS. The examples of STE: an electric generator, a microphone, a bulb, etc[33]. The number of physical NCE is estimated from several thousand to tens of thousands, and physical STE — in hundreds of thousands.

Figure 22. A flat wedge. Changing the direction of movement and the magnitude of transmitted force by a flat wedge

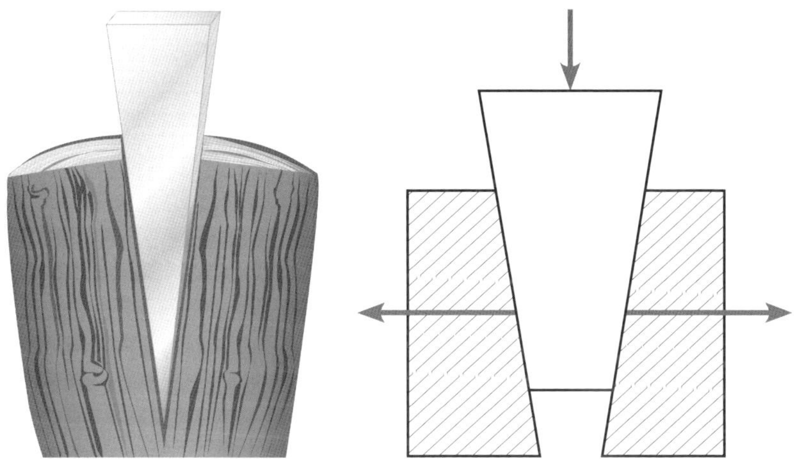

Figure 23. A screw wedge

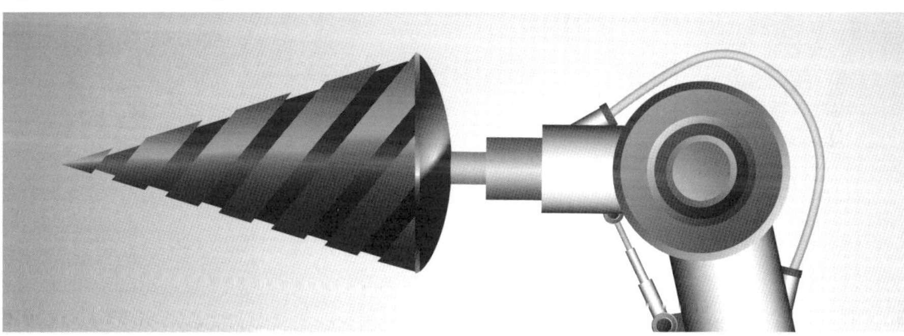

The amount of chemical and biological effects and phenomena estimates are approximately the same. There are significantly fewer geometric effects.

As examples of chemical effects, one can provide a group of common effects used, for example, for the transformation of a substance (changes in mass, concentration, shape, phase state, chemical properties, etc.) — the chemical transport reactions accompanied by the transfer of an initial solid or liquid substance from one temperature zone to another as a result of the formation and decomposition of intermediate gaseous compounds[34].

The example of a biological effect: the acceleration of cheese ripening is achieved by adding ready-made enzymes — the organic substances of a protein nature, which are the catalysts or mutant bacteria that produce the large amount of these enzymes.

Figure 24. A simple lever

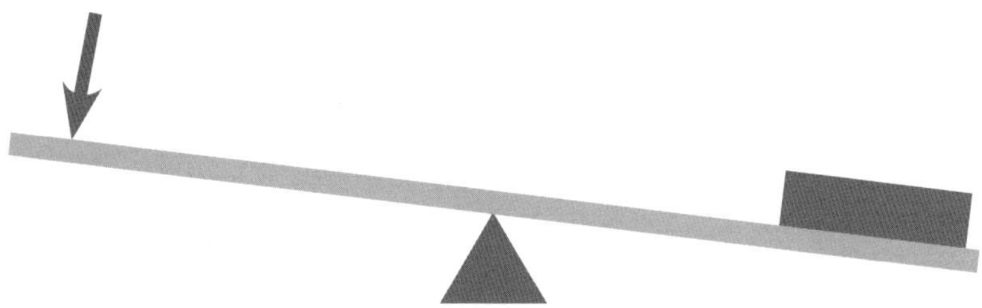

Figure 25. Archimedes spiral: a — right (counterclockwise rotation); b — left (clockwise rotation)

a) b)

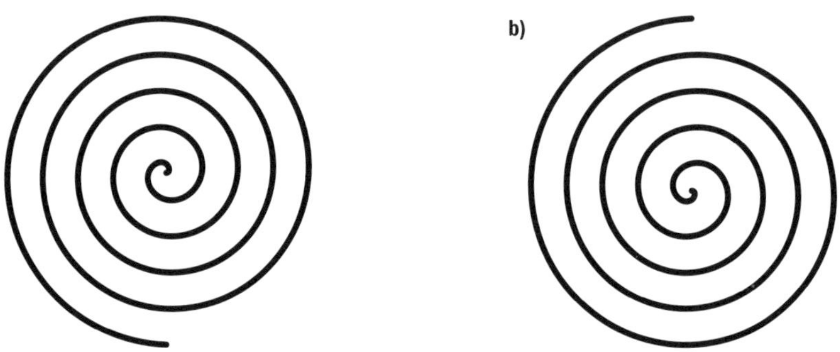

The combinations of objects geometric characteristics that generate new functions and TS properties are called *geometric effects*. So, a flat wedge can change the direction of movement and the magnitude of transmitted force (Fig. 22), and a screw wedge (screw) can additionally convert rotary motion into translational and vice versa (Fig. 23). With the help of a simple lever, you can effectively change the magnitude of transmitted force and movement (Fig. 24). Many people know the aphorism attributed to Archimedes: "Give me a fulcrum and I shall move the Earth", which characterizes high efficiency of the lever.

The Archimedes spiral, in which the normal distance between adjacent loops is constant, is well known (Fig. 25). Goethe called the spiral "Curve of life". Oddly enough, but in nature, most sea and river shells have the shape of the Archimedes spiral, a hurricane is twisted in a spiral, eyewitnesses say that a deer herd is scattering in a spiral, a DNA molecule is twisted in a double spiral, the geometry of objects arrangement in the Galaxy is also a spiral[35]. The examples of the Archimedes spiral use in the technical field are given in Fig. 26.

Figure 26. Examples of Archimedes spiral use in the technical field: a — a spiral staircase; b - heating element of the electric stove; c — clock spring

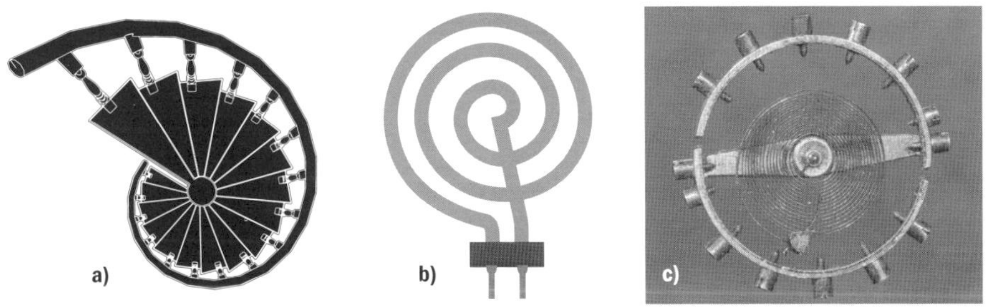

Figure 27. Mobius strip

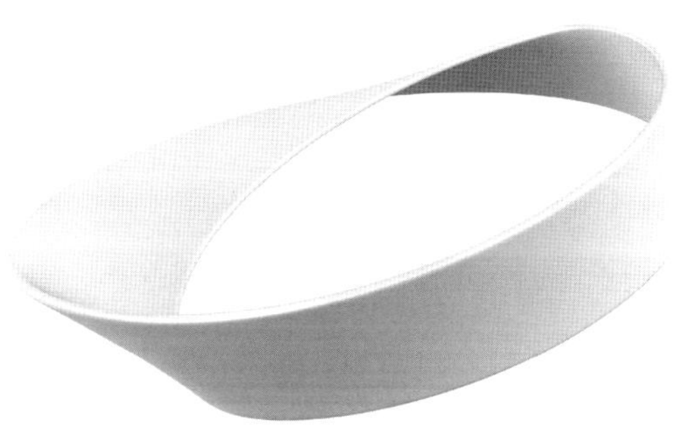

Another striking example of a geometric effect is the Mobius strip (one of the objects with a one-sided surface). Such objects have many practical applications in engineering, architecture and other fields (Fig. 27). It is easy to make a Mobius strip oneself: you need to take a paper strip, long enough, and glue the opposite ends, having previously turned one of them. As the examples of the use of the Mobius strip geometric effect, one can see a conveyor belt for transporting abrasive bulk material, a grinding belt, a band saw (Fig. 28), the service life of which is doubled. Other one-sided surfaces are known as well.

Of course, all geometric effects are realized in the physical principles of action.

The principles of TS operation can be specified in a variety of technical decisions (TD), which can be improved by optimizing TD's parameters and properties.

All TS go through the phases of inception (creation), accelerated development, deceleration of development, and withering away.

Figure 28. Examples of the Mobius strip geometric effect using: a — conveyor belt; b — sanding belt; c — band saw

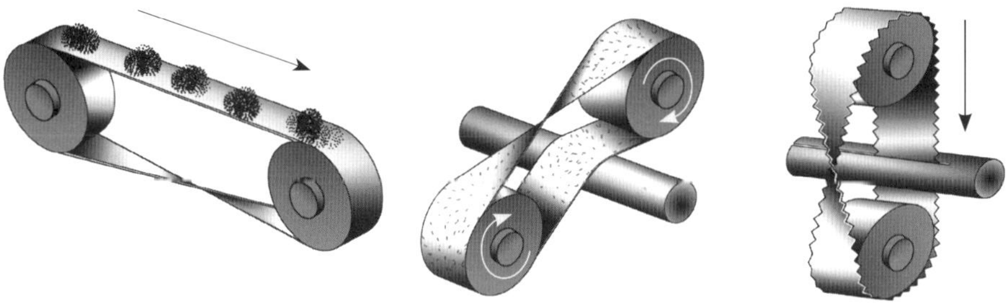

a) b) c)

The creation of conceptually new technologies and generations of technology requires development and use of new action principles. In this case, the most significant practical results are recorded when the structure of action principle includes effects that have not been previously used in this subject area. For example, electric lighting emerged on the basis of the operating principle that included new (compared to the principle of a candle) effects and their new combinations.

The task of constructing an operating principle is of great combinatorial complexity. As a result, even having information about the significant number of effects, a specialist is not able to combine them "manually" (without using computers) effectively, choosing optimal combinations from the point of view of the given criteria[36].

Under my leadership, the automated system for searching and synthesizing TS operation principles was developed, based on the logic, the essence of which is disclosed in Section 6.3.6.

1.4.5 Technical decisions formation

The principle of TS operation can be specified in a variety of technical decisions (TD).

In the Big Encyclopedic Polytechnic Dictionary, TD is defined as "a device, a structure, a product that is the structural element or a set of structural elements that are in functional and constructive unity; the method, the process of performing interrelated actions on a material object and with the help of material objects; a substance, an artificially created material formation, which is a set of interrelated elements, ingredients (the substances include, for example, the materials for manufacturing objects, structures, used for coatings, insulation, depreciation, used as the conductors of energy, medicinal, cosmetic, food, flavoring substances, fodder products, chemical reagents, substances-emitters and substances-absorbers of radiation, surfactants, biologically active substances, including pesticides, growth stimulants)[37]". With great respect for the Big Encyclopedic Polytechnic Dictionary and accepting almost completely this definition, I, together with the authors[38], consider it necessary to emphasize that TD is not a device, a structure, a product, etc., but it is its **description**.

Technical decisions can be divided into three main groups: constructive, technological and other TD, while the latter includes substances, biological objects, the topology of integrated circuits, etc.

The description of TD is usually supplemented with a graphical representation. About 95% of all inventions are TD, less than 5% are descriptions of TS actions principles.

Although the descriptions of many TD do not have any quantitative characteristics, a specialist can usually estimate the difference in potential effectiveness of TD from one area.

Below is the example of a constructive TD.

Figure 29. Bicycle

TD — bicycle (fig. 29): "A bicycle consists of a frame (1), an articulated steering fork (2) with a handlebar (3) and a front wheel (4), a rear wheel (5), a saddle (6) on a seat tube (7), crank pedals (8), chain drive (9) and brakes (10). Additionally, a bike can be equipped with one or two shock absorbers, fenders (to protect against splashes), a trunk, lighting equipment (lamps and / or reflectors), a sound signal, a "footboard", etc."[39]

A typical example of a technological TD is a recipe for making borscht (Fig. 30).

TD — a recipe for making Moskovsky borsch: "Ham bones are boiled in a meat broth, the broth is filtered, boiled. A cabbage, chopped into strips, is placed in the boiling broth, boiled for 8–10 minutes; add browned vegetables and boil again. Then add stewed beets, season with white sauce or browned flour, diluted with the broth or water. At the end of cooking, add salt, pepper, bay leaf, vinegar, sugar and cook until it's ready. Prepare a set of meat products; it includes ham, meat, sausages. Use one piece of each type per serving. The meat products are cut into slices. Pour in a small amount of the broth and boil.[40]"

Figure 30. Borscht cooking technology

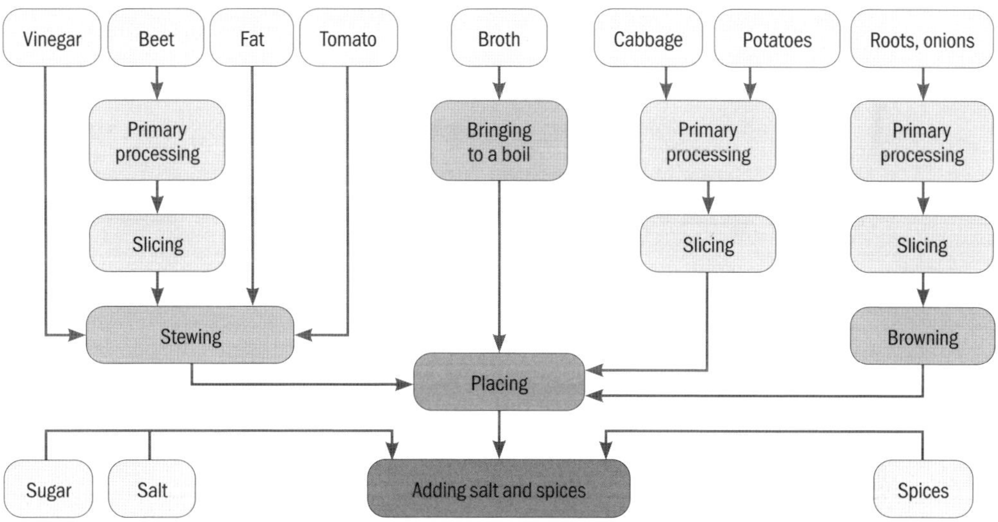

The example of TD — the description of a biological object. TD — Lactobacillus acidophilus bacterial strain (Fig. 31): "the cells size of an 18-hour culture in milk is 3–6 microns, thin rods, the cells are evenly colored in methylene blue, located singly or in the form of the chain of 2–3 segments, on agar with hydrolyzed milk form colorless colonies with the diameter of 1–1.5 mm, deep, dark in the form of spiders. The minimum growth temperature is 20 °C, the optimum is 36–39 °C, and the maximum is 60–65 °C. Relation to carbohydrates — ferments glucose, lactose, mannitol, does not ferment arabinose, weakly ferments rhamnose, cellobiose. The milk is curdled at the optimum temperature and application of 2% of ferments in 4.5–5 hours, 4% of ferments in 3.5 hours, after 24 hours acidity reaches 138 ± 10 ° T (pH 3.95 ± 0.1), after 48 hours — 170 ± 10 ° T (pH 3.8 ± 0.1), after 3 days — 190 ± 10 ° T (pH 3.7 ± 0.1). The strain is homofermentative, racemate, produces DL-isomer of lactic acid, has a pronounced antagonistic activity against pathogenic bacteria. The culture has a high proteolytic activity and differs from the known acidophilic strains by a lower limit of acid formation"[41].

A technical decision (TD) embodies technical system (TS) functions, properties, functional structure and a principle of operation, and with a high efficiency of TD maker's creative activity, it can be the outstripping reflection of a technically realizable human need, which has not yet become the subject of public consciousness. History knows a lot of TD ahead of their time.

Figure 31. Lactobacillus acidophilus bacteria

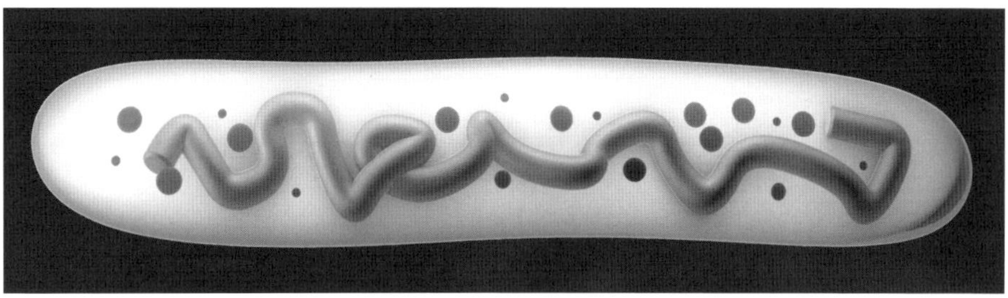

TD can be protected as a result of intellectual activity, improved by optimizing quantitative parameters (see more on this in the next section), and also used in design documentation for its implementation (sometimes even without optimizing quantitative parameters).

1.4.6 Optimization of property parameters in technical decisions

A technical decision can be significantly improved by optimizing TD properties parameters. This can be done by choosing the best combination (according to the selected criterion) of properties quantitative characteristics that directly affect the utility (consumer value) of TS being developed, and the properties that determine the utility charge of TS being developed.

To analyze the quantitative characteristics of properties and select the best combination of them, the mathematical and / or other models of TS being developed are used, on which quantitative changes in some characteristics of TS model are studied when other characteristics change. For example, by changing only the quantitative characteristics of automobile body geometry, it is possible to improve its aerodynamics significantly. At the same time, the potential maximum speed of a car (with the same quality parameters of properties) can increase by 10–30%, and the fuel consumption can also decrease by 10–30%. It will be recalled that the same effect was obtained in TS for towing barges, but due to a change in the functional structure of TS (see Fig. 20 and 21).

1.4.7 Creation of a technical system and its life cycle

A technical decision can be the basis for TS creation or its experimental sample, which must successfully pass tests in order to proceed to next stages, ending with the mass production of TS.

The concept of "the life cycle of a technical system" is widely spread, which characterizes the process stages covering various TS states, starting from the moment of the need arising for such TS and ending with its complete decommissioning, including utilization or burial. However, there is no generally accepted list of these stages. The following simplified list can serve as an example:

- The definition of the functions and consumer properties of TS (which corresponds to working-out of technical specifications in the process of TS designing; for more details about design, see Chapter 3);
- The selection or developing of a functional structure, principle of operation and technical decision of TS (which corresponds to developing of technical specifications, draft design and / or technical design);
- A detailed design related to calculation and optimization of TS parameters, selection or developing of manufacturing technology, drawing up design documentation;
- TS manufacturing and testing;
- TS transportation and storage;
- TS operation, malfunction diagnostics and repair;
- TS utilization or burial as a result of its physical or moral deterioration.

The life cycle of TS tends to decrease in duration. For example, it took almost 400 years to introduce massively the printing press invented by Johannes Gutenberg at the beginning of the 15th century, a common telephone was being introduced for 50 years, a mobile phone became widespread in 7 years, and social networks — in 3 years. In the modern conditions of intense competition in the markets, requiring the constant renewal of assortment and improving the consumer properties of goods, the tendency to shorten the life cycle of TS is even more expressed. For example, Moore's Law — one of the most famous empirical (based on experience, facts study) rules in the world — states that every two years the number of transistors, placed on a computer chip, doubles.

The knowledge and understanding of the main stages of TS life cycle makes it necessary to create such TS in which functions and properties will be implemented ensuring not only its efficient operation, but also transportation, repair (maintainability), utilization and burial of TS. These stages (steps) are described in detail in Chapter 3 "Scientific and technical creativity as formation and implementation of designs."

1.4.8 Characteristic phases of a technical system inception (creation), development and withering away

Pierre François Verhulst, a Belgian mathematician showed in the middle of the 19th century that the growth of biological populations occurred along an S-shaped, or, as it is also called, logistic curve. Subsequently, many scientists, based on the results of large-scale studies, have proved that the life cycle of any technical systems (creation, development and withering away)[42] is proceeding according to that pattern. The qualitative characteristics of the phases of inception (creation), development and withering away of any newly created TS on an S-shaped curve (Fig. 32) can be interpreted as follows.

Phase 1 — the inception (creation) of TS is characterized by a slow growth in the consumer properties of TS, since almost all properties of TS require research, experimental verification, as well as the assessment of consumer demand.

Figure 32. Characteristic phases of TS inception, development and withering away

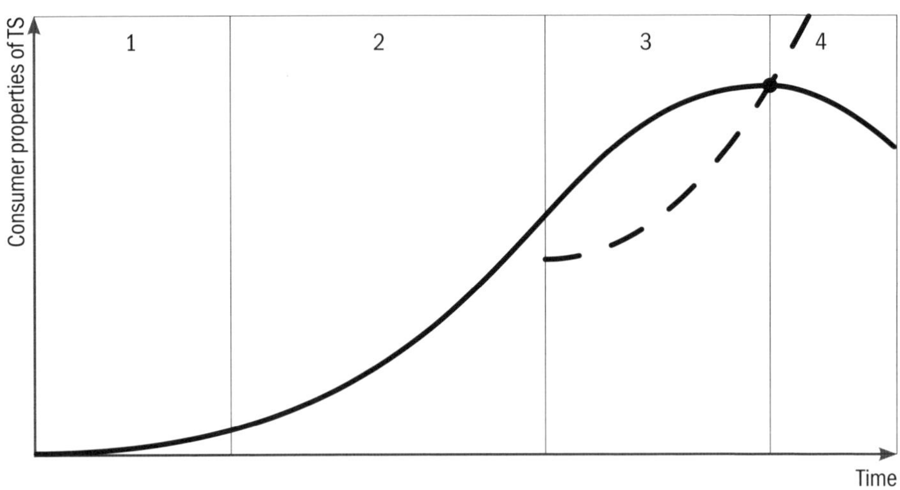

Phase 2 — accelerated development of TS, which begins and continues with high consumer demand. It is this phase that is characterized by development and implementation of a variety of technical decisions of TS based on the previous or new functional structures and principles of TS operation or existence.

In the absence or very low consumer demand, the life of TS ends immediately after the first phase, or at least stops indefinitely until the need for TS is recognized and formed (for example, in case of accidental discoveries) or until the necessary materials are invented and / or technology. For this last reason, many inventions of the genius Italian Leonardo da Vinci were ahead of their time (for example, ball and roller bearings, a parachute, Fig. 33 and 34). In general, in technical creativity, one should follow the advice of the outstanding American inventor Thomas Edison: "I find out what the world needs. Then I go ahead and try to invent it."

Figure 33. Ball bearing: a — proposed by Leonardo da Vinci; b — modern

a) b)

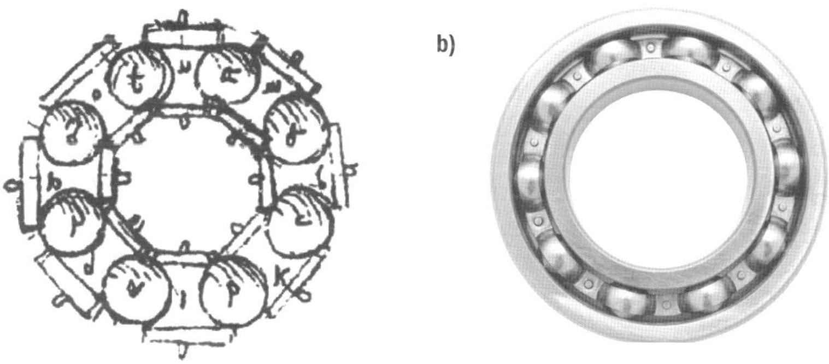

Phase 3 — a slowdown in the development of TS, which is associated with the emergence of new human needs, which can no longer be fully satisfied by technical decisions within existing TS.

Phase 4 — actually characterizes TS dying off due to a contradiction that has arisen: consumer demand for TS is falling, and it is necessary to develop the consumer properties of TS, but with the existing functional structure and the current principle of operation or existence of TS, its consumer properties cannot continue to develop significantly (contradictions in technical and other spheres are discussed more detailed in chapter 4).

To meet the growing human needs, it is required to switch to new generation TS based on a new functional structure and / or another principle of the operation or existence of TS (this is a dashed line in Fig. 32). However, creation of new generation TS should be started in advance (as a rule, this is the beginning of phase 3 in the "old" TS life), so that by the end of phase 3 of the "old" TS life, the consumer properties of new generation TS will be no worse than maximum values of the

Figure 34. Parachute: a – proposed by Leonardo da Vinci; b – modern

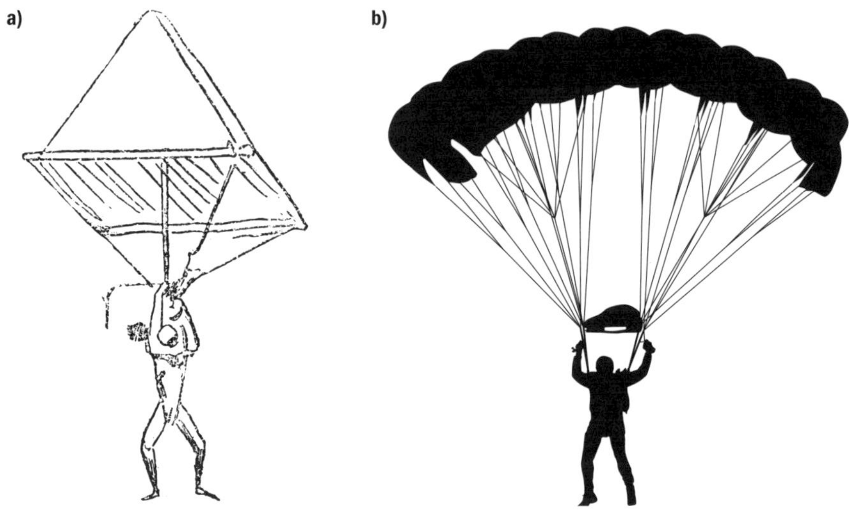

"old" TS consumer properties. Then transition to new generation TS will take place without negative consequences and this TS will continue its life along its S-shaped curve, after it new generation TS will slow down its development as well. It is necessary to develop in advance next generation TS, that is, the phases shown in Fig. 32 will be repeated.

It should be noted that there are many examples when TS do not "die" after phase 3, but continue to "live" and "live" indefinitely without significant improvement in consumer properties. For example, a scooter has been used for more than a century without fundamental changes, and a motorcycle and a car have still not displaced it.

Chapter 2
Realization of human needs — a creative process

2.1 Basic concepts of the creative process and scientific-technical creativity

In a general sense, **a creative process (creativity)** is aimed at creating qualitatively new material or spiritual values. Also, creativity is understood as a result of creating something subjectively new (for a man or a group of people), which, when checked, may turn out to be objectively new (never existed before).

According to the goals, there are the following types of creativity (very con ditionally): artistic — the creation of new aesthetic values that cause the new emotional state of a person; scientific — the creation of new knowledge; technical — the creation of new means of labor; sports — the creation (achievement) of new sports results. Then scientific and technical creativity is the creation of new knowledge and means of labor.

Human needs realization at first can be carried out within the framework of reproductive activity (reproduction of already known), that is, practically without any elements of creativity. But then this activity, as a rule, turns into the creative process for the creation of technical systems (TS) to meet the same needs, but in a new quality.

For example, in ancient times the organic need of a man to quench his thirst under natural conditions (moderate temperature, humidity and air velocity) was carried out in the simplest way: a man was lapping, he was drinking from his hands, a wide leaf, a fruits shell, plant seeds, mollusk shells, a mail, a horn and suitable animals bones. Soon drinking vessels appeared at first simple ones, and then with a lid, mouthpiece and spout, double bottom, with functions for heating or maintaining the temperature of the liquid (which they began to saturate, for example, with gas), and then they began to come up with functional and design attributes (for example, drinking straws or umbrellas for cocktail glasses), etc.

However, there are many cases when even simple technical solutions have not appeared for a long time inexplicably. For example, the first can opener was

Human needs realization
is a creative process.

Creativity can be artistic, scientific, technical, scientific-technical and sporting.

There are 6 levels of novelty created as a result of TS creative activity.

It is necessary to combine all knowledge into meta-knowledge (mega-science) for creative activity to be as effective as possible.

invented only 45 years after the creation of a can (for almost half a century people suffered by opening cans without special means).

Creativity is necessarily required (sometimes of a very high level) when the conditions for satisfying a need are changing. For example, it is not easy to realize the need to drink, I mentioned, when there is no drinking vessel or you need to drink in space, where the temperature is extremely low and there is practically no gravity. By the way, due to the structural complexity of the toilet on the International Space Station, it costs $19 million, which is more expensive than the gold bar of the same mass (Fig. 35).

Figure 35. Toilet for astronauts, developed by SPE "Zvezda"

Unfortunately, the overwhelming majority of creative process models that are still used in educational institutions and industrial enterprises to train specialists and improve their qualifications, determine the overall low efficiency of a creative process, since they are based on passive stages — the stage of ideas "incubation" (maturation) which is recommended to be intensified due to "diversion from the task and switching to another object", as well as the stage of unpredictable waiting time of "enlightenment" and sudden "insight", which may not arise. Such models are suitable only for people who are naturally endowed with outstanding creative abilities (there are very few of them). Their incubation time is minimal, and insight is almost guaranteed. Although sometimes even geniuses had incubation cycles that lasted for years, less often for decades. Isaac Newton was nursing the law of universal gravitation for over 20 years. William Harvey, his compatriot, spent the same amount of time nourishing the thought of the Body's Circulatory System.

The overwhelming majority of people need the special methods and tools (this book is just about them), which minimize the time of the *incubation* process and dramatically increase the probability of *insight* at every stage of creativity: setting a task, solving a problem (implementing a task), commercializing and / or other using of results.

The intangible results of a creative process (the results of intellectual activity — RIA), which represent TD descriptions, for example, in the form of *inventions, utility models, industrial designs, trademarks*, computer programs and *databases* can be protected by the titles of protection. For more information on the stages of creativity and protection of creative activity results, see Chapter 6.

2.2 The concept of "new" in scientific and technical creativity

The results of scientific and technical creativity — scientific knowledge materialization — consist in the formation of a "new" one that has not previously existed or has been repeated in other conditions, connections and relationships, with a special functional meaning and effect.

The ancient Roman philosopher Lucius Annei Seneca said: "Novelty is often admired more than greatness." Many people, creating "new", do not know that their "new" already exists. For this reason, this newly created "new" is called "subjectively new".

The result of creating an objectively "new" depends on the combinations of the following factors as the components of novelty: the problem of scientific and technical creativity, the methods and means used, the existing conditions and limitations (Table 3).

The level of TS novelty depending on the changes in TS characteristics is shown in Table 4.

Table 4 shows that the creation of a new one in TS occurs when new technically realizable human needs arise. At the same time, the lowest level of novelty (1) is formed when only quantitative parameters (size, weight, speed, color, etc.) are changed in TS. In some cases, changing even one parameter — size — was considered revolutionary (for example, the transition from a long-length dress

Table 3. Variants of factor structures as components of novelty

Characteristics			
Tasks	Methods used and means of solving the problem	Conditions and restrictions	A finding
New	Known	Former	New
New	Known	New	New
Former	New	Former	New or former (but achievable easier, simpler, cheaper, faster)
Former	Known	New	New
New	New	Former	New
Former	Known (but new for this task)	Former	New
New	New	New	New
Former	New	New	New or former (but achievable easier, simpler, cheaper, faster)

Table 4. Dependence of TS novelty level on the changes in TS characteristics

Technically realizable needs	Characteristics TS						TS novelty level
	Functions	Properties	Functional structure	Principle of action (functioning)	Technical decision	Quantitative parameters	
New	Former	Former	Former	Former	Former	New	1
	Former	Former	Former	Former	New	Former, new	2
	Former	Former	Former	New	New	Former, new	3
	Former	Former	New	New	New	Former, new	4
	Former	New	New	New	New	Former, new	5
	New	New	New	New	New	Former, new	6

to a midi in the 40s and to a mini in the 60s of the last century). It can be assumed that primitive women, when they made clothes from skins, probably preferred the mini style, which did not hinder movement. The highest level of novelty (6) is when all characteristics of TS change. At all levels of novelty from 2 to 6, the quantitative parameters can remain the same, but the changes in other characteristics of TS (from a technical decision to functions) form the necessary novelty and satisfy any technically realizable need (for example, installations for the production of fresh water can have the same quantitative parameter — productivity, but one of them can be based on the desalination of sea water, and the other — on the production of water from the air).

The potential efficiency of TS increases in proportion to the level of novelty, because of changes in its characteristics (for more details, see Section 3.2.1).

2.3 Factors affecting the effectiveness of scientific and technical creativity

2.3.1 The efficiency level of knowledge convergence and the basal principle of heuristics

The formation of something new in the technical sphere occurs in the process of convergence, the transfer of knowledge about existing TS and the objects of animate and inanimate nature to TS under formation. This process, which is most important for scientific and technical creativity, is called *the convergence* of knowledge. The level of its effectiveness is proportional to the amount and *relevance* (adequacy) of knowledge used to develop new TS. The complexity of this process lies in the fact that the relevance of knowledge about known TS and the objects of animate and inanimate nature relative to TS under formation can be potentially very high, but implicit. For example, it is psychologically difficult to see the possibility of using knowledge about the peristaltic mechanisms of the mammal that provide transport of products in the gastrointestinal tract or the circulatory system when creating highly efficient peristaltic pumps and even the simplest tube with cream or paste (Fig. 36), although the relevance of this knowledge is very high. That is why experts believe that the convergence of traditionally non-overlapping knowledge has implicit, but great *heuristic* capabilities and such capabilities should be carefully identified and widely used.

In order to increase the efficiency of knowledge convergence when creating new TS, ideally, it is necessary to consider all the knowledge about living and inanimate nature that exists in the world. This is not an easy task, and quite difficult to comprehend. But if one can manage it, then the number of variants of new TS will be maximum. Twice Nobel Prize winner Linus Karl Pauling once said: "The way to get good ideas is to get lots of ideas." This approach is shared by the prominent American scientist Percy Hill. According to him, at all (except for the regulated) stages of creating a new product (TS) and its commercialization (use), starting with the formation of an important need, it is necessary to generate at least 60 fundamentally different variants of ideas with optimal quantitative properties parameters, and then to choose (using the known methods and means of decision-making) the best of them, which is most likely to be competitive[1].

Figure 36. Examples of peristaltic TS: a — tube with paste; b — pump

a) b)

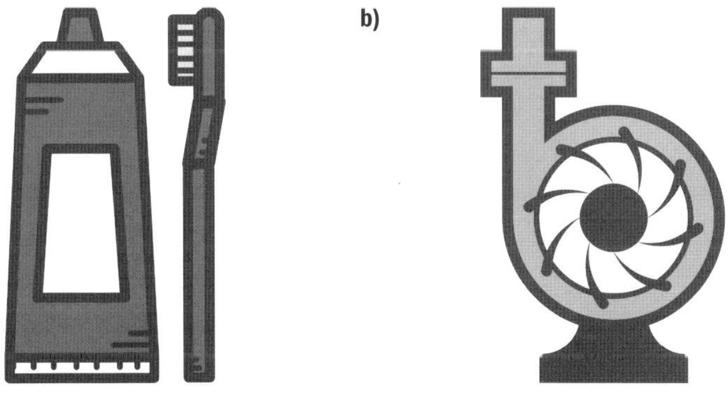

The understanding of high importance of combining knowledge about animate and inanimate nature for active involvement of all available knowledge in the economic circulation, including for the convergence of knowledge in scientific and technical creativity, led to the fact that in the poll taken by *New Scientist* magazine in 2005, scientists ranked fourth on the idea of creating a science about everything — *metaknowledge, or megascience* in the top ten greatest ideas of mankind. At the same time, Albert Einstein's theory of relativity was only in seventh place[2].

Meanwhile, the processes of *differentiation* of sciences and knowledge are intensifying in the world. Universities have not trained "general" physicists, chemists, biologists, economists for a long time, but ultimately train narrow specialists, for example: "opticians", "mechanics", "nuclear scientists", etc. Several dozen independent sciences have emerged from chemistry, biology, economy and others. On the other hand, many disciplines have been integrated into one general science. For example, chemical physics, physical chemistry, biochemistry, etc. have arisen. All this undoubtedly complicates the unification of knowledge.

In Russia, the process of creating megascience is being carried out by the National Research Center "Kurchatov Institute", where the scientific direction related to the convergence of nanoscience, bioscience, information science, cognitive and socio-humanitarian sciences and technologies (NBICS) is actively developing. At the same time, we must realize that ultimately all sciences and branches of knowledge should be combined into megascience.

However, long before the poll was taken by *New Scientist* magazine, Soviet creative methodologists focused on the key role of metaknowledge (megascience) in the concept of the basal (fundamental) principle of heuristics, the essence of which was that in order to solve a specific problem, it was necessary to go beyond knowledge about it (and even a section of science) into the area of metaknowledge or megascience and with the information introduced, return to the problem being solved. In fact, this is the convergence of knowledge. It is clear that the narrow specialization of scientists and engineers, and most importantly, information resources distributed across the fields of science impede the basal principle implementation of heuristics, therefore, to resolve a problem situation, a chemist will most often be dominated by chemical methods, a mechanic — by mechanical methods and so on, although a more effective solution can be obtained based on knowledge from another area (or other areas). The diagram explaining the basal principle of the heuristic is shown in Fig. 37.

Due to the low efficiency of knowledge convergence in Russia, the conversion is also ineffective (full or partial transfer of defense industry enterprises to the production of civilian products and consumer goods). The most odious example of ineffective conversion: a military plant that had produced high technology products switched to the production of pans, pots, and shovels.

Figure 37. Scheme explaining the basal principle of the heuristic

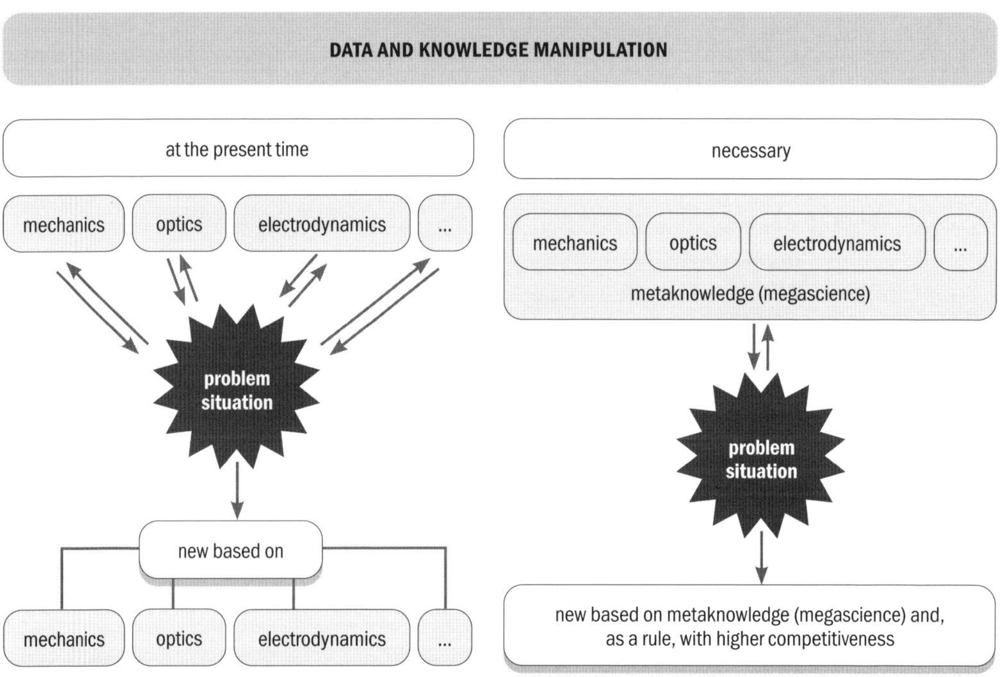

So far, the convergence of knowledge is not sufficiently promoted by modern information and communication systems that manipulate TS descriptions of different sectors of the economy, in which, for example, functionally identical transport TS can be called differently: "cargo carrier", "conveyor", "batcher", "feeder" and others. Due to the objective indexing of TS descriptions, a computer will find only those analogs, the names of which are specified in a search query by a user. At the same time, perhaps the most effective technical decisions, descriptions of which could be used with great efficiency when creating new transport TS, will not be found.

The way out is to index the descriptions of each TS with a set of functions (useful, harmful, neutral), as well as generalizing functions (see 6.2.5.1) that combine the TS functions of different economy sectors. Then, by the name of any (even harmful function), it is possible to provide a search for the TS analogs of any economy branch available in electronic resources, regardless of analogs name.

In general, the methodology for creating metaknowledge (megascience) and convergence of knowledge is not yet sufficiently developed, and this process of creation itself (taking into account constant emergence of new knowledge) is endless. An original approach to combining and using knowledge about human needs, functions, properties, functional structures, principles of operation, TS technical decisions of different branches of the economy, based on knowledge of all sections of physics, chemistry, biology, economics, is being developed under my leadership. The results are presented in Chapter 6.

2.3.2 Reproductive and productive knowledge

Insufficient efficiency in the search for TS analogs in the various sources of information is not the only thing that reduces the efficiency of the convergence of knowledge and scientific and technical creativity in general. An additional factor leading to the decrease in the effectiveness of scientific and technical creativity is the insufficient *heuristicity* of descriptions (the potential for knowledge to grow) of TS analogues and other related knowledge, which in most cases are presented in the form of reproductive descriptions, reflecting mainly the actual state of TS analogues (with different specificity and display methods — texts, tables, graphics, animations, etc.). Such descriptions can be successfully used as reference manuals — for training, reproducing (reproduction, manufacturing) according to a model and / or instructions, but due to their low heuristicity, they rarely serve as a clue for generating the relevant new knowledge leading to the creation of new TS.

Reproductive TS descriptions have a weak potential for knowledge to grow; therefore, the productive knowledge hidden in TS, in living and inanimate nature, in systematized reproductive knowledge should be used.

The productive knowledge, provided with the methods for its use, is called heuristics.

Such opportunities are possessed by productive knowledge hidden:

1)in TS; 2) in living and inanimate nature; 3) in systematized, ranked and selected according to a certain criterion, reproductive knowledge. Of course, there must be methods for using productive knowledge. Then this knowledge turns into heuristics (from Greek ευριστικω — I seek, find, discover): principles, laws and patterns of TS development, strategies, tactics, methods, techniques. Productive knowledge can (should) be identified in a variety of ways[3].

In the first case, productive knowledge (and the heuristics formed on its basis) arise when new TS is created. It is the heuristics that lead a developer to new TD, which has potentially better characteristics compared to previous TS. Very often, such heuristics are not understood by a developer and, accordingly, are not described, although their value sometimes exceeds the value of TS developed on the basis of such heuristics (see the wise statements of K.A. Helvetius, G.V. Leibniz, L.D. Landau on this issue in the preface). For example, it is unlikely that the creator of the halberd (pole arms with a combined tip made of a spear point and a battle ax — Fig. 38) formulated a heuristic — realization of the the most natural need to "provide multifunctionality", which was the methodological basis for creating a halberd from analogues (spear and ax).

Figure 38. One of the first examples of multifunctionality implementation: a — spear; b — battle ax; c — halberd

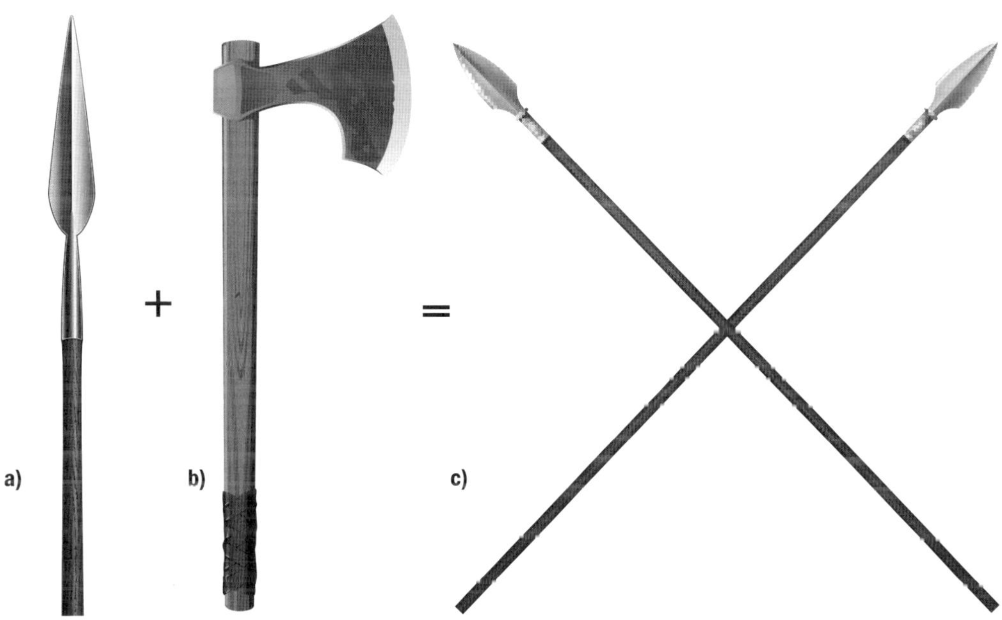

This heuristic was formulated by another person many years later; it is now widely used in the world. For example, the Wenger Giant Swiss Army Knife has 87 built-in tools and 141 functions (fig. 39).

Figure 39. Wenger Giant Swiss Army Knife – record holder

In general, in the first case, the productive knowledge is revealed during functional-physical (chemical, biological) analysis and synthesis of TS (FPA), which consists in the multifaceted comparison of the last TS with the previous TS (analogue) and formulation of heuristics, due to which new properties or new property parameters in the last TS (Fig. 40) have been obtained. Along the way, the productive knowledge is revealed, aimed at the further development of the last TS.

It is important to note that FPA can be carried out without developers of the last TS and its previous analogues. The FPA methodology is described in Appendix 2.

In the second case, the productive knowledge is identified in the objects of animate and inanimate nature, considered as the potential *analogs* of TS under development, based on *analogies* between them. Upon that, analogies are described in the form of needs, functions, properties, functional structures, principles of operation, and even technical decisions, and these descriptions are used to transfer knowledge about analogs to TS being developed. For more information on inferences by analogy (the method of analogies) see Section 5.1.4.

The formation of productive knowledge from animate or inanimate nature objects can be carried out with the separate analysis of objects (Fig. 41) or within the framework of FPA.

Figure 40. Identification and formation of productive knowledge during functional physical (chemical, biological) TS analysis

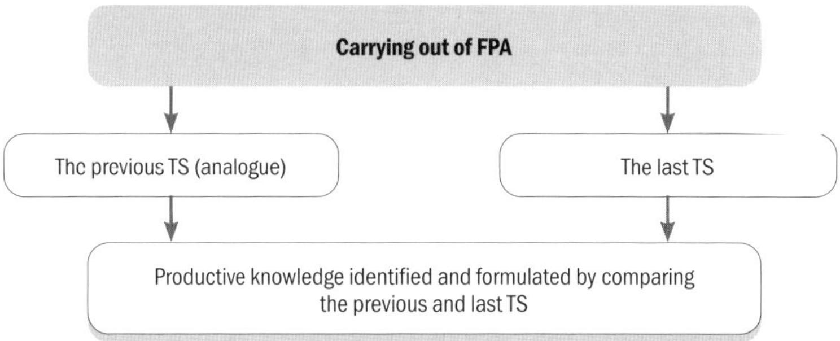

Figure 41. Identifying of productive knowledge from descriptions of animate and inanimate nature objects in the form of the analogies and analogs of TS

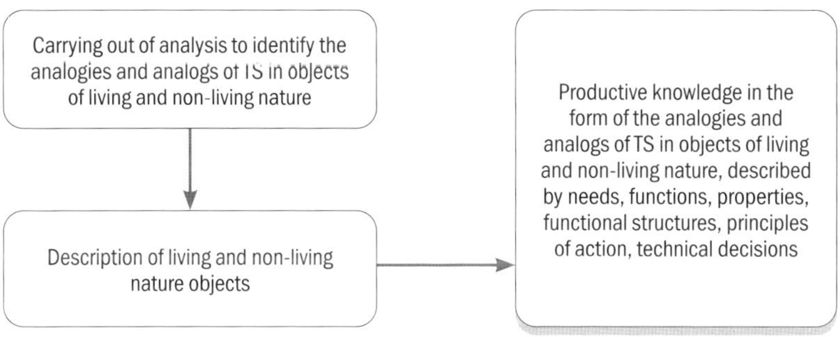

In the third case, the productive knowledge is identified during accumulation, systematization, ranking and selection according to the specific criterion of reproductive knowledge, which leads to the emergence of heuristic opportunities that contribute to new knowledge generation. In fact, the productive knowledge arises. For example, the descriptions of chemical elements separately — the reproductive knowledge, but their accumulation, systematization (by groups and periods) and ranking of elements (by their atomic masses) led D. I. Mendeleev to the formulation of the periodic law (heuristics) in 1869, the logic of which made it possible to predict the discovery of new (absent in the periodic table) elements.

In addition, by ranking the reproductive knowledge, it is possible to identify the most demanded ones, the use of which in new TS development with a high probability can lead to a significant improvement in the consumer properties of TS being created. For example, when ranking many thousands of human needs according to the frequency of their manifestation (occurrence), I have formed

The outstanding methodologist of scientific and technical creativity G.Ya.Bush has collected 3000 heuristics, I recommend 24.

Psychological barriers hinder effective creativity. The most harmful of these is the term barrier.

the group of the most urgent needs, called the *most important human needs.* A list of them and recommendations for use on creating new TS are presented in section 6.2.2.

Exclusively due to the ranking of reproductive verbs and adjectives characterizing TS functions and properties, I have formed lists of "good" and "bad" verbs and adjectives with productive knowledge capabilities (presented in Sections 6.2.8 and 6.2.9).

The general scheme of the opportunities formation of the productive knowledge into the reproductive knowledge is shown in Fig.42.

Figure 42. Opportunities formation of the productive knowledge in the reproductive knowledge

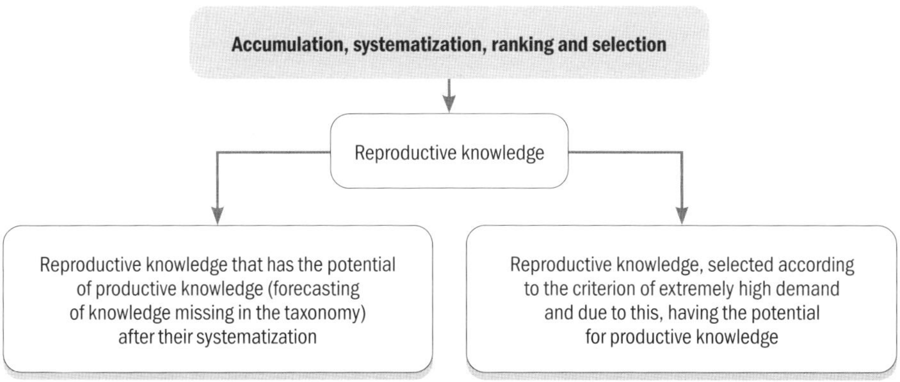

It is important to note that the so-called contemplative-explanatory type of presentation still prevails in the literature on creativity, when the information about the features of scientific and technical creativity is presented mainly in the stories of great creatures generating, which, of course, is useful, but does little to develop one's own creative potential and increasing the efficiency of creative activity. That is why the ability to identify and formulate the productive knowledge independently (as well as the methods of their use — to get, in total, the heuristics) from TS, the objects of animate and inanimate nature and the reproductive knowledge not only allows the deeper understanding of TS essence, the surrounding world, but also to increase own creativity significantly (for many, many times).

Let me remind you that the famous Russian scientist in the field of psychology of creative activity V.A. Molyako[4] called heuristics "mental karate", because heuristics act in creativity as techniques (and most importantly — ideology) of karate. Everyone knows that a person who masters karate will always defeat another, even physically stronger opponent who does not master karate. This explains the title of this book.

Table 5. Recommended list of heuristics[5]

The name of heuristic	Purpose or scope of use
The basal heuristic principle	Elimination or leveling of harmful consequences of narrow specialization of scientific and technical creativity subjects and ensuring a sharp increase in the efficiency of the convergence of knowledge mechanism
The method of using the most human needs	Formation of new urgent needs during TS development , the implementation of which is likely to lead to significant improvement in TS consumer properties
Stages, strategies and basic design tactics	Formation of the rational sequence of designer's actions (innovator) for various design purposes and in various situations: in the presence of TS analogs' multitude, which ensures the choice of an analogue (prototype) and the realization of a need (with possible parametric optimization); in the presence of TS analogs that require improvement that goes beyond parametric optimization; in the absence of TS analogues (prototype), which necessitates pioneer TS creation
The model of scientific and technical creativity using heuristics	Imaginative representation of creativity subject actions with the use of heuristics, contributing to the deeper understanding of the methodological foundations of effective creative activity
The reproductive knowledge of human needs, functions (including "generalizing" functions) and properties of technical systems (TS) and methods of their use, accumulated and systematized in the form of thesauri	Search and generation of new human needs, TS functions and properties with the creation of pioneer TS, TS improvement and creation of new TS generations
Elementary operations of R. Koller and the method of their use	Formation of TS functional structures at a high level of abstraction, preventing the emergence of psychological barriers to thinking, including the most harmful – the barrier to terms
The method of synthesis of physical, chemical, biological, economic principles of action (existence) TS	Search and synthesis of operating principles automation (existence) TS
The criteria for progressive development of TS and methods of their use	Determination of the most probable tendencies of changes in the properties parameters of TS being developed, leading to an increase in TS competitiveness
The concept of intensified contradiction, the methodology for the formation of intensified contradictions and standards for resolving contradictions	Elimination or leveling of psychological barriers to thinking when developing TS; intensification of productive ideas generation for the creation of new competitive TS
The heuristic techniques, methods of their identification and use	The use of effective technical solutions implemented in TS, formulated in the form of methodological advice, when creating new competitive TS
Analogs and analogies of TS in the technosphere, in the objects of animate and inanimate nature, the method of inference by analogy	Convergence of knowledge of analogues and analogies in the development of new competitive TS

The name of heuristic	Purpose or scope of use
Taxonomy of ready-made, modified and transformed resources hidden in TS and the environment, methods of modification, transformation and use of resources	Use of ready-made, modified and transformed resources hidden in TS and the environment to reduce the cost of operating existing TS or to develop new competitive TS
The laws and patterns of TS development and methods of their use	Determination of the most promising areas for new competitive TS development or new TS generations creation
Unusualness criteria for the properties of materials and methods for their use	Search and / or creation of materials with unusual properties for new competitive TS development
The concept of "ideal final result" and the method of its formation	Development of new competitive technical decisions close to ideal
"Good" ("strong" or "controlling") and "bad" verbs, their taxonomy and recommendations for use	Using "good" verbs in the development of competitive TS to form new useful functions, and "bad" verbs to identify possible harmful functions
"Good" and "bad" adjectives, including "smart" and "clean", their taxonomy and recommendations for use	Formation of new competitive TS of new "good" TS properties during their development that must be implemented, or the identification of "bad" TS properties, the appearance of which must be prevented, as well as ideas generation for the creation of new classes of competitive TS based on the adjectives "smart" and "clean"
Functional-physical (chemical, biological) analysis and synthesis of TS	Identification, systematization and use of productive knowledge for the synthesis of new functions, properties, functional structures, principles of operation and technical decisions of TS
Functional cost analysis and synthesis of TS	Identification of opportunities to reduce the cost of goods production and provision of services due to revaluation of the utility and charge for utility of the functions and properties, implemented in TS
Morphological analysis and synthesis of TS	Formation of a multitude of new functions, properties, functional structures, operating principles, technical decisions of TS
The Method of Focal Object	Overcoming existing psychological barriers to thinking or preventing the emergence of such barriers in innovation activity
Constructive mathematics, or STC operator	Formation of potential feasibility abstraction in thought experiments for generation of technical decisions, when they are distracted from the practical limitations of design, technological capabilities in size, time, cost, weight, material and others, which helps to eliminate psychological barriers to thinking
Metaphors and methods for their formation and use	Transfer of one object properties to another according to the principle of their similarity in any ratio or by contrast, which contributes to the deeper understanding of TS essence, phenomenon and can serve as an impetus for the emergence of many interrelated productive ideas in innovation activity
"The ascent-descent " method	Formation of a variety of opportunities for using the results of innovative activity for their commercialization and / or other use, further development

2.3.3 Creativity barriers

Methodologists of creativity have the widespread concept of the barrier to creativity as a psychologically conditioned obstacle or motive that reduces the effectiveness of creative activity at different stages or even prevents its occurrence. The consequence of such barriers is unproductive thinking in a creative sense, which is called "stereotyped", "stereotypical", "rigid" (difficulties in perceptual restructuring and representations in a changed situation), "vertical" (straightforward logical thinking), "reproductive", etc. A plenty of obstacles also arise in collective creative activity, mainly due to *communication barriers*, therefore it is useful to conduct periodically an easy introspection (or collective analysis) of the psychological state of creative activity subjects in order to identify barriers and overcome them, as well as prevent their occurrence. No special knowledge is required to prevent or overcome most of these barriers.

The most common barriers types of creativity (some are groups of barriers) and their characteristics are given in Table 6.

Oddly enough, but the most harmful (in my opinion) barrier to creativity — the barrier to terms — has been paid attention not by professional psychologists, but by methodologists of scientific and technical creativity.

The barrier to terms arises immediately in the object problem formulation (for example, "to develop a filter"). At the same time, firstly, because of the term "filter", the traditional image (conceptual model) of a filter appears immediately in the brain as a kind of obstacle that lets something through, but detains something (Fig. 43). Secondly, a traditional conceptual filter model connects involuntarily the subject of creativity with corresponding analogs, which should be forgotten for a while, since, on the one hand, the technical decisions implemented in these analogs do not provide the necessary characteristics (therefore, the task was set to "develop a filter"), on the other hand, psychologically these analogs impede new technical decisions generation based on the new principles of action and / or functional structures.

It is unlikely that the subject of creativity, bound by the barrier to terms, will be able to come up with an electrostatic filter, which does not have a barrier in its traditional form (Fig. 44), but it is the electrostatic filter that retains fine dust better than any filters made of cotton wool, material, etc.

Table 6. Types of barriers to creativity

Types of barriers	Barrier characteristic
The cognitive and psychological barrier	This barrier is believed to have two main functions. The purpose of the first one is to slow down the movement of creative thought (in order to exhaust completely the achieved stage of development), and the purpose of the second one is to stimulate the process of transition to a higher stage of development
The value-orientation barrier	Difficulties in motivating and determining the meaning of creative activity
The uncertainty barrier	Associated with a lack of information about the necessary results of creative activity
The functional thinking barrier	Causes difficulties in creative activity associated with stereotyped thinking, the use of ineffective mechanisms for the knowledge convergence
The semantic barrier	It occurs during the interaction of creative activity subjects, when they use different semantics, that is, different semantic meanings of the same words
The character accentuation barrier	Exaggerated development of certain character traits to the detriment of others, as a result of which interaction within the framework of collective creativity worsens
The barrier to negative emotions (fear, anger, resentment, irritation, apathy, depression, suffering, grief, shame and guilt, contempt, disgust, squeamishness, etc.)	It is associated with the emotional and mental state of an individual, prejudices, psychophysiological characteristics, repulsive habits, etc.
Intellectual snobbery	Excessive self-confidence, based on superficial abundance of knowledge, which "does not teach the mind" (Heraclitus), the identification of knowledge and understanding
The barrier to a wrong mindset	This is a lack of readiness to act in accordance with a communication situation. A wrong mindset arises due to stereotypes of thinking, bias, rigidity, conformism, lack of attention and interest, neglect of facts
The barrier to terms	Occurs in the object (instead of functional) problem formulation

Figure 43. The simplest filter: 1 — housing; 2 — filter baffle plate; 3 — layer of captured particles

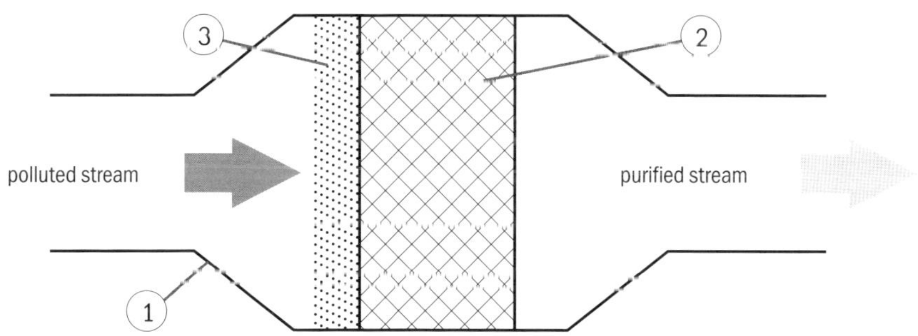

The ability to overcome psychological barriers to creativity effectively and go beyond stereotypes is formed on the basis of high qualities and competencies of a creative person.

"They say that one can comprehend everything with just one mind. Do not believe it! One thread is the mind, the other is the feeling, and they always touch each other in creativity."

V.I. Vernadsky

Figure 44. Electrostatic aerosol filter

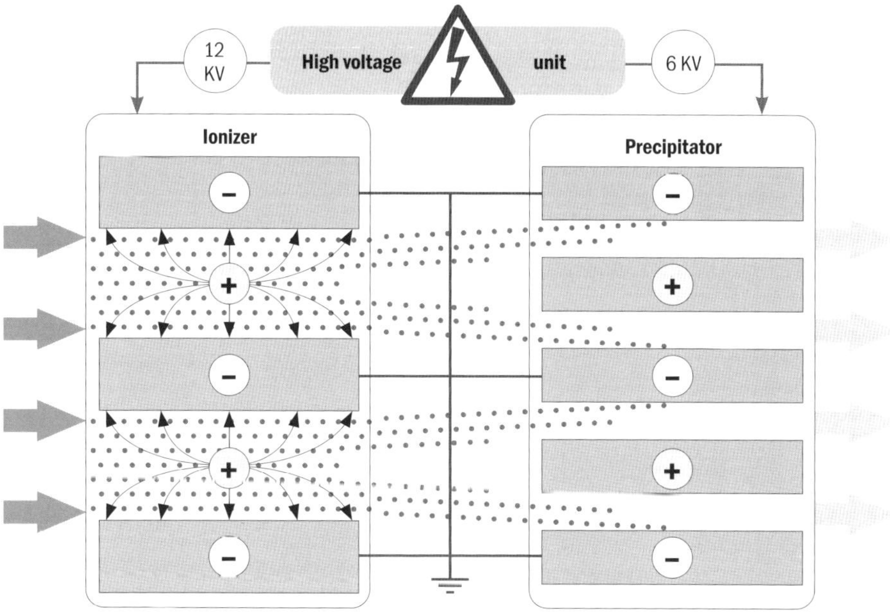

2.3.4 Qualities of a creative person

The famous American psychologist Alice Paul Torrance believed that, in general, creativity was generated by the strong need for a person to relieve tension that arose in a situation of uncertainty or incompleteness. The Soviet (Russian) academician V.A. Engelhardt was agreeable to that, he believed that the innate instinctive force of creativity was the desire to reduce the degree of ignorance about the world around us.

In the presence of TS analogue, scientific and technical creativity requires mental detachment from the imperfect analogue, which, in essence, has become a stereotype, which most often turns into a barrier on the way to creating a new one, to the transition to a creative level of problem understanding.

The ability to overcome such barriers and go beyond stereotypes is essential qualities of a creative person.

Many researchers, having studied such characteristics of lots of geniuses, have come to almost the same results. In detailed and supplemented form, they are presented in Table 7[6]

The logic of common sense shows that all the positive person's qualities in general can be attributed to the qualities of a creative person. I have counted more than 200[7] of these qualities.

Many creative individuals are inconvenient in communication, everyday life and are often lonely due to creative work — often stressful, unregulated and requiring distraction from everyday problems and everyday trifles. "Loneliness is the fate of all outstanding minds" — this is how Arthur Schopenhauer, a German philosopher, summed up his observations. His compatriot Jean Paul, a writer, agrees with him: "Solitude is the attribute of the strong. The weak is always found in the crowd." — and many others.

Table 7. List of qualities and competencies that ensured the successful creative activity of geniuses

Geniuses	Qualities and competencies		
Leonardo da Vinci, Rembrandt, Galileo, Newton, Hegel, Napoleon, Darwin, Einstein and others	Psychological readiness for creativity		
	Orientation to the new	Broad outlook	
	Attitude to transform the existing world		
	Openness to the perception of new information		
	The ability to wonder	Keenness	
	Readiness for continuous adjustment and revision of own positions		
	Readiness to overcome psychological barriers to creativity		
	Ability to work in a situation of uncertainty		
	Gnosiological readiness (associated with the process of cognition)		
	Flexible and fluid thinking	Thinking imagery	
	Understanding the relativity of available information		
	Ability to analyze, summarize and integrate information		
	Aspiration to realize "one's own" approach "to the end"		
	Experience in solving various types of creative problems		
	Absorption in the process of problem solving		
	High interest in the result	High motivation for creativity	
	Orientation to complex problems	Ability to work in a team	
	Methodological and methodical preparation		
	Awareness of the social significance of creative activity		
	Awareness of the need for creativity		
	Ability to criticize and perceive criticism correctly		
	The need for self-expression	Altruism	Optimism
	Ability to organize your own creative activity and the activity of a creative team		
	Respect for another point of view	Benevolence	

However, it is precisely because of their keenness to creativity that they do not suffer much from loneliness, everyday inconveniences. All this, of course, does not mean that a harmonious family and a well-established life reduce the effectiveness of creativity.

Creative skills are inherent in everyone, but most people lose them under the influence of the environment. However, they can be developed through the formation of special competencies and motivations for creativity.

2.3.5 Motivation for creativity

Motivation depends both on the object to which the action is directed and on the need satisfied as a result of its commission. Many experts believe[8] that the motivation of creative activity is a system of motives and value orientations that initiate creativity, goals and intentions aimed at their implementation, as well as ways of responding to difficulties and failures that arise in the process of creative activity.

In general, all types of motivation are divided into groups: external, internal, positive, negative. For convenience, many recognized types of motivation, due to various reasons, activities, the implementation of which can be well creative, are summarized in Table 8.

Human emotions can enhance the motivation for creativity: both, the lower ones, based on instincts and being their expression (for example, the emotions of hunger, thirst, fear), and the higher ones — feelings (for example, duty, love, companionship, curiosity). The most stable feelings are personality traits (for example, honesty, humanity)[9].

The prominent Russian scientist V. I. Vernadsky believed that emotions play an important role in the creative process. He wrote: "They say that one can comprehend everything just with one mind. Do not believe it! One thread is the mind, the other is the feeling, and they always touch each other in creativity"[10]. However, scientists have experimentally found[11] that there is a certain optimal level of motivation at which any activity is performed best. In other words, very high motivation can sometimes impair the effectiveness of creative activity[12]. "Burnt out" — so they say about such conditions. In general, needs, motivations, motives, emotions, feelings lead to a phenomenon that makes our thought return to the same topic again and again. This phenomenon, which determines the mechanisms of active behavior of a person, his spirituality, the desire for knowledge, creativity, overcoming difficulties, is called the *dominant* in physiology. The dominant phenomenon was discovered by A. A. Ukhtomsky, one of the outstanding thinkers of the twentieth century, and formulated by him as "a fairly persistent excitement flowing in the centers (of the brain) at the moment, which had the significance of the dominant factor in the work of other centers: accumulating excitement from distant sources, but inhibiting the ability of other centers to respond to impulses directly related to them"[13]. This phenomenon is rightly called "the dominant of Ukhtomsky". On the basis of the dominant focus of excitement, a specific adaptive activity is formed, focused on achieving useful results.

For the formation of the positive dominant of scientific and technical creativity and the inhibition of negative dominant foci in the human brain, it is necessary to follow consciously the needs, motivations, motives, emotions, feelings, ultimately aimed at the effective achievement of creative results.

Table 8. Types of motivation for creativity

Name of the motivation type	Motivation characteristic	Specification of motivation characteristic or examples of its manifestation
Extrinsic motivation	*Extrinsic* motivation, not related to the content of a particular activity, but associated with external circumstances with respect to the subject[14]	Seeking for material gain, ensuring own position[15]. With the dominance of external motivation, less educability, more impatience, preference for simplified tasks, less creativity are observed[16]
Intrinsic motivation	*Intrinsic* motivation, which is mainly associated with the content of creative activity	The desire for self-expression, accompanied by the pleasure of the creative process itself and aesthetic satisfaction. Intrinsic motivation is very conducive to creativity
Positive motivation	Motivation based on positive incentives[17]	"If I finish the year without C marks, then I'll get a tablet"
Negative motivation	Motivation based on negative incentives[18]	"If I am disciplined at work, I will not be deprived of the premium"
Affiliation motivation	Motivation based on joining individuals, groups[19]	Aspiration to establish or maintain relationships with other people. Sometimes the strength of affiliation is so high that a person goes to work in order to communicate[20]
Power motivation	The totality of person's aspirations to gain influence on individuals, groups of people and the development of a situation with the help of power means	To realize such aspirations, for example, coercion, privilege, positive and negative reinforcement in the form of approval and punishment are used
Achievement motivation	Associated with the need to succeed and avoid failure	To determine the level of motivation, experts identify four main factors: the significance of success; hope for success; subjectively assessed probability of success; subjective standards of achievement. Each person has a dominant tendency to be guided either by the motive of achievement or by the motive of avoiding failure[21]

Name of the motivation type	Motivation characteristic	Specification of motivation characteristic or examples of its manifestation
Motivation of identification with another person	One person's desire to be like another	Identification with another person (hero, idol, authoritative creative personality) leads to an increase in the energy potential of the individual due to the symbolic "borrowing" of the energy of the hero, idol, which increases the power of motivation. However, it should be borne in mind that sometimes, with a lack of upbringing, a person can take an example from asocial personalities, even creative ones, but with false needs[22]
Self-development motivation	The totality of person's aspirations, independently or in cooperation, to develop their qualities or to form new ones	Self-development can be physical, mental, emotional and spiritual. Some experts believe that emotional development of a person has an even greater impact on success in life than the development of intelligence. For this, they form "smart" emotions, correct shortcomings in the development of the intellectual sphere. Self-development motivation can be blocked by the desire for safety and self-preservation
Self-affirmation motivation	It is characterized by an orientation towards proving own worth to others (society) and to oneself	The desire for self-affirmation is often equated with the motivation of prestige. In general, the motivation for self-affirmation, aimed at increasing one's status in society, leads to an increase in self-esteem, can be realized in scientific and technical or artistic creativity, communication activity, social career, aspiration for leadership and provides additional incentives for further self-development
Prosocial motivation	Characterized by an orientation towards actions related to understanding the social significance of activities, the sense of duty and responsibility to people	To a large extent, pro-social motivation has given rise to a powerful volunteer movement in various fields. This motivation can also be manifested in scientific and technical creativity, for example, aimed at creating technologies, equipment and products that facilitate or improve the life of people with disabilities
Procedural and content motivation	The desire for any activity caused by the very content of this activity (close to intrinsic motivation)[23]	If a person likes to do something, he does it. In a series of studies tracing the development of people from primary school to adulthood, E.P. Torrance found that people who did what they loved to do were more creative. However, the long absence of positive results, for example, in scientific and technical creativity (even if the process of creativity itself is pleasant), can restrain the desire for creativity and even turn away from it. The use of technologies and means that ensure the intensification of creative activity makes it possible to multiply the likelihood of obtaining many positive results of scientific and technical creativity

The high level of positive dominant forms great additional human capabilities in all types of creative activity. The life and work of many outstanding people can serve as examples. According to legend, Homer, who created the Odyssey and the Iliad, was blind. Johann Sebastian Bach began to lose his sight at the age of 14 and by the end of his life became completely blind. Ludwig van Beethoven lost his hearing at the age of 28 and for almost 30 years, being completely deaf, wrote music, played and conducted. Leonardo da Vinci suffered from cerebral palsy. Due to a spinal cord tumor, Boris Kustodiev was bedridden for the last 15 years of his life; he painted his best pictures lying down.

The high level of the positive dominant of scientific and technical creativity often led to a sharp change in the lives of many people who achieved outstanding results in that ("second") life. For example, Antoine Lavoisier (18th century), the founder of modern chemistry, served as a tax collector; at the age of 16 Alexander Graham Bell (early 20th century), the inventor of the telephone, got an appointment of a music and oratory teacher; George Agricola, a German physician, laid down the foundations of mineralogy, mining and metallurgy in the 16th century; in declining years academician A. I. Berg, who achieved high recognition in the development of radio electronics, was actively engaged in cybernetics, which in the 50s of the last century was considered a pseudoscience in the USSR.

The dominant mechanism of the brain's work ensures effectively not only the formation of the most important natural need for new knowledge, but also the receipt of significant scientific and practical results in the learning process.

Such a mechanism is implemented in the creative pedagogy developed under my leadership, which, in particular, is based on the fact that the goal of teaching is not only mastering a certain educational material (subject), but also achieving a creative goal using heuristic strategies, tactics, methods, techniques — "Mental karate". Read more about creative pedagogy in Chapter 7.

Chapter 3
Scientific and technical (and other) creativity as the formation and implementation of designs

3.1 Basic design concepts

Human activity is reproductive (the reproduction of known results) and / or productive (the creation of a new one). Scientific and technical (and other) creativity, in essence, consists of the formation and implementation of projects. In the Modern Russian Explanatory Dictionary, a project (from *Lat.* Projectus literally "thrown forward") is defined as: 1) a set of documents (calculations, drawings, etc.) for the creation of any structure or product (I add from myself — technology, as well as others TS included in the TS variety; see Fig. 1); 2) the preliminary text of any document; 3) concept, plan[1]. In turn, the process of creating a project is called design (from the *English* design). However, if in Russia the concept of "design" is still very widespread only as a process or result of the artistic decoration of an object, then abroad, since the 70s of the last century, the concept of "design" has been widely used — as a "third culture", which has traditionally united non-overlapping scientific-technical and humanitarian-artistic cultures[2]. That is why "social", "political", "educational" and other "designs" have appeared, in which industry technologies are developed on the base of design activities methodology.

For example, in Russia, there are norms and rules (GOSTs) which are used to regulate the stage by stage development of a design (a set of documents) and carrying out of other related work.

The stages provide for the development of a technical proposal, a draft design, a technical design, working design documentation for the manufacture and testing of a prototype (pilot batch) and for series (mass) production[3].

It should be noted that in many cases (especially in everyday life) the design as a set of documents is not drawn up, although in accordance with the need for the human brain, a technical or another solution is consciously or unconsciously formed, which is implemented then.

3.2 Systematic methodology of design activities

Sometimes, in the designing, technical decisions are used without changes, implemented in TS analogs. Quite often, well-known technical decisions require "binding" to specific conditions, which is expressed in changing (optimizing) the parameters of properties in the design being developed. When there are no analogues of technical decisions (which is extremely rare) or when the preliminary analysis of known technical decisions shows that the desired design result will not be achieved only by changing the properties parameters of technical decisions, it is necessary to conduct additional research, and then form a new technical decision in a creative way in accordance with the phased implementation of human needs (see Section 1.4).

To rationalize design processes, taking into account the research of many specialists on the effectiveness of creative activity, the methodology for the hierarchical choice of design decisions developed by A. I. Polovinkin[4], the need to use metknowledge (megascience) to implement the basic principle of heuristics (see Section 2.3.1), I have developed the *systematic methodology for design activities*, which includes eight stages, three strategies and three basic tactics of design activities of high creativity level (higher than development, for example, calculation and cost estimate documentation, preparation of drawings, production and testing of prototypes, etc.)[5].

3.2.1 Stages of design activities and their effectiveness

The stages of design activities are shown in Fig. 45 in accordance with the systematic methodology. The first four stages are related to *external* design. These stages, in essence, are technical design tasks. Stages 5 through 8 are referred to as *internal* design.

The design (formation of a technical decision for TS being designed) proceeds from the abstract to the concrete (in Fig. 45 from top to bottom). However, the practice shows that the degree of contribution of each stage to the overall result (respectively, and the significance of the stages) increases from stage 8 to stage 1.

Indeed, it is possible, for example, in a car, without changing the engine, suspension, control system, etc., to optimize the technical decision of the body (for example, dimensions), which will improve its aerodynamic properties (stage 8). With the implementation of such a technical decision, the maximum vehicle speed can increase by an average of 10–30%, and fuel consumption can also decrease by an average of 10–30%.

Figure 45. Stages of design activities

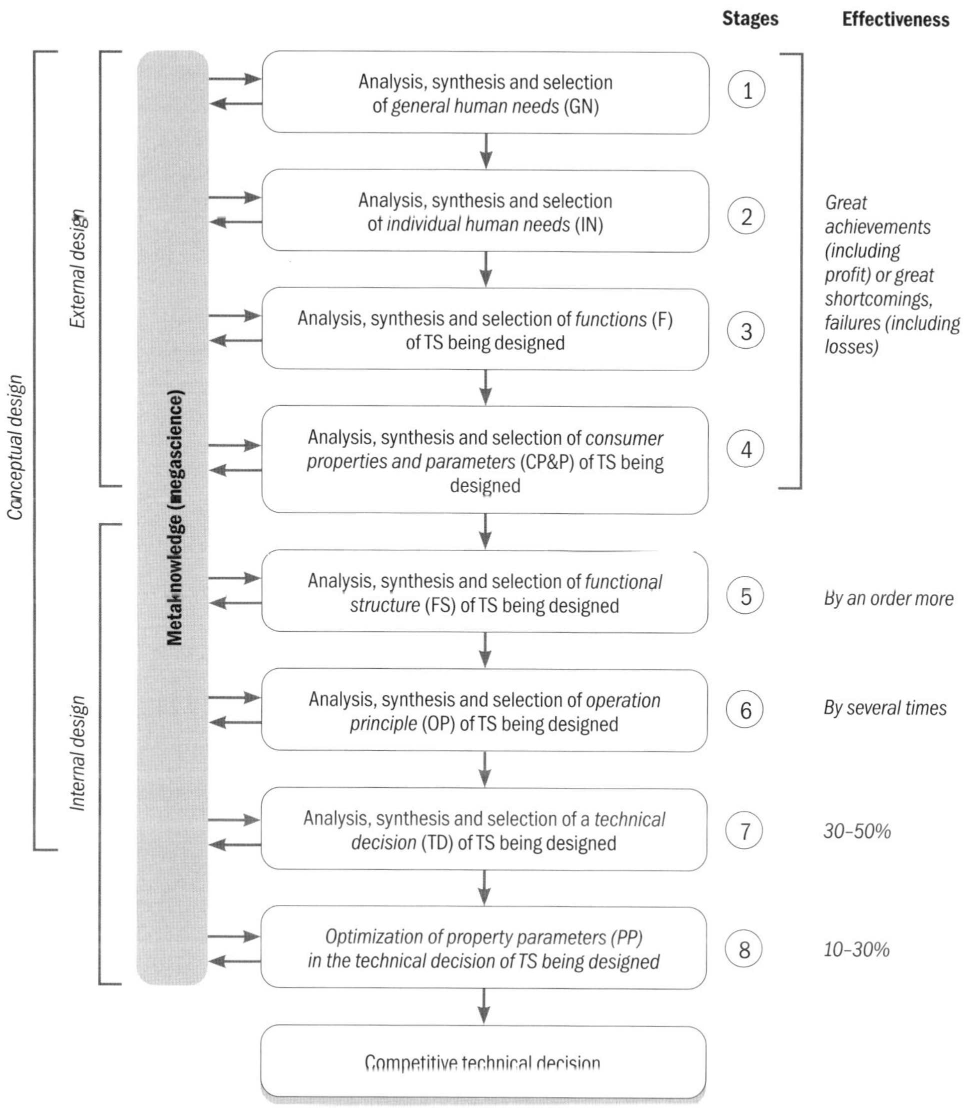

A significantly greater effect (on average 30–50%) can be obtained due to changes in the technical decisions of TS analogs (prototype) (stage 7). About 95% of inventions in the world are the results of design object transformations at this stage.

Several times greater effect, as a rule, can be obtained when implementing a new operating principle in TS based on the use of new physical, chemical,

Scientific and technical creativity is the formation and implementation of designs.

The systematic methodology of design activities includes eight stages, three strategies, three basic tactics.

The first seven stages are the most creative and relate to the stages of conceptual design, at which the qualitative characteristics of TS being designed are formed, which determine the success or failure of the design as a whole.

biological, geometric and other effects and phenomena and / or their combinations (stage 6)[6].

For example, the maximum speed of the aircraft with the propulsion system based on a jet engine is several times higher than that of the aircraft with the propulsor based on propeller piston engine. In other words, these are aircrafts with fundamentally different propulsion systems.

The increase in efficiency by an order of magnitude or more is possible with the change in the functional structure (FS) of TS being designed (stage 5). A striking example of this can be the comparison of two FS of underground sulfur mining: FS that existed before the beginning of the twentieth century (traditional mines, pits and development workings for breaking and delivering sulfur to the surface), and FS that is widespread now, which implements the technology of underground smelting of sulfur through wells (geotechnology with manless mining)[7].

However, the greatest achievements (including profits) or major shortcomings, failures (including losses) may occur during the implementation of the stages related to a technical proposal, otherwise — a task (from 1st to 4th). Indeed, the synthesis and / or selection of a new need (for analogs or prototype of TS being designed) can be the main idea of creating a competitive technology, equipment, and product. Recall that the classic of modern management Peter F. Drucker says: "It is the want that is unique and not the means to satisfy it"[8]. And vice versa, the choice of a false need for any implementation will, sooner or later, certainly lead to negative consequences.

Thus, at stages 1–7, the qualitative characteristics of TS being designed are formed, which determine the success or failure of the design as a whole. At different times those stages were called "engineering, or research, design", "exploratory (sometimes artistic) design", "initial stages of design", and in the last 15–20 years — *conceptual design*. This term is the most accurate as it emphasizes the high level of creativity. The results of conceptual design at each stage are conceptual models of technologies, equipment, products with varying degrees of abstractness in TS description. If necessary, at each design stage, design documentation is prepared, including justifications, calculations, drawings.

Thus, if we correlate conceptual design (CD) with the traditional (regulated) stages of design development, then CD includes the development of a technical proposal, draft design, technical design. The latter, in essence, is a technical decision (TD). Traditionally, the design subject in the formation of TD had previous design stages which were implicit and "blurred", and the design result (TD) embodied the results of mental analysis, synthesis and choice (see Fig. 45) of general and individual human needs, functions, consumer properties, functional structure, operation principle of TS being designed — and appeared in the mind immediately in the form of a certain TD appearance (for these reasons, it is rarely competitive). Splitting this unified process into design stages has allowed me to find (and develop) heuristics that include methods, technologies and means that can dramatically increase the efficiency of design in general and each of its stages in particular.

It is advisable to consider almost any design object, at least temporarily, as a technical system (TS), since the number of characteristics used in the design of TS is much greater than the number of characteristics used in the designs of any non-technical systems, and the methodological support for the design of TS has been developed more than for non-technical systems.

Then people consciously develop very creative opportunities (especially in non-technical areas, including art, architecture, music, literature and other areas) to generate new human needs, functions, properties, functional structures, principles of action, technical solutions aimed at achieving a new aesthetic reflection of the world.

For example, kinetic (dynamic) art and architecture arose on this basis. Characters, structural elements of paintings, sculptures, furniture, architectural structures can change quantitative parameters, properties, states (rest, movement), positions in space, and more of these objects under the influence of natural or artificial forces. An example of the implementation of new physical principles of action in music is color music, proposed at the end of the 19th century by the composer Alexander Nikolaevich Scriabin or the creation in 1919 by the acoustic physicist Lev Sergeyevich Termen of a fundamentally new instrument, the thereminvox, which is the forefather of electronic music.

One should consider performances in all genres of art (temporarily) as technologies, which include processes and operations. It contributes to the generation of competitive ideas in art.

Once again, we note that it is necessary to use all available knowledge (metaknowledge, megascience, see Fig. 45) at each design stage, rather than just knowledge from a narrow problem area.

3.2.2 Strategy of technical systems design

3.2.2.1 Overview of design strategies

Design strategies are long-term, non-detailed plans for the implementation of complex ideas of scientific and technical (and other) creativity. I have suggested three design strategies.

The essence of the first strategy is most vividly reflected in the ironic and lyrical performance of the Moscow State Academic Central Puppet Theater named after S. V. Obraztsov "Divine Comedy" based on the play by I. V. Stock (not to be confused with Dante Alighieri's Divine Comedy). In this performance, after the events when God creates the sky, earth, sea, plants, fish, birds, animals, reptiles and man in seven days, a voice behind the scene says: "And when God saw that everything he had created was good, he decided to come down to the earth to see if it was good."

Improving the efficiency of using the design results (developed TS) can be implemented at low cost by expanding the initial list of objects and situations in which the results can be used. For example, one has developed the mobile phone which generates high-frequency vibrations that repel mosquitoes in order to fulfill the important human need "to protect man from pain and discomfort." The generation of different frequency oscillations can be added at minimal cost (optimization of parameters, stage 8, see Fig. 45), and that will allow to scare away dogs using the phone. Or you can install this device in doors or windows, using the same operation principle (the generation of high-frequency oscillations) and even the same oscillator with the changed TS functional structure in this case. Then you can leave them open without anti-mosquito nets, since mosquitoes can not fly into the house. It will be shown in section 6.2.14 how to expand the list of objects of influence (use), functional structures, principles of action, functions, TS consumer properties, important human needs through the use of the "ascent-descent" method.

In practice, the second strategy is most often used — improvement of TS analogs or creation of new TS generations. For example, the creation of new TS generations — unmanned vehicles on the roads, on the water, in the air, in space (drones) — became an important human need not just for a long time, but forever. Moreover, these drones will gradually combine movement capabilities in all environments.

The third strategy is aimed at creating TS that has no analogues in the world. Such a strategy can be used, for example, to strengthen the defense of country's capability (sometimes it is enough to notify potential opponents that such TS are being developed to form a deterrent factor), its rapid economic development, the generation of new, important human needs, etc.

3.2.2.2 Strategy for increasing the results efficiency of technical systems' design

Once again, I draw the attention of readers to the fact that this is the lowest cost strategy, and one should start scientific and technical creativity exactly with it, that is, consider the question: is it possible to expand the initial list of objects for applying the results by changing the parameters of developed TS and / or its functional structure, to increase the excess of "benefit" over "charge-for-benefit" which is already perfect?

This strategy, exampled by the mobile phone (see the previous section), is shown in Fig. 46.

Figure 46. Changes at the design stages in the strategy for increasing the results efficiency of TS design: ◯ – the same (no changes); ◯ – new one (changes take place); ◯ ◯ – the same or new one

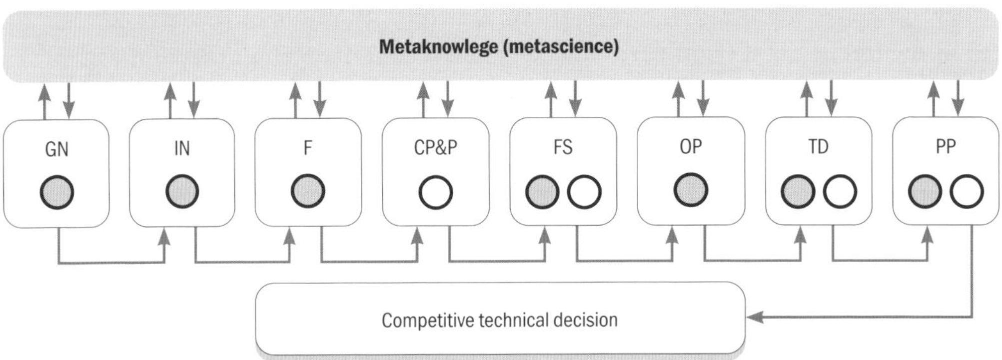

3.2.2.3 Strategy for improving analogs of technical systems or creating new generations of technical systems

Improvement of analogs can occur with changes at the design stages shown in Fig.47

Figure 47. Changes at the design stages in the strategy of improving TS analogs or creating new generations of TS: ◯ – the same (no changes); ◯ – new one (changes take place); ◯ ◯ – the same or new one

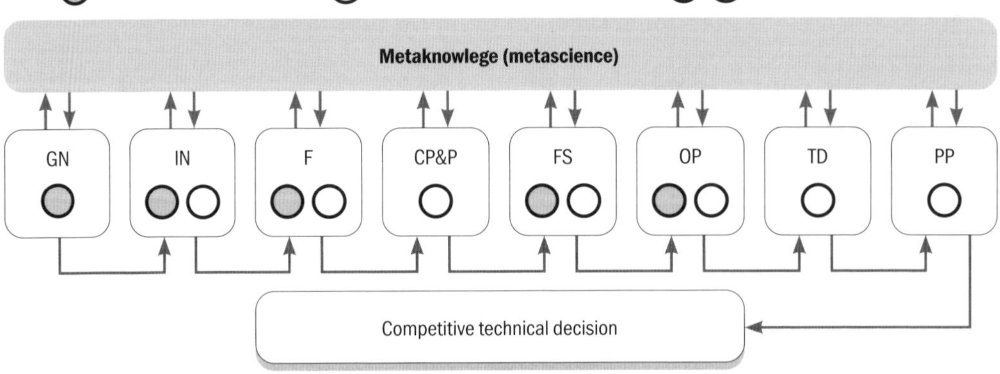

Developing the example with the mobile phone that repels mosquitoes, as well as the generator of sound vibrations installed on windows and doors, we can formulate the new individual need (IN) to create TS that scares away other insects, animals; the new individual need for people using another FS (for example, an oscillator clothing), other principles of action (for example, the generation of bright light flashes, loud sounds, directed emissions of liquids with repelling odors), etc.

3.2.2.4 Strategy for creating pioneering technical systems

When developing pioneer TS, all the results, obtained at each (from the 1st to the 8th) design stage, are new (Fig. 48).

Figure 48. Changes at the design stages in the strategy for creating pioneer TS:
○ – new one (changes take place)

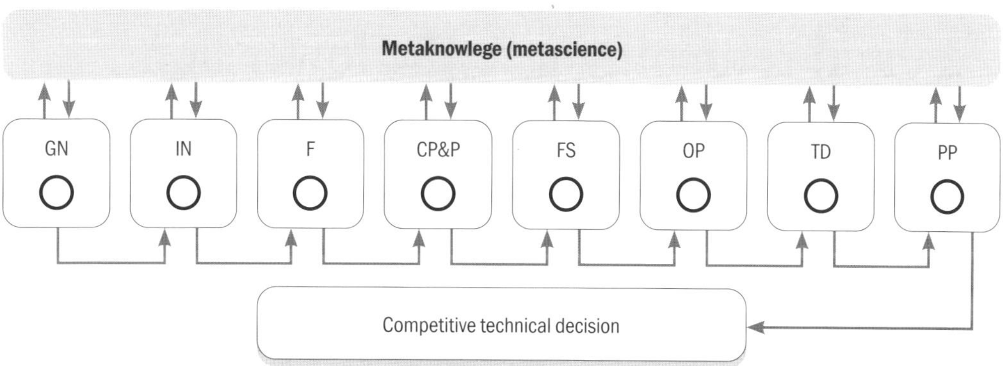

3.2.3 Basic design tactics

3.2.3.1 Direct analogy tactics

The important component of the systematic methodology of design activity is the basic tactics of design activity, which include the basic tactics of direct analogy, the basic tactics of iterative approximations, and the basic tactics of creating pioneer TS.

It should be noted that almost all of the heuristics shown in Table 5 can be attributed to design tactics. However, they should be used after choosing one of the basic (main) design tactics (hereinafter we will call them simply "design tactics"): direct analogy, iterative approximations, and the creation of pioneer TS.

The cheapest design tactics is the direct analogy tactics. Even a baby unknowingly uses it.

The tactics of iterative approximations resembles lifting an elevator and returning it to the lower floor, with the search, selection, and the synthesis of a solution at each stage.

The tactics of creating pioneer TS is based on the sequential passage of all stages and the adjustment of the technical task, if stage 6 does not give a result.

Direct analogy tactics is the most common and cheapest one (Fig. 49). After the formation of the technical task (sequential passage of the first four stages), a designer immediately turns to technical decisions (TD) in the known analogues (prototype) (stage 7), and if the quantitative parameters of TD in the found prototype meet the requirements of the technical task, then the design ends off. If it is necessary to optimize the parameters of the found solution, stage 8 is to be passed through.

Figure 49. Direct analogy tactics

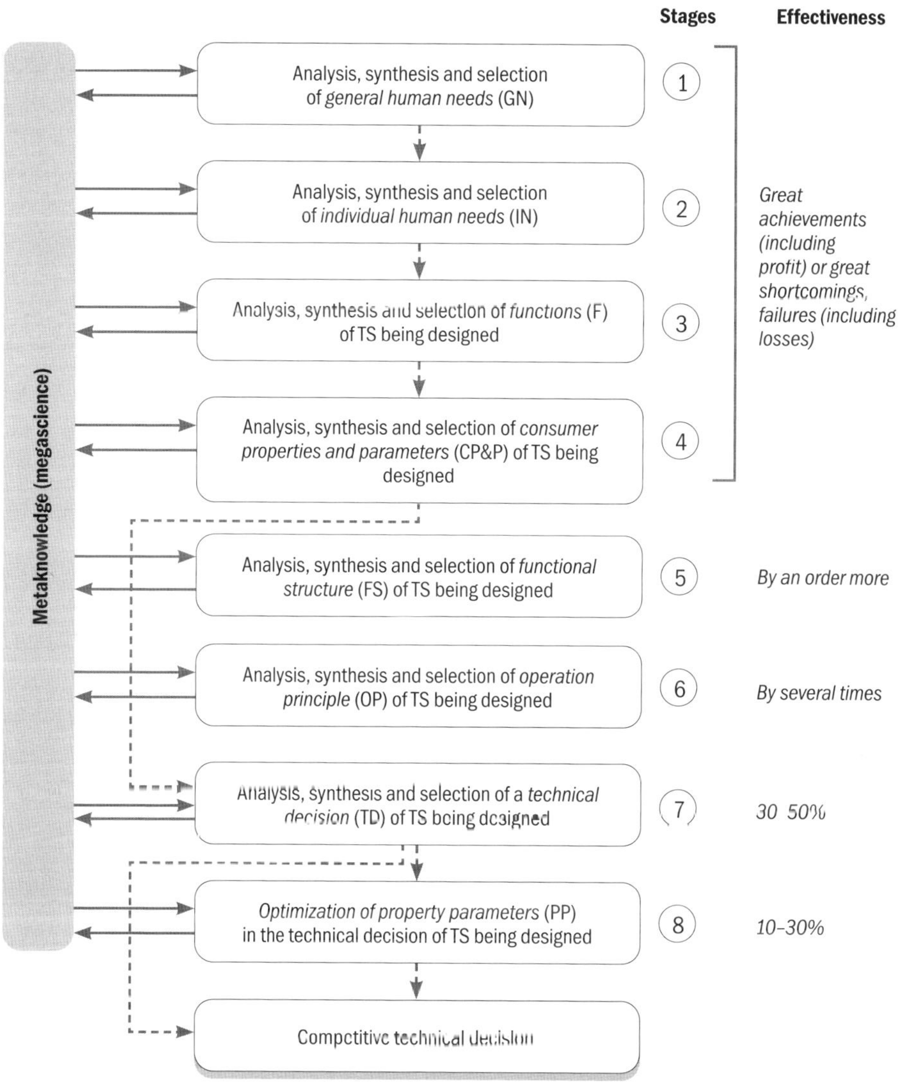

In fact, each of us, consciously or unconsciously, has repeatedly used the tactics of direct analogy in life. Even an infant, if accidentally waving his hand in front of his eyes, feels fear and instantly, unconsciously, due to genetic memory, closes his eyes and maybe even makes a defensive movement with his hand, in fact, acting according to the tactics of direct analogy. Unconsciously and instantly the child forms the first four stages (technical design task) and immediately (reflexively) implements a technical decision (closed eyes, raised hand), which protects from danger (by the way, animals, birds, fish, insects protect themselves from sudden extraneous movements in about the same way or somehow in their own way).

Let us consider in more detail another example from life, when the tactics of direct analogy is performed step by step (from top to bottom).

Stage 1. We have the general need for protection from the influence of the environment (from cold, sunlight), from the immodest looks of people, as well as the additional need to decorate our appearance.

Stage 2. We determine the individual needs that you can implement with the help of technical systems (technologies, equipment, products). For example, with the help of clothes, shoes, hats (in the context of this book, these ones are various TS). Let me remind you that some needs cannot be technically realized. For example, the need for respect can hardly be realized by means of expensive and fashionable clothes, because in the end ..."they are greeted according to their clothes, but seen off according to their minds".

Stage 3. We determine TS functions: for example, "regulate heat exchange" (a man with the external environment), "restrict the movement of air, water" (to the human body from the external environment and from the human body to the external environment). For (sports) footwear, this can be the function of "converting (mechanical) energy" of walking into mechanical energy that makes it easier for you to walk, run, for example, due to the elastic sole material properties.

Stage 4. We determine the properties of clothing, footwear that we need (for example, properties of fabric, fur, leather that provide the protection at a specific temperature range, as well as waterproofness, windproofness, elasticity, crease resistance, etc.). The important property of TS is its cost. In addition, in this case, the parameters (sizes of clothes, shoes) are important. Thus, you have actually formed a technical task (quite often in everyday life there are a lot of such tasks prepared — for many occasions).

According to the tactics of direct analogy, it is not necessary to develop the functional structure and operation principles of TS we need, but we should (bearing in mind the functional structures and operation principles of clothes and shoes known to us) go directly to stage 7 (see Fig. 49) and look for ready-made analogs (prototypes), in which technical decisions are implemented that correspond to your technical task (albeit to varying degrees). In everyday life, this means that you need to go to the store, to the market and look for ready-made clothes, shoes, and hats there.

In this case, two options are possible. The first option, if all the properties in the found analogs: model, material, color, size, price — are completely suitable for us, then you buy it (end of design). The second option is if, for example, the size does not suit you. Then you try to adjust this parameter: take in, stretch (if possible).

If, coming home with the purchase, you are disappointed or the optimization of the purchased items (taking in, unstitching, stretching) does not satisfy our needs, then you get rid of these products somehow and move on to the tactics of iterative approximations.

3.2.3.2 Iterative approximations tactics

The tactics resembles the movement of the elevator from the lower floor right to the upper one (in our case, this is the stage with higher efficiency) and return to the lower floor with the search, selection or synthesis of solutions on each floor (Fig. 50).

With regard to the above example of satisfying the need for protection from the environment and for hiding nudity with clothes, shoes, and a headdress, the tactics of iterative approximations should be used (in case of failure with the tactics of direct analogy) as follows.

In order not to confuse the sequence of actions in the tactics of iterative approximations with the stages of design activities, the elements of this sequence will be called steps.

Step 1. It is necessary to "rise" to stage 7 "Analysis, synthesis and selection of a technical decision for TS being designed ", look again in the store (other stores) and select other models of clothing, footwear, headwear (in general, conduct the search and selection of TS analogs (prototype) in other industries). After that, two options are possible: 1) if everything suits the price, color, size, then you buy (the design is over); 2) if the sizes do not fit, you buy and "go down" to stage 8 (see Fig. 50) — you take in, stretch. The designe ends off again. If these actions do not bring the desired result (do not satisfy our needs), you proceed to the next step.

Step 2. It is necessary to "rise" mentally to stage 6 and try to find in other industries and / or form new action principles of clothes, shoes, headwear, based on new physical (chemical, biological, geometric) effects and phenomena and / or their combinations ... For example, you can find and / or form principles of action

Figure 50. Iterative approximations tactics

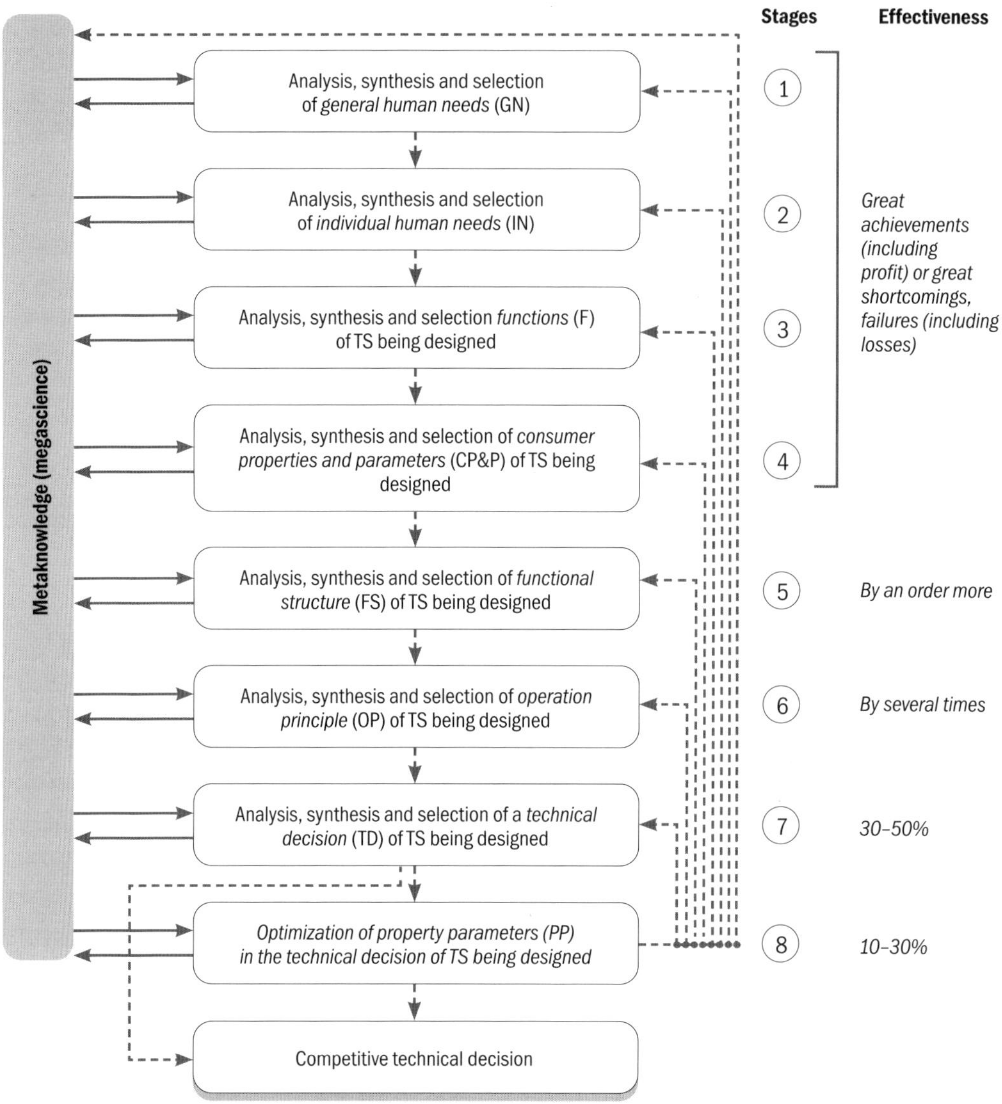

so that when the conditions of contact with a person and / or with the environment change, clothes change their color, temperature, be water-repellent, release protective substances on human skin (such types of clothes are already in sales), shoes would be equipped with a pedometer, and a hat would be equipped with a solar battery (to recharge the phone).

Step 3. After choosing (or forming) new action (existence) principles of clothes, shoes, headgear at stage 6, one should "go down" to stage 7 and choose new analogues (prototype) of technical decisions for clothes, shoes, headgear based on new principles of their operation (in general, select or develop new technical decisions based on new operation principles).

Step 4. You should analyze the technical decisions found and / or developed on new operation principles. Further, as in the tactics of direct analogy, two options are possible. If the new technical decisions satisfy you, then this is the end of the design. If not, then you need to go to stage 8 "Optimization of property parameters in the technical decision of TS being designed".

If the described actions have not satisfied your needs, proceed to the next step.

Step 5. It is necessary to "rise" mentally to stage 5 (see Fig. 50) and look for, choose or come up with a new functional structure of clothes, shoes, headgear. For example, instead of the dress — a suit of two or three items (jacket, vest, skirt or trousers), boots and a hat made of integral (removable) parts. On the basis of new functional structures of clothing, footwear, headgear, we should search, select or synthesize new principles of action for them (stage 6) and then, "going down" to stages 7 and 8, consistently come to the end of tactics.

If these actions have not satisfied your needs, proceed to the next step.

Step 6. It is necessary to "rise" mentally to stage 4 in order to change some properties, the implementation of which you have planned (for example, to abandon expensive items in clothes, shoes, headwear), and which can dramatically change, for example, the price of products. After that, on the basis of new properties, it is necessary to select or form a new functional structure (stage 5), on its basis — new action principles (stage 6), and on the basis of new action principles — new technical decisions (stage 7). If necessary, you can optimize the properties of clothing, footwear, headgear (take in, unstitch, stretch).

If these actions have not satisfied your needs, proceed to the next step.

Step 7. It is necessary to "rise" mentally to stage 3 and form the new list of functions of clothes, shoes, headgear. Some functions (and features) may have to be abandoned altogether, since they do not "fit" into the available budget. For example, give up the properties of clothes to change color, a pedometer on shoes and a solar battery on a headdress. After that, it is necessary to go through all stages consistently (from 4th to 7th) and try to find solutions for the implementation of new functions (to form new properties, functional structures, principles of action, technical decisions). Next, you should perform the actions provided for in step 4. If the actions specified in step 7 have not satisfied your needs, then proceed to the next step.

Step 8. We "rise" mentally to stages 2 and 1 and review our needs. Perhaps some will have to be abandoned, since plainly there is not enough money for everything. For example, you will have to limit yourself only to the need for shoes. Then we perform steps 1–7 again.

3.2.3.3 Creating pioneer technical systems tactics

If we know for sure that in the technical field there is no prototype (closest analogue) of TS, which implements the technical decisions we need, we will use the tactics of creating pioneer TS (Fig. 51). For example, we need an astronaut's clothing (spacesuit) for a spacewalk. When it was developed for the first time, they acted according to the tactics of creating pioneering technical systems — immediately sequentially through all stages from 1 to 7 (if necessary, 8). If the goals were achieved, then the design ended. Although, strictly speaking, in this case, the analogue was a diving suit.

This tactic's feature is that the need to return to the step-by-step adjustment of the technical task may arise already at stage 6 (analysis, synthesis and selection of the operation principles of TS being designed), when from the effects and phenomena known to science (in essence, determining the level of modern knowledge) it is not possible to form the principle of TS operation — the basis for the development of the technical decision corresponding to the terms of reference (stages 1-4) for design. Then, as in the tactics of iterative approximations, it is necessary to go through all the steps, starting from step 5 (see Section 3.2.3.2.).

3.3 Design result assessment and selection

At each design stage (Fig. 51), several, and sometimes (using special methods and automated means), hundreds of options for needs, functions, properties, functional structures, principles of operation, and technical decisions of TS being designed are sequentially formed. These options need to be assessed, ranked and the best ones should be chosen which will subsequently be implemented. For this, methods are used for quantitative assessments of new objects' quality, processes, labor products, developed within the framework of the scientific discipline — qualimetry.

For example, the variants of design results at each stage can be ranked according to the maximum or minimum value of the property (criterion) or according to the integral (complex) criterion that summarizes the values of several properties. Moreover, quantitative values can be assigned to any property by experts (for example, beauty or other aesthetic characteristics of a technical decision can be evaluated in points).

The choice of the best option can be influenced by political, social, economic and other situations, depending on which, expertly, higher weight values for individual properties can be established. For example, in the case of increased danger, the best (highest value) may be the technical decision that quickly provides the highest degree of protection, regardless of the cost of its implementation.

Many systems of computer support for decision-making (assessment and selection of the best variant of the design result) have been developed.

Figure 51. Creating pioneer technical systems tactics

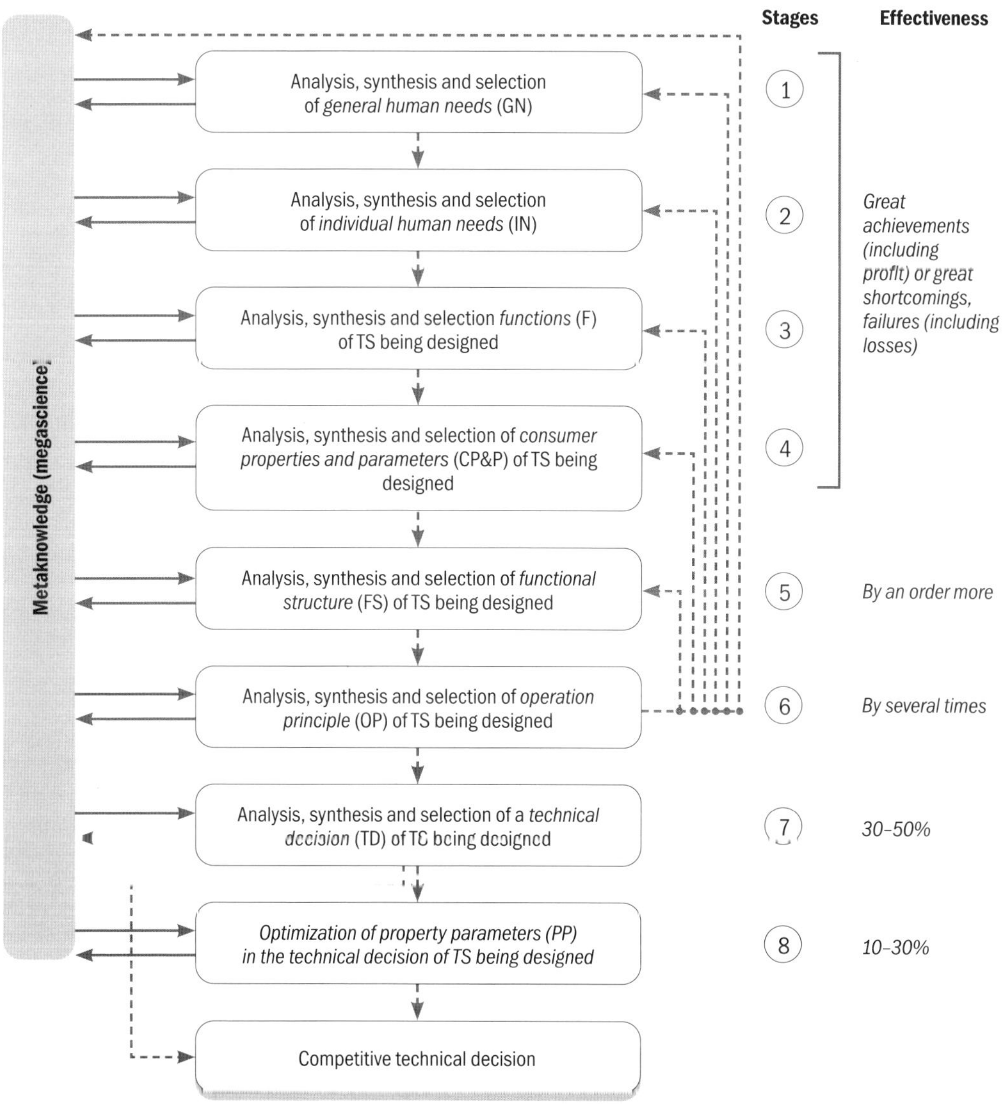

In some cases (especially when solving social problems), the so-called "Descartes square" can help to comprehend the problem, select the solution and calculate the consequences of its adoption, which in essence represents the grouped answers to four simple questions:

1) What will happen if this happens?
2) What will happen if this does not happen?
3) What will not happen if this happens?
4) What will not happen if it does not happen?

Chapter 4
Contradictions in fulfillment of human needs

4.1 Basic concepts related to contradictions and types of contradictions

Fulfillment of human needs is associated with the resolution of contradictions existing in all spheres of life, animate and inanimate nature, society, cognition.

According to the Great Soviet Encyclopedia, *dialectical contradiction* is the interaction of opposite, mutually exclusive sides and tendencies, objects and phenomena, which at the same time are in internal unity and interpenetration, acting as the source of self-movement and development of the objective world and knowledge[1].

Formal opposites are such properties of objects (phenomena, processes) that occupy limiting, extreme places (day — night, good — evil, top — bottom, right — left, dry — wet, hot — cold, truth — delusion, analysis — synthesis, life — death).

The unity of opposites forms a contradiction, and the struggle of opposites leads to the resolution of contradictions, which is the source of the development of all spheres of life, living and inanimate nature, society, knowledge (for example, universalization — specialization). At the same time, the meaning of opposites can be different for development: one of the parties can play a progressive role, the other can inhibit development; one is to be a leading force, the second is a follower; both opposites can perform a creative function, but they do it in different ways. The process of resolving a contradiction is a leap in the development of an object, a phenomenon, the result of which is a new quality. The main forms of contradictions resolution are[2]:

- Preservation of opposites on the basis of their transition into each other and, accordingly, into more developed forms[3]. The transition from the sailing vessel that uses the force of the wind for movement to new generation TS — the ship with the engine that provides movement in wind and calm would be an example. However, in recent decades, very large ships have been actively built, which, along with the engine, have the opportunity, with a tailwind, using the latest advances in mechanization

Fulfillment of human needs is associated with the resolution of contradictions existing in all spheres of life, in society, knowledge, in living and inanimate nature.

In the technical sphere, such a contradiction is most common: with the improvement of TS, some of its properties improve, while others deteriorate.

and automation, to deploy sails, but, of course, not made of canvas anymore (Fig. 52);

— "Victory" of one of the opposites and "death" of the other (for example, the predisposition to left-handedness can defeat right-handedness even the forcibly implanted one);

— The disappearance (death) of both opposites during a radical transformation of the system (everyone should be aware that the resolution of contradictions with the use of nuclear weapons can lead to irreversible consequences and the death of civilization)[4].

Figure 52. Modern sailing yacht Maltese Falcon with two diesel engines and 15 straight sails (five on each mast), retractable inside the mast. 80m yacht was built in 2006

There are many classifications of contradictions types, but there is not a single one generally accepted, therefore, Table 9 presents examples from various classifications.

People, having no theoretical training, but proceeding solely from life experience and intuition, have formulated one of the most important examples of the unity and struggle of opposites (the improvement of some TS properties and the deterioration of others during its improvement simultaneously) in the form of the saying "There is no evil without good, there is no good without evil".

Taking this into account, the creation of competitive TS or its further development (if this task does not contradict the laws of nature and is potentially achievable), in essence, comes down to the fact that the positive effect of improving some TS properties significantly exceeds the negative effect with the deterioration of other properties.

Table 9. Examples of contradictions types

The name of the industry or area in which contradiction is manifested	Name and definition of contradiction	An example of a contradiction or comment
Social sphere	Social contradiction. The contradiction between the individual and society	For example, when the innovation offered by the individual collides with the conservatism of the society. At the extreme degree of contradiction aggravation, a conflict may arise
Chemistry	Chemical contradiction. Association and dissociation	In parallel with the decomposition of molecules into ions (dissociation), the process of combining ions (association) proceeds. Ions are atoms that have a positive (cations) or negative (anions) charge[5]
Physics	Physical contradiction. Mutual attraction and repulsion of molecules	With the decrease in the distance between molecules, the repulsion increases, with the increase in the distance, the attraction increases
Biology	Biological contradiction. Dissimilation and assimilation	Dissimilation and assimilation are mutually opposite processes that ensure, in unity, a continuous process of vital activity of organisms, which constitutes metabolism. The contradiction between dissimilation and assimilation determines the dynamic balance of a living body[6]
Pedagogy	Pedagogical contradiction. The contradiction between the amount of knowledge accumulated by mankind and the amount of knowledge adopted by the student	Although the amount of knowledge accumulated by humanity is enormous and no one will ever master it fully, each student needs to master the basic methods of cognition in order to assimilate knowledge in the chosen field effectively, independently as well
Living and inanimate nature	The contradiction between animate and inanimate nature. The process of synthesis and destruction of organic matter	The contradiction between animate and inanimate nature is the development source of the biosphere due to self-organization. However, man often has a great negative impact on this process, therefore, it is necessary to develop environmental education and upbringing of a man, society

The name of the industry or area in which contradiction is manifested	Name and definition of contradiction	An example of a contradiction or comment
Technical area	Technical contradiction: — between the properties of TS being designed and the laws of nature, which must be observed when implementing these properties;	The creation of a perpetual motion machine is impossible, since it contradicts the first and second laws of thermodynamics
	— between the properties of TS elements and processes, which must ensure implementation of properties of other elements and processes of this TS;	The energy reserve of the flywheel (inertial accumulator of kinetic energy) is proportional to the square of the angular velocity of its rotation. However, insufficient material strength can prevent the flywheel from spinning at extremely high speeds. Improving the protective properties of a tank by increasing armor thickness inevitably increases its weight, which (while maintaining engine power) inevitably leads to the loss of speed
	— between the properties of TS, which improve with its perfecting, and the properties of this TS, which simultaneously deteriorate	Improving the protective properties of a tank by increasing armor thickness inevitably increases its weight, which (while maintaining engine power) inevitably leads to the loss of speed

4.2 Typical and "physical" ("intensified") contradictions in technical systems

A variety of human needs and TS implies the presence of the large number of new technical contradictions and their emergence, which requires their effective resolution.

The outstanding Soviet (and then Russian) methodologist and organizer of the mass movement in scientific and technical creativity, author of The Theory of Inventive Problem Solving (TRIZ) G. S. Altshuller[7] (more about him and his theory is discussed in Chapter 8) analyzed more than 40 thousand patents for inventions in various industries and formed the compact list of significant typical contradictions (typical for most of the considered TS), and also formed the list of 40 typical heuristic techniques — proven by practice, rational creative actions leading to the effective resolution of contradictions. These actions are formulated in the form of methodological advice[8].

To exclude or neutralize the negative effect of improving TS (deterioration of some TS properties while improving others), G. S. Altshuller proposed an effective methodological technique — hyperbolization (excessive exaggeration or understatement of a contradiction), which I called an "intensified" contradiction.

However, for many developers, the awareness of the possibility of a negative effect psychologically interferes with finding a revolutionary solution when improving TS or at least an effective compromise.

The elimination or leveling of this drawback can be facilitated by the methodology for the formation and use of the new type of contradictions called physical proposed by G. S. Altshuller. These contradictions are extremely aggravated technical contradictions, which on the whole can be characterized as the implementation of the *imperative* "There is good without evil!" At the same time, with the improvement of some TS properties, other properties may practically not deteriorate.

Noting the high value of this approach, it is worth pointing out that the name of this contradiction type — physical contradiction — is not quite suitable, as it often leads to confusion. First, most of the "ordinary" technical contradictions in physical TS are physical; secondly, it is not clear what one should do with biological or chemical TS, where the contradictions, respectively, are biological and chemical.

In essence, physical (according to G. S. Altshuller) contradictions are hyperbolized (too exaggerated) or exaggerated contradictions. However both terms are discordant and difficult to pronounce, therefore, hereinafter, I propose to use the term "intensified" contradiction instead of the term "physical" contradiction.

The general approach to the formation of an intensified contradiction in the technical sphere is that the first part of the contradiction (human need, property or state of TS or its environment), reflecting the positive effect or possibility (condition) of the functioning or existence of TS or its environment, is formulated as well as in the usual technical contradiction, and the second part is formulated as contradictory, but without negative effect. There are several examples of the formation of intensified contradictions on the basis of "ordinary" technical contradictions. Among them, there is an example, that is a classic one, of the technical contradiction, when there is the need to move quickly along the roads, including overcoming intersections (this is the reflection of the desired positive effect), but at the same time traffic safety at intersections decreases (a possible negative effect). Then, on the basis of this technical contradiction, the intensified contradiction is formulated as follows: "Cars must move to transfer people and goods and cars must not move for the sake of complete safety."

Another example. To raise water level in a river or to create a reservoir, a spillway dam is built, which ensures the overflow and flow of water in the river after passing the reservoir (the simplest design of a spillway dam is shown in Fig. 53). However, with large flood waters, the dam may not be able to cope with the overflow of water, which leads to water spills in front of the dam and even floods. The intensified contradiction in this case: "Due to the flood, the inflow of water to the spillway dam increases, leading to flooding of the area in front of the dam, but there should not be the increased inflow of water to the dam in order for the dam to cope with the drainage of water."

Figure 53. The simplest spillway dam design

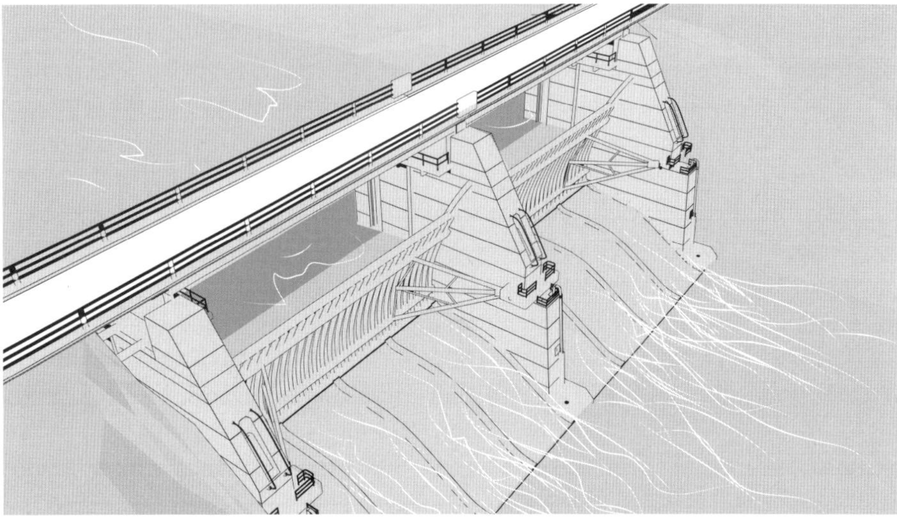

An example from the history of aviation. As soon as the aircraft with wings perpendicular to the fuselage reached supersonic speed, a jump in air pressure and density appeared at the leading edge of the wings, from which the wings collapsed (Fig. 54). The intensified contradiction in this case: "When the speed of sound is exceeded by the aircraft with wings perpendicular to the fuselage, there should be a jump in pressure and air density at the leading edge of the wings (due to the laws of physics), and there should not be such a jump in pressure and air density to keep the wings from collapsing".

Let me give you an example from mining. In the underground mining of solid ore (see Fig. 55), a technology is used where the ore is separated from the ore mass (2) by blasting fan-shaped boreholes (3), fragmenting the ore massif (2) into pieces (4) for further transportation. The problem is that if all the rings of the boreholes are blasted simultaneously (which is desirable for high productivity of ore breaking), only the explosion of the boreholes ring, the first from the exposed surface (5) of the ore massif, will separate the ore from the ore massif (2) and crush the ore.

At the same time, the simultaneous explosion of the remaining rings will not separate the ore and crush it. These holes will only be blown up by the explosion, practically without disturbing the ore massif around them. And if the

Figure 54. Increased force action on the leading edge of the wings perpendicular to the fuselage with a jump in pressure and air density at supersonic speed

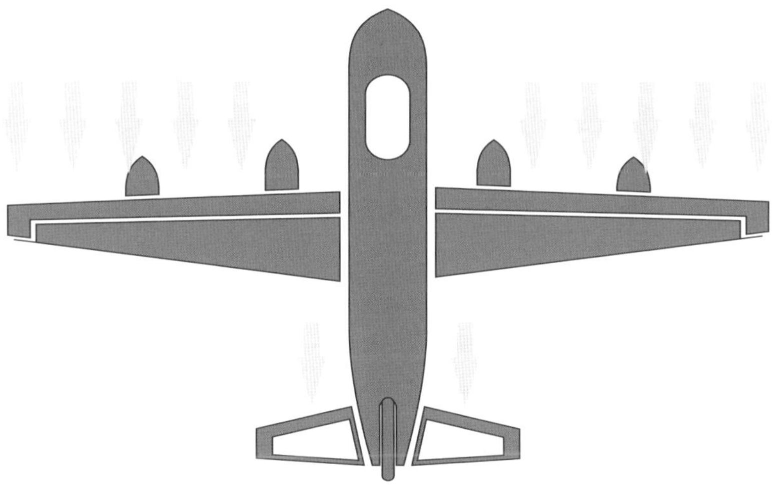

rings of the holes are blown up alternately from the exposed surface (5) of the ore massif, then the entire ore massif, limited by the rings of the holes, can be crushed. However, this process is unproductive, since it is required to charge the ring of boreholes closest to the exposed surface (5) of the ore massif with explosives, to detonate the explosives remotely, and then to ventilate the mined-out space (6) and the borehole (1) for a long time. Then all work must be repeated for each subsequent explosion of the boreholes ring. The intensified contradiction in this case: "To ensure high productivity of ore blasting, all rings of blast holes should be blasted simultaneously, however, due to the fact that ore can only be crushed by the explosion of the holes ring, the first from the exposed surface of the ore massif, all rings of holes should not be blasted simultaneously."

Or an example from everyday life. It is known that most people drink tea hot or very chilled. However, if you pour normally hot tea (about 60 °C) into an ordinary faceted glass (Fig. 56), the drink will quickly cool down and become simply warm, which, in the opinion of many, dramatically worsens its properties. And if you pour boiling tea into an ordinary glass (so that it does not cool down longer) and immediately start drinking it, you can burn yourself. The intensified contradiction in this case: "The tea in the glass must be very hot (so that it remains at a high temperature during tea drinking) and at the same time not very hot (so as not to burn you)."

Thus, in the general case, when an intensified contradiction is formed, mutually exclusive requirements are simultaneously applied to TS properties (or its

When an intensified contradiction is formulated, mutually exclusive requirements are simultaneously applied to TS properties or the environment surrounding TS: the object or phenomenon must be and not be, move and stand, be hard and soft, cold and hot, long and short, etc.

element, or process) or the environment surrounding TS: the object or phenomenon must be and not be, move and stand, be hard and soft, cold and hot, long and short, etc.

Figure 55. Technology of underground ore mining using ore breaking: 1 - drilling room; 2 — ore massif; 3 — rings of blast holes; 4 — crushed ore; 5 — exposed surface of the ore massif; 6 — goaf

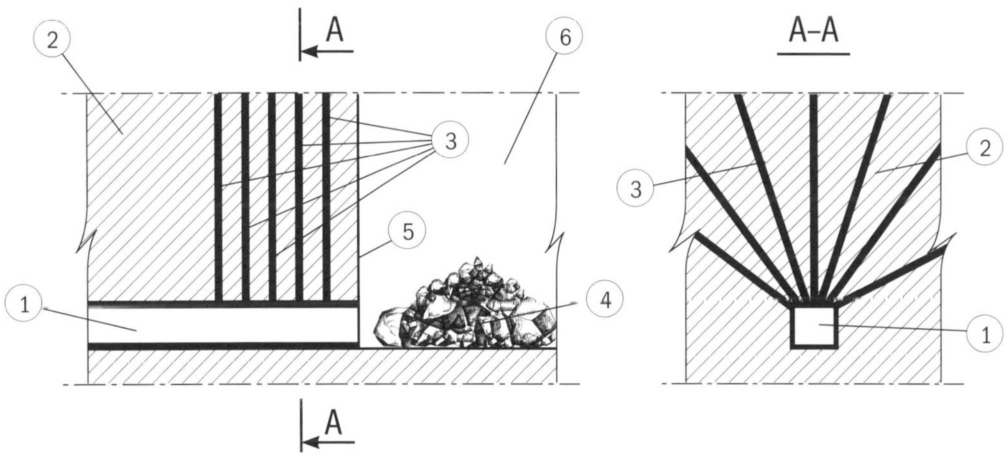

Figure 56. Faceted glass: a — ordinary; b — with a cup holder

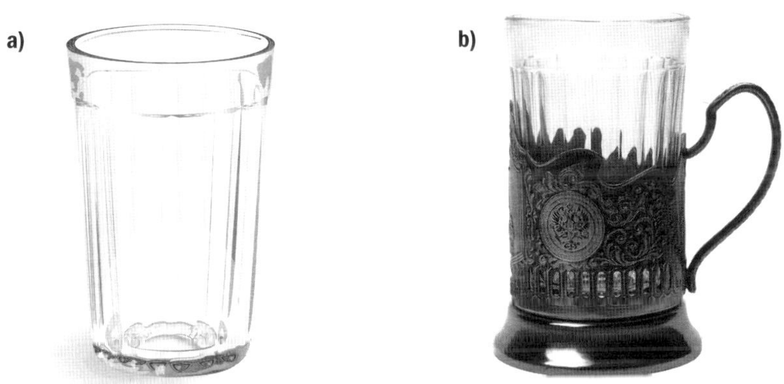

The formation of an intensified contradiction in TS or in the interaction of TS with the environment completely or significantly removes the developer's thinking barrier that prevents him from finding a revolutionary solution (or at least an effective compromise), and also prevents the emergence of such a barrier.

Methodological recommendations on the formulation of intensified contradictions in TS, standards for their resolution, as well as results of the application of these standards for the effective resolution of the above intensified contradictions are given in Section 6.3.3.

Chapter 5
General information on methods and means for resolving contradictions in technical systems

5.1 Traditional methods and means based on the use of reproductive knowledge

5.1.1 "Trial and error" method

Resolution of contradictions in TS is a creative process. The most widespread method of any creativity in the world, including scientific and technical, is the "trial and error" method. The method definition and explanation are given in the preface to this book.

The method of "trial and error" in the creativity of different levels of complexity is used by people of different ages, social status and education: a preschooler, a schoolboy, a housewife, a worker, a pensioner, an academician.

The "trial and error" method is one of the main elements of the "old" innovation culture and, in general, is characterized by the low efficiency and great unpredictability of the time to obtain the result.

The models of creativity using trial and error method and the alternative model are presented in section 5.2.3.

5.1.2 Intuition and insight

In the history of scientific and technical creativity, there are many examples of brilliant guesses, discoveries, ideas made by people of various categories, occupations and ages based on intuition.

In the Big Encyclopedic Dictionary "intuition" (*medieval Lat.* Intuitio — from intueor — I look intently) — comprehension of the truth by direct perception of it without substantiation with the help of proof, the subjective ability to go beyond the limits of experience by mental grasping ("enlightenment") or generalizations in the figurative form of unknown connections, patterns[1].

The most widespread creative method in the world is the "trial and error" method. It is ineffective and unpredictable in terms of the time it takes to get the result.

One of the most effective methods of convergence (transfer) of knowledge in creativity is the inference by analogy.

There are analogies:
functional, structural, substrate,
of external form, of relationships, etc.

English psychologist Guy Claxton[2] identified the following types of intuition (table 10).

Table 10. Types of intuition[3]

Type of intuition	Characteristics of intuition type
Expertise	Non-reflexive performance of complex highly skilled actions
Implicit learning	The way of acquiring knowledge through unconscious, non-conceptual means
Judgment	Making an accurate decision and categorizing when it is impossible to explain or prove it
Sensual	Increased conscious or unconscious attention to detail
Incubation	The process characterized by the phase of creativity - solving a problem, when the conscious, deliberate activity to find a solution is suspended and a person is engaged in affairs that are not related to the problem
Thinking	The process of "chewing" the experience over and over again in order to extract new knowledge from it

Strictly speaking, any human thinking is the unity of intuitive and analytical types of thinking. Moreover, it is believed that creative process is bipolar, its psychological mechanism consists of "distant", often opposite, polar elements: original + model; creation + *reflection*; involuntary + arbitrary; stimulation + motivation; impulsive + volitional; unconscious + conscious; congenital + acquired[4].

Extreme positions: purely intuitive — the psychobiological instinct of animals, which, for example, when they feel unwell, almost unmistakably choose plants and minerals for treatment; purely analytical (logical) — modern computer programs.

It is generally accepted that women are more intuitive than men.

Intuition is associated with insight (enlightenment), which is considered a desirable stage in solving any problem — the phenomenon of thinking, which is expressed in the sudden and quick, meaningful solution of a problem (or comprehension of the truth) on the basis of intuitive enlightenment, internal "discovery". Understanding of essential relationships during insight is not directly related to the past experience, learning of the subject and is the product of a sharp change in the flow of thought[5]. Insight is also called "aha-reaction", meaning the exclamations that a person says when he catches the essence of the problem and sees the way out of it.

Many people know examples of bright enlightenments from school years and lucky people to whom these enlightenments came in a bath, under an apple tree (you may remember), in other places and during various activities, for example, cleaning shoes (as the composer Brahms claimed).

Many books (domestic and foreign) offer to develop intuition (insight).

From the point of view of psychology, intuition is the opposition of *reflection* (the principle of human thinking, which directs it to comprehend and realize its own forms and patterns), therefore the main tips for developing intuition in these books are extremely trivial — for example, how to learn to relax and not concentrate on exciting problems, to break an obsessive chain of thoughts, emotions (watch a movie, read, take a walk), train the ability to guess simple events (for example, who will enter first — a man or a woman, etc.), increase the readiness of fixing sudden enlightenments (for example, to have a notebook and pencil with you everywhere).

However, most psychologists and philosophers agree that focused and persistent work on a specific problem can accelerate the emergence of insight, including obtaining maximum information about it and its attendant factors, studying the problem from all sides, with the inevitable alternation of work on the problem and complete distraction from it with the subsequent return to work with even higher motivation.

And I will add from my personal experience that insight often comes during activities that distract from minor everyday problems and do not require concentration (swimming, walking, skiing or cycling on safe tracks). At the same time, mental attempts to use productive knowledge to solve the problem are very helpful (see section 2.3.2). Of course, one should not lose concentration (think deeply) even for the sake of the possibility of insight in many other places and during the actions of increased danger (for example, skiing, mountain tourism, driving a car).

5.1.3 Combinatorics

Combinatorics is widely used in many types of activity, including scientific and technical creativity.

In the Great Soviet Encyclopedia, combinatorics are interpreted as follows: 1) the same as mathematical combinatorial analysis; 2) the section of elementary mathematics related to the study of the number of combinations subject to certain conditions that can be composed of the given set of objects (no matter what nature, it can be letters, numbers, any objects, etc.)[6].

However, few people have thought about the fact that all English literature written in modern English consists of combinations of words — patterns (pattern — model, sample, template, technique, stable configuration of connections) formed by only 26 letters, and all literature written in modern Russian Cyrillic consists of words combinations — patterns formed by 33 letters. All patterns written by artists are brushstrokes-patterns from the combinations of only three primary colors (red, green, blue). All the music composed is various combinations of only 12 notes (including semitones). All numeric expressions (in Arabic numerals) consist of only 10 characters. For any computer calculations, the patterns of only two components (1 and 0) are used. The same number of components is in Morse code, where the alphabet, numbers, punctuation marks and other service characters are represented by combinations of two components — dots and dashes (Fig. 57), in the sound version — the combinations of short (like a dot) and long (about three times longer) sounds of the same tone. And finally, Braille — a relief-point tactile font designed for reading by blind and visually impaired people — in which different alphabets and some symbols represent the combinations of dots (Fig. 58, 59).

Figure 57. Morse code (Cyrillic alphabet)

Morse code				
А · —	К — · —	Ф · · — ·	1 · — — — —	. · · · · · ·
Б — · · ·	Л · — · ·	Х · · · ·	2 · · — — —	, · — · — · —
В · — —	М — —	Ц — · — ·	3 · · · — —	; — · — · — ·
Г — — ·	Н — ·	Ч — — — ·	4 · · · · —	: — — — · · ·
Д — · ·	О — — —	Ш — — — —	5 · · · · ·	? · · — — · ·
Е ·	П · — — ·	Щ — — · —	6 — · · · ·	! — — · · — —
Ж · · · —	Р · — ·	Ь, Ъ — · · —	7 — — · · ·	- — · · · · —
З — — · ·	С · · ·	Ы — · — —	8 — — — · ·	" · — · · — ·
И · ·	Т —	Э · · — · ·	9 — — — — ·	(— · — — · —
Й · — — —	У · · —	Ю · · — —	0 — — — — —	/ — · · — ·
		Я · — · —		

Combinatorics is based on a rather complex mathematical apparatus that includes many rules and principles for determining mainly the number of placements, permutations and combinations of objects.

Since in the preface of this book it was promised not to use any formula, we will restrict ourselves to a simple example of combinatorics. For example, an architect designes three buildings (Fig. 60). Let's consider the problem of determining the number of permutations: in how many ways can you set the order of three different objects one after the other? The options for solving this problem are shown in Fig. 61.

Combinatorics is the basis for the morphological analysis and synthesis of technical systems, which will be discussed in Section 6.3.4. In addition, combinatorics is closely related to the realization of the most important human need to "regularize". In general, the most important human needs are presented in section 6.2.2.

Figure 58. Cyrillic Braille

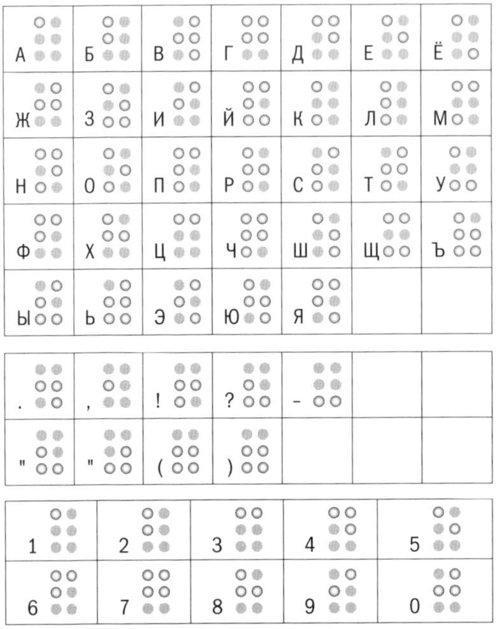

Figure 59. Latin Braille alphabet

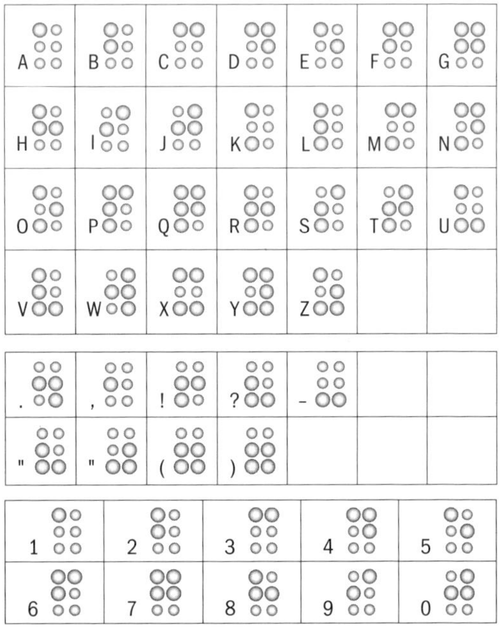

Figure 60. Designed structures: structure A, structure B, structure C

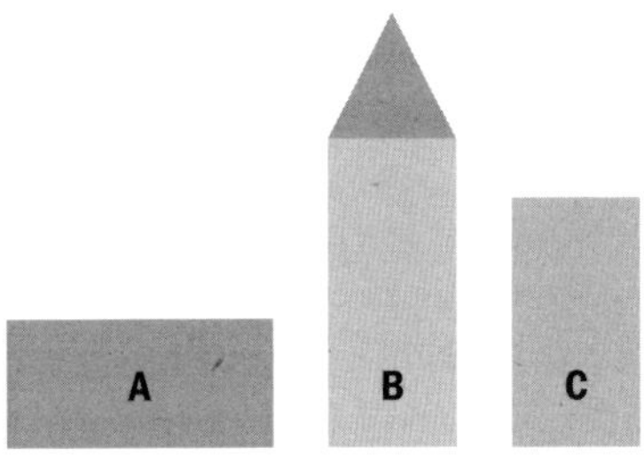

Figure 61. Permutations number of three objects

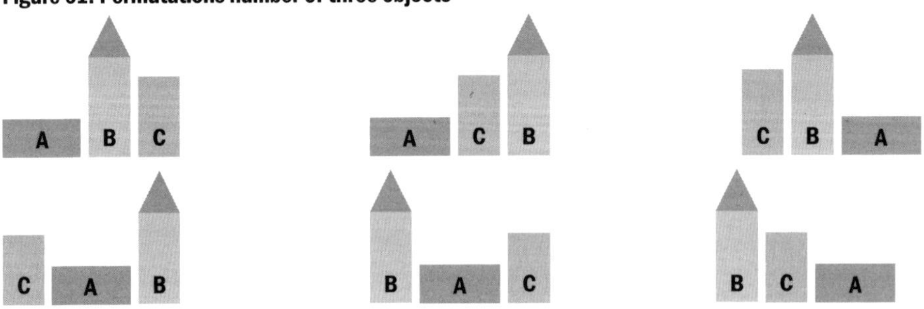

5.1.4 Inference by analogy (method of analogies)

In creative (as in any other practical) activity, inferences by analogy, also called *the method* of *analogies*, are widely used. Analogy (*ancient Greek* analogia — correspondence, similarity) — similarity, likeness of objects or phenomena in any properties or relationships[7]. Accordingly, inference by analogy is the logical action, the result of which is the conclusion that the object being investigated or developed, called a *model*, has some characteristics that are also inherent in the similar object called a prototype. In scientific and technical creativity, the model is TS being investigated or developed.

Inference by analogy is, perhaps, one of the most effective methods of convergence (transfer) of knowledge when improving TS or creating new TS by searching for analogues (prototypes) of TS being developed in animate and inanimate nature, as well as in the technical sphere based on the similarity of analogues (prototypes) and TS.

At the same time, to search for analogues (prototypes) of TS being developed, *the analogy of properties and the analogy of relations* are used (sometimes not fully consciously).

The analogy of properties is an inference in which the similarity feature of analogs (prototype) of TS being developed is their properties.

In a simplified way, the analogy of properties in scientific and technical creativity can be described as follows.

TS being developed has features *a, b, c*. Object A in the considered subject area has features *a, b, c*. In addition, object A has feature *d*. It is likely that the feature *d* can be used in TS being developed. For example, when developing a highchair (armchair), when searching for analogues, the following features can be used: the presence of a seat, legs, back (features *a, b, c*), which can be used to find different types of chairs — throne, country, palace, boudoir, theatrical, baby and others; then one can choose, for example, the features *d* of the throne chair (characterizing belonging to the throne) and use this feature (for example, a crown) with probable success when creating an armchair for a little princess (prince).

The analogy of relations is the inference in which analogues similarity features (prototype) of TS being developed are the relations planned in TS being developed (relations between the elements of TS, TS and TS, TS and man, TS and the environment) and implemented in analogs (prototype).

In a simplified way, the analogy of relations in scientific and technical creativity can be described as follows. TS being developed (A) is similar to another TS (C). Object B in the considered subject area is similar to object D. There are relations R between TS (C) and object D. Probably, there are relations R between

TS being developed (A) and object B.

Let us consider the option of using the analogy of relations in the development of a high-mobility car on the basis of a car (characterized by low cross-country ability on a natural surface). So, TS being developed (A) — a vehicle with high cross-country ability — is similar to another TS (C) — a car. Object B in the considered subject area — a natural surface — is similar (in terms of the presence of properties) to object D — an artificial road surface. There is the following relationship between TS (C) — the car — and the object D — the artificial surface:

— R_1 is the dependence of the specific pressure of the wheels (propulsion system element) on the road surface, determined by the force value of the pressure of the wheels on the contact area of the wheels with the surface;

— R_2 is the dependence of the deformation value of the surface, determined by the value of the specific pressure of the wheels on the road surface and the strength characteristics of the coating material affecting the car's passing ability.

Unambiguously there are the same relations between TS being developed (A) — a vehicle with high cross-country ability on a natural surface — and the natural surface itself (B).

An analysis of these relationships could lead to the conclusion that the specific value of the pressure of the propulsion element in TS being developed on the natural surface and, accordingly, the value of deformation of the natural surface can be reduced with the increase in the contact area of the propulsion element with the natural surface. This could serve as a hint to use a caterpillar instead of a wheel as a propulsor element, since the caterpillar has a much larger contact area with the surface, which dramatically increases the vehicle's cross-country ability.

The properties and relationships that characterize the features of analogy are classified in the following five types of objects analogy in animate and inanimate nature, as well as in the *technical sphere* of TS being developed (Table 11).

Examples of analogy types are shown in Table 12 and in Fig. 62-71.

The effectiveness of the analogy method increases immeasurably with its computerization. The description of the subject area with a variety of properties and their parameters is the most laborious (Fig. 72).

It should be borne in mind that the decrease in the likelihood of conclusions by analogy, as a rule, is associated with the increase in the heuristic possibilities of the analogy. This allows the analogy method to be used to generate new ideas efficiently.

In general, the use of the method in scientific-technical and other creativity is carried out according to the following scheme (Fig. 73) with the implementation of 4 stages.

Table 11. Types of analogy and their characteristics

Types of analogy	Characteristics of analogy type
Functional	The analogy is determined by the similarity of the main function of the compared objects and phenomena.
Structural	The analogy is determined by the similarity of the functional structure of the compared objects and phenomena.
Substrate	The analogy is determined by the totality of the material basis of the compared objects and phenomena. *Substrates* of all known *physical processes* are elementary particles and fields, the fundamental interactions of which determine the physicochemical forms of motion. *The substrate of a chemical reaction* is atoms that remain stable during the formation and transformation of substances. *The substrate of biological processes* in living organisms are molecules of nucleic acids (DNA and RNA) and protein substances, which act as elementary "units" of life as the form of matter motion. *Man is the substrate of social forms of movement*[8].
Of external form	TS being developed uses a variety of shapes, as well as the characteristic features of analogs surface
Of relationships	The analogy is determined by the similarity of relations (connections) between the structural elements of TS, TS and TS, TS and man, TS and the environment, man and man (society) and the objects of analogy

Table 12. Examples of types of analogy

Type of analogy	Developed TS	Object of a possibly used analogy
Functional	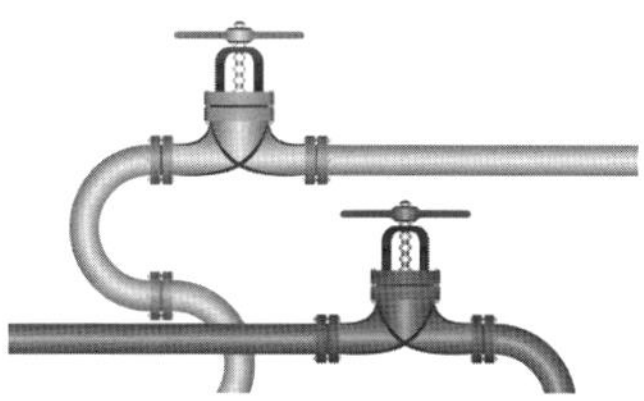 **Figure 62. Pipeline**	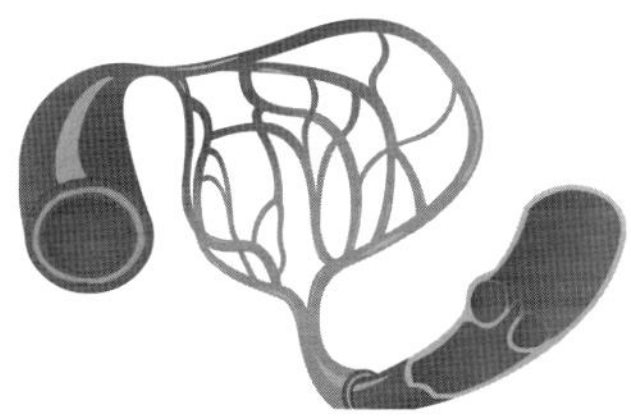 **Figure 63. The circulatory system of the animal**
Structural	 **Figure 64. Cellular architecture of buildings**	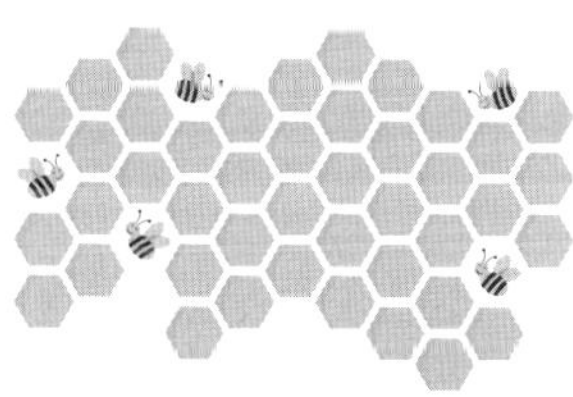 **Figure 65. Honeycomb**

Type of analogy	Developed TS	Object of a possibly used analogy
Substrate	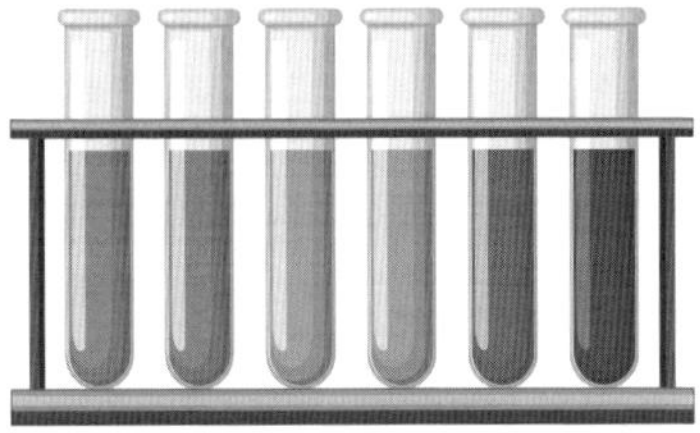 **Figure 66. Artificial colors**	 **Figure 67. Plant sap**
Of external form	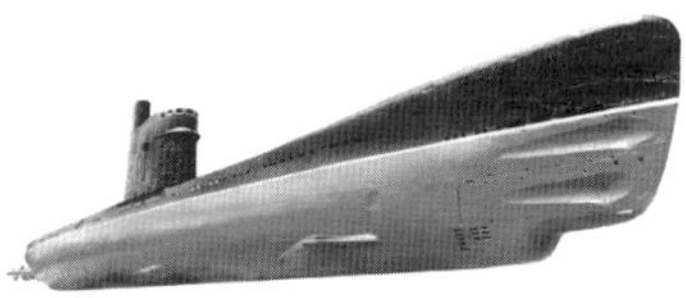 **Figure 68. Geometry of the submarine head**	 **Figure 69. Geometry of the sperm whale head**
Of relationships	 **Figure 70. Planetary gear train**	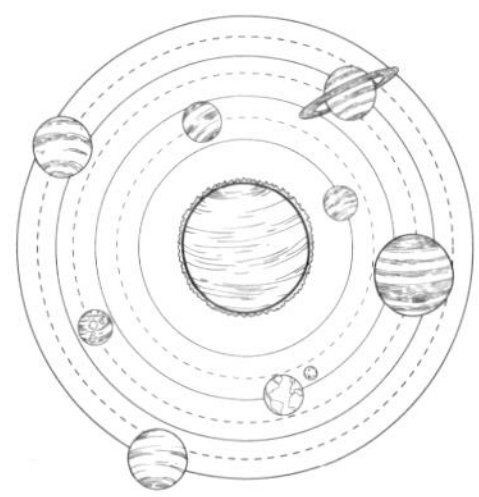 **Figure 71. Heliocentric (planetary) model of the Solar System**

Figure 72. Description of the subject area

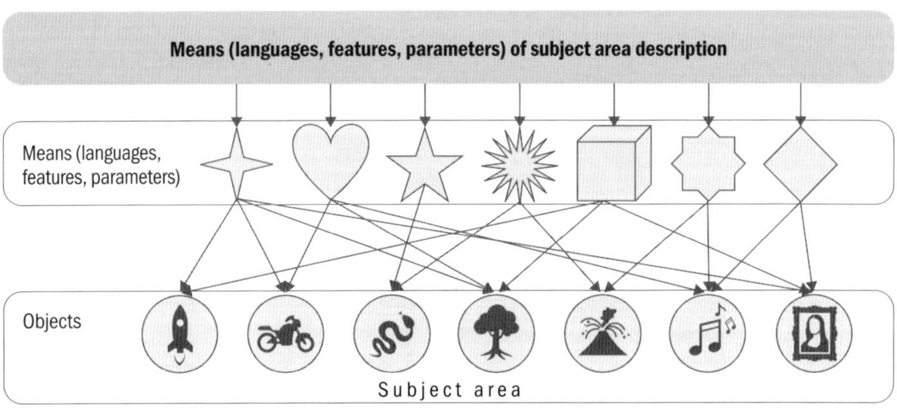

Figure 73. Scheme of using the method of analogies in scientific and technical creativity

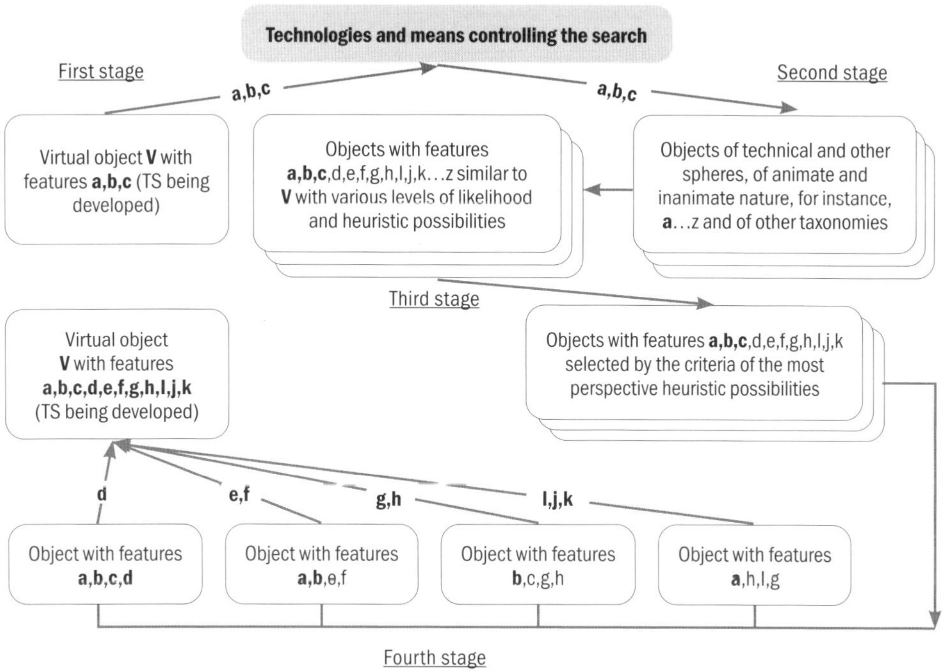

At the first stage, for the technologies and means that control the search, the task is formed to search for analogs for a virtual object **V** (TS being developed) with features, for example, **a, b, c**.

At the second stage, objects similar to object **V** with different levels of likelihood and heuristic capabilities are selected in objects of technical and other spheres, living and inanimate nature.

At the third stage, from the objects selected at the second stage, objects are selected that meet the criterion of the most promising heuristic capabilities.

At the fourth stage, in the objects selected at the third stage, the features are selected that are advisable to implement additionally in object V.

The thesis about the interdependence of the likelihood levels of conclusions by analogy and its heuristic possibilities is confirmed by the paradoxical postulates of analogy formulated by G. Ya. Bush:

1) "The more complete the analogy between the compared objects, the less complete the analogy between the compared objects, the more original the conclusion is.
2) The closer the analogy between the compared objects, the more trivial

"Imagination is more important than knowledge. For knowledge is limited, whereas imagination embraces the entire world, stimulating progress, giving birth to evolution."

Albert Einstein

The likelihood of conclusions by analogy, as a rule, decreases with the increase in the heuristic capabilities of the analogy.

There are a lot of associations: by contiguity in space, by contiguity in time, by similarity, by contrast, cause-and-effect, etc. Any association can accelerate creative insight.

the conclusion is, the more distant the analogy between the compared objects, the more original the conclusion is.

3) The analogy of the secondary and side, at first glance, features and relations of objects often has the high degree of heuristicity, which is explained by the disclosure of the previously unnoticed significance of the analogy in the process of analogization.
4) The more general features and relationships are used in the analogization process, the more space for searching for new ideas can be used.
5) The analogy between the special and the general case is more heuristic than the analogy between the special cases.
6) The greater the heterogeneity of the classes to which the analogs belong, the more original the conclusion is."[9]

The complete analogy is characterized by the similarity of almost all the features of the compared objects. At the same time, it is very difficult to identify a new (not available in TS being developed) feature in the analogue. However, if this succeeds, then there is a high probability that such a feature will be realized in the original technical solution in TS being developed.

The close analogy between the compared objects is characterized by the complete similarity of functional features inherent in technical systems of one class or the object of animate or inanimate nature and TS, the characteristics of which are described by the knowledge from one sphere. In this case, it is unlikely to form an original conclusion by analogy. I will add, in practice, the last postulates mean that descriptions of the subject area should, if possible, include the broad knowledge that goes beyond the narrow area of knowledge associated with TS being developed. Then the likelihood of generating non-trivial new ideas increases.

It is necessary to take into account, along with the similarity of the objects of analogy, their difference. It is not unusual when one difference points to invalidity of analogy, despite many similarities. For example, there are many similar features between the Earth and the Moon, but the absence of atmosphere on the Moon does not allow one to draw conclusions about the presence of life on the Moon on the basis of analogy with the Earth[10]

Based on G. Ya. Bush, it should be noted that analogies can be effectively used to solve the following problems:

- Identifying trends in the development of technical objects and branches of technology;
- Forecasting the development of social and personal needs and technical means to meet them,
- Studies of the genesis of similar problem situations;
- Setting a problematic technical task by comparing different prototypes

of the problem itself and known means of solving it;
- Research on the model of the essential qualities of the prototype or ideal idea (concept) of an inventor or designer;
- Verification and demonstration of the concept of solving a problem task on the model, etc[11].

In conclusion, it can be noted that other scientific methods are closely connected with analogy as the general scientific method of cognition and creativity, with conclusions by analogy: *modeling, interpretation, extrapolation, induction, deduction, generalization, concretization, classification*. The reader will become more familiar with some of these methods in other sections of the book.

5.1.5 Inference caused by associations

The psychological phenomenon can shorten the appearance time and increase the likelihood of insight in scientific and technical creativity — the *association* characterized as "the connection between two contents of experience (sensations, images, thoughts, feelings, etc.), which is expressed in the fact that the appearance in the mind of one of the contents entails the appearance of another. The more often such a connection arises in experience, the stronger the association becomes"[12].

The action of this connection — the *actualization* of the association — consists in the fact that the appearance of one member of the association regularly leads to the appearance of another (others).

In contrast to inference by analogy, when in scientific and technical creativity as a clue for creating a new (or improved) object or phenomenon (model)[13], the characteristics of the analogue (prototype) of the model being developed are used, deliberately identified on the basis of the direct similarity of the prototype and the model, for inference caused by association, such a hint can be the characteristics of the images of analogs (prototype) of the model being developed, which have arisen involuntarily in the human brain on the basis of mental connections (sometimes, at first glance, very distant) of the analog (prototype) and the model as a result of a random event or identified as a result of a purposeful semantic search or *conditioned reflex*.

In this case, the following types of associations may arise (the links between mental formations, table 13), the characteristics of which can be used in scientific and technical creativity, including for teaching, education, for memorization (Fig. 74).

The emergence of random associations can be expected for an unpredictable long time and it may not happen, although the history of scientific and technical creativity knows quite a lot of bright inventions based on random associations.

For example, there is a legend that the construction of a suspension bridge was prompted to the English engineer Samuel Brown by the random association

Table 13. Types of associations

Type of associations	Characteristics	Example
By adjacency in space	The connection between mental formations arises due to their simultaneous appearance in consciousness	Sea – coast, mountains – valleys, beehive – bees, oil – gasoline, car – road
By adjacency in time	The connection between mental formations arises as a result of their appearance one after another	For most people, the word "Friday" evokes the association of "Saturday", "lesson" – "change", etc.
By similarity	The connection between mental formations is based on the similarity of their characteristics (needs, functions, features, functional structures, principles of action or existence, technical solutions)	Chameleon – color change – disguise, cobweb – fishing net, watermelon – ball, circulatory system – plumbing, bat – radar, seashell – superglue
By contrast	The connection between mental formations is based on their opposite characteristics	Drought – flood, black – white, good – evil, acceleration – deceleration, strong – weak, poison – a medicine
Cause-and-effect	The connection between mental formations arises under certain conditions (reasons) and disappears (or changes) in their absence. The reason for the connection can be signals: visual, auditory, olfactory, gustatory, tactile, temperature; consequence – conditioned reflex	The bright flash of light – pain, strong wind – chills, grinding – destruction, breakage; command by sound, gesture, tactile – a change in the state or behavior of an animal

caused by a cobweb under a bush. Or the example that is already classic: the structure of the benzene molecule, an aromatic substance (Fig. 75), was prompted to the German chemist August Kekule by an association — a scene of monkeys grappling each other by their paws and tails accidentally seen in the zoo (there are other legends of this discovery, including one that comes in a dream).

Figure 74. The use of associations in scientific and technical creativity

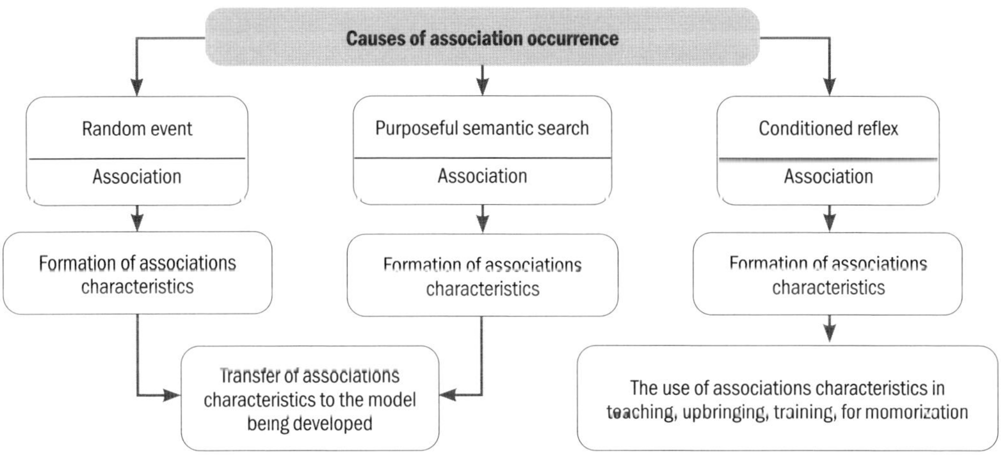

It is much more reliable to form a set of semantic associations (garlands of associations) and then use their characteristics for the development or improvement of TS models. Fig. 76 shows many associations for the word "bridge" that I have formed in a short time.

It is highly likely that the formed set of objects characteristics that are associations, after some thought, will suggest the original design of the bridge or elements of the adjacent infrastructure.

Various means of computer support for the formation of word associations are being developed. For example, on the Internet you can find the project "network of word associations", as well as many electronic dictionaries of synonyms (for example, there are 204 associations and synonyms for the word "development"), automated thesauri. A query for the word "mnemonics" suggests methods of development based on the associations of memorizing information in any areas of science, technology and culture.

Figure 75. The structure of the benzene molecule

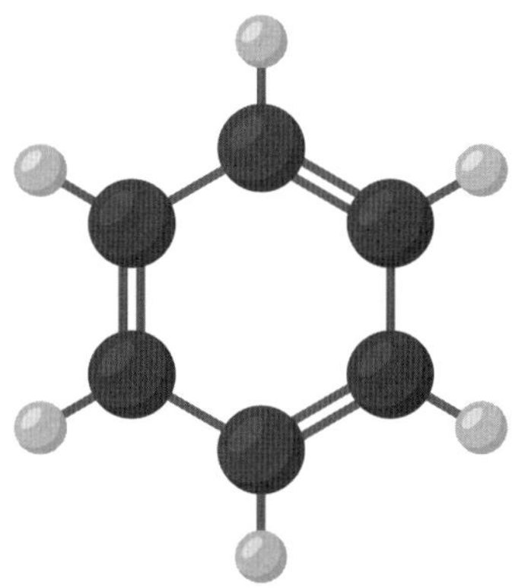

For the emergence of random associations, methodologists advise walking with the problem to various museums, flea markets, and even garbage dumps.

Developed associative thinking helps to assimilate new information, correlate it with previous experience and previously obtained data, and also contributes to the formation of the holistic picture of the world and the creative perception of reality.

Association garlands can be effectively used in the focal object method, which will be presented in Section 6.2.13.

Figure 76. Example of semantic associations to the word "bridge"

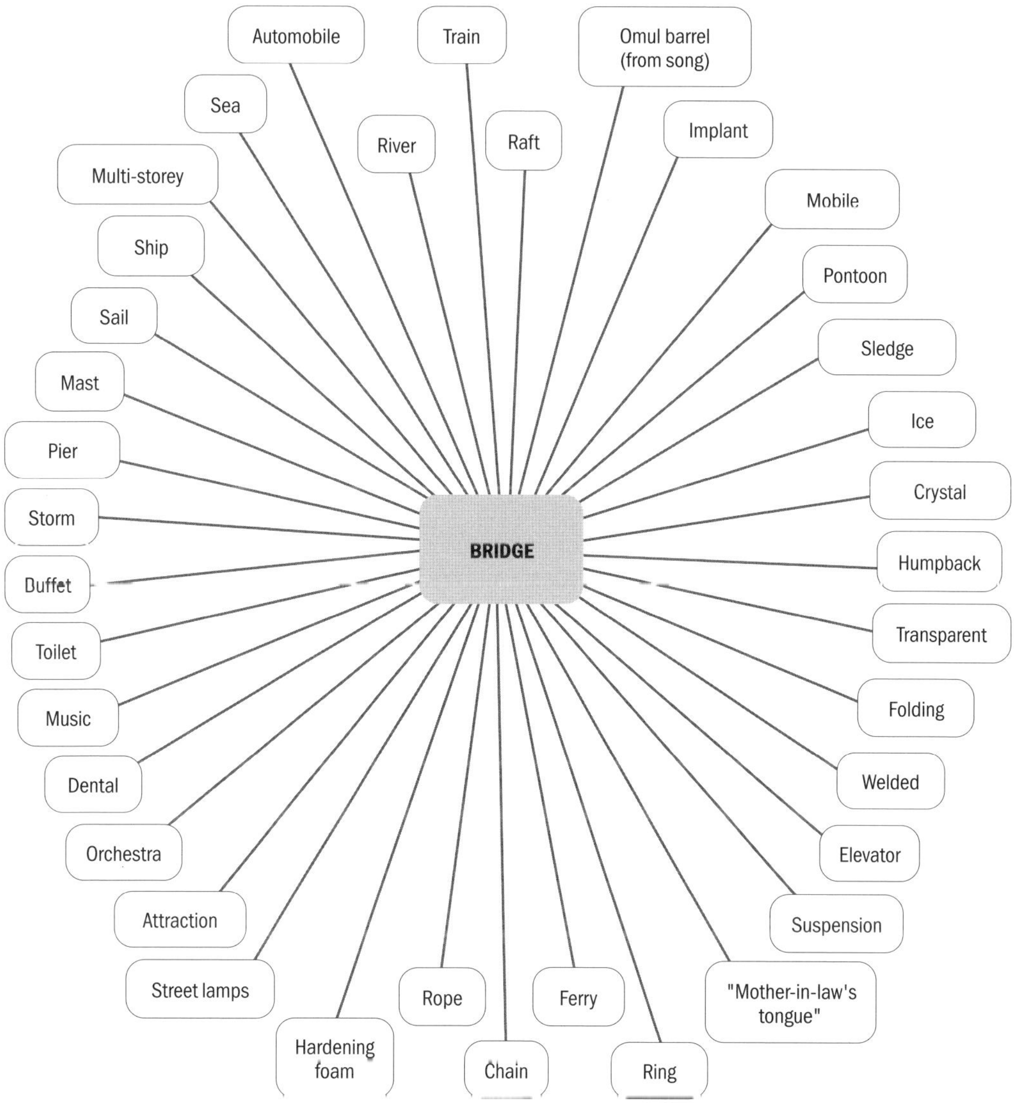

5.1.6 Imagination (fantasizing)

Any new knowledge is the product of creative imagination, since new is not the mechanical sum of old knowledge, but the synthesis of empirical (based on experience, real facts) data.

Imagination (fantasy) is the mental activity consisting in the creation of representations and mental situations that have never been generally perceived by a person in reality[14].

There are many types of imagination (fantasy) distinguished by the character, the degree of volitional efforts activity, the degree of reality transformation. They are shown in Fig. 77.

Figure 77. Types of imagination (fantasy)

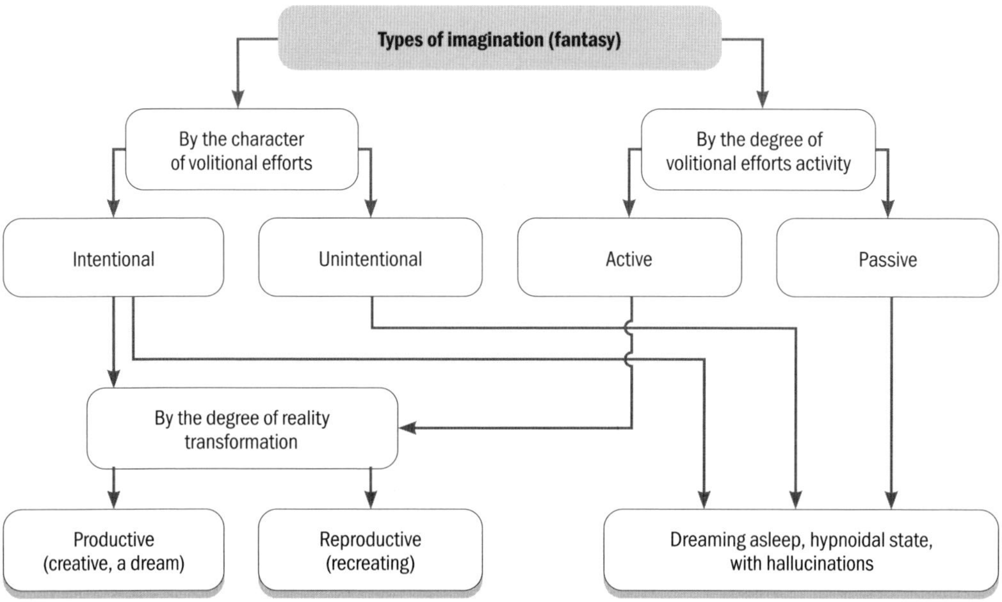

There is also emotional, figurative, verbal-logical, abstract and concrete imagination.

A special type of creative imagination is a dream, desired future images creation, which are not directly embodied in the certain products of activity[15].

Imaginings close to a dream are imaginary visions, the play of imagination in a dream, in a fever or in reality.

The famous psychologist L.S. Vygotsky identified the following functions of imagination (Table 14)[16].

The physiological basis of imagination is analytic-synthetic activity[22] (mental decomposition of initial ideas into component parts — analysis and their subsequent connection in new combinations — synthesis) using the following methods of new images forming (Table 15).

The word "fantasy", derived from the ancient Greek φαντασία — imagination, fantasy, whim, is similar in meaning to the concept of "imagination". A person

Table 14. Imagination functions

Function name	Brief description or comment
Cognitive	The ability to fancy, to imagine plays a huge role in man's mastering (especially during school years) of new forms of human experience for him[17]
Emotional	Imagination (fantasy) "opens" the doors to the magical world, where a child and an adult can realize their needs in a fairy tale, play, fiction, while experiencing many excitements
Educational or formative	In the process of playing, a child prepares for future encounters with social reality, for the upcoming role situations, actions and decisions. An active person uses the cognitive and educational functions of imagination all his life[18]
Anticipating	Provides the ability of a person to anticipate the possible consequences of his actions, to foresee the development and outcome of any situation, to imagine the result of the activity, being carried out, in advance. In essence, a thought experiment in scientific and technical creativity is based on anticipation[19]
Creative	Imagination (inextricably linked with thinking) actively participates not only in the formation of new artistic images and technical solutions, but also in the generation of such an integral formation as the "picture of the world"[20]
Regulatory	Using the imagination, a person is able to control the functional states of his own body and psyche, minimize the negative impact of stressful environmental factors, maintain volitional effort with insufficient (reduced) motivation[21]

with a developed imagination (fantasy) can predict the creation or occurrence of many unusual (for a given time) objects or phenomena (events) in the scientific and technical sphere.

They say that the one who "sees only what he sees" is devoid of fantasy. The example of the maximum possibility of fantasy manifestation is the white dragon known from Chinese mythology, which is "depicted" with a blank white sheet of paper, which allows each person to imagine "his own" dragon.

The outstanding Soviet methodologist G. S. Altshuller analyzed the predictions of science fiction classics[27] more than half a century ago, and it turned out that Jules Verne had 98 predictions out of 108, Herbert Wells had 77 out of 86, and Alexander Belyaev made a mistake only three times in his 50 scientific and technical ideas.

Other examples of amazing fantasies that have come true are shown in Table 16.

How to develop imagination (fantasy) in scientific and technical creativity? A good developmental workout is to come up with unusual uses of various objects (for example, a brick, a cabinet, a bottle (glass and plastic), a button, a paper clip, a thin strong thread, bearing balls, ice, etc.).

New images in human consciousness can arise due to agglutination, accentuation, hyperbolization, schematization, typification, etc.

Most of the predictions of science fiction classics came true:
98 out of 108 for Jules Verne,
77 out of 86 for Herbert Wells,
47 out of 50 for Alexander Belyaev.

Table 15. New images formation techniques in scientific-technical and artistic creativity imagination

Technique name	Technique characteristic (from various encyclopedias and dictionaries)	Example of using the technique or comment
Agglutination	The name comes from lat. agglutinatio (gluing) is an elementary mechanism for transforming person's ideas in the process of imagination. It consists in the synthesis, combination ("gluing") of various objects or parts, as well as events or processes, as a result of which the images of objects are created that do not occur in reality. It can be observed in acute schizophrenia[23]	Multi-barrel mortar; amphibious aircraft; aircraft with vertical takeoff and landing; flying car; jet (rocket) pack (fig. 78); multifunction folding knife (see fig. 39). In artistic creation: a centaur, a mermaid, a many-headed flying kite, a sphinx, a hut on chicken legs
Accentuation	One of the mechanisms for processing representations into imaginary images, which consists in highlighting a characteristic detail in image being created; focusing is a common synonym for accentuation	The cumulative effect (Munroe effect) is the increase in the effect of an explosion by its concentration (accentuation) in a given direction, achieved by the special design of the charge (with a notch facing the target; Fig. 79); in artistic creation (in the circus): "snake woman", "gutta-percha boy"[24]
Hyperbolization	One of the mechanisms for the processing of representations into imaginary images, which consists in a significant increase or decrease in the object as a whole or its individual parts[25]; in addition, the separation over long distances of object parts connected by flexible or rigid ties is used (for example, a prototype of a rope elevator, a mine cage can be hundreds of meters away from the counterweight)	Skyscraper; space elevator; giant movie screen; nanorobot (Fig. 80); in artistic creation: Gulliver, Lilliputian, Thumbelina, Tom Thumb; artistic understatement called litotes
Schematization	Consideration of an object or phenomenon in general terms, smoothing out differences, simplifying at the schema level	This allows you to fully display even complex structures, technologies, phenomena and not "drown" in structural details and processes (see, for example, the radio circuit in Fig. 81); in artistic creation, schematization is used, for example, to develop ornaments from elements of the plant world (leaves, branches, roots, seeds) and objects of inanimate nature (snowflakes, crystals, Fig. 82)
Typing	Development of standard designs or technological processes based on technical characteristics common to the number of products (processes)[26]. Typing is one of the methods of standardization	There is a very serious attitude to typing, for example, in industrialized Germany, in which products are manufactured on the basis of high-quality reference (typical) samples; deviations towards individual exclusivity are not supported; it is believed that only in this way it is possible to develop an all-powerful, firm, genuine taste of the nation; in fiction, typing is used to form images endowed with the features characteristic of their era (for example, Onegin, Pechorin, Oblomov)

Figure 78. Jet (rocket) pack

Figure 79. Cumulative charge

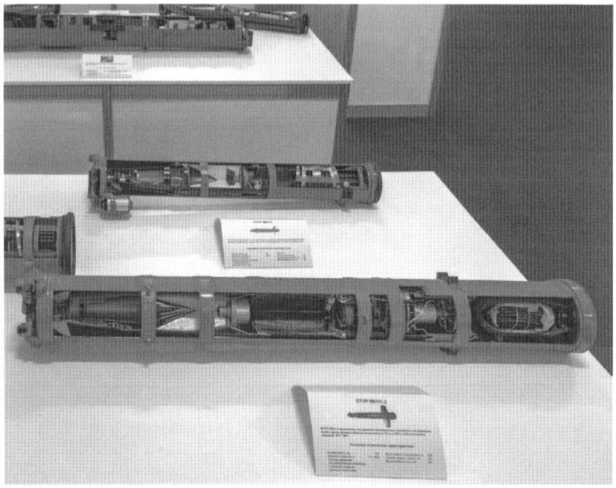

Figure 80. Nanorobot

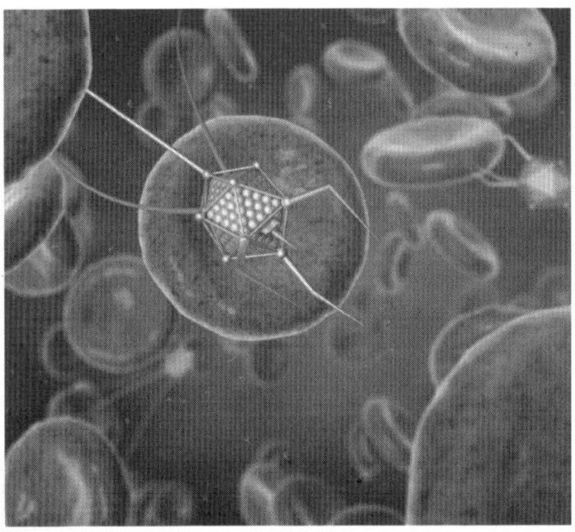

Figure 81. Schematic diagram of the detector receiver: 1 – antenna; 2 – variable capacitor; 3 – inductance coil; 4 – diode; 5 – low-pass filter; 6 – earpiece; 7 – grounding

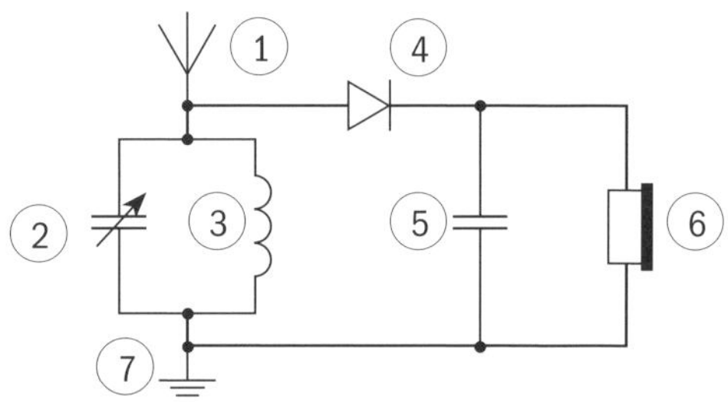

Figure 82. Examples of ornaments from elements of the flora and objects of inanimate nature

Table 16. Coming true predictions of science fiction writers

Science fiction writer, years of life	Name of the edition where the prediction was published, year of publication	Name or essence of a come true or realized prediction
Karel Chapek (1890-1938)	Script "Rossum Universal Robots" (1920)	The term "robot"
Edwards "Doc" Smith (1890-1965)	The novel "Triplanetary" (1934)	Metal coated discs as storage media
Aldous Huxley (1894-1963)	The novel "Brave New World" (1932)	Some Aspects of Genetic Engineering: test tube babies, cloning development
Clifford Simak (1904-1988)	The novel "Cosmic Engineers" (1939)	Rocket pack on a spacesuit, magnetic soles on boots
Robert Heinlein (1907-1988)	Book "Stranger in a Strange Land" (1961)	Pocket phone, money changer, GPS, waterbed (the first waterbed appeared seven years after the book was published)
Frank Herbert (1920-1986)	Book "The Dosadi Experiment" (1977)	Alarm clock using aromas instead of sound. These alarms can be purchased as a fun gadget
Ray Bradbury (1920-2012)	The story "There will be gentle rain" (1950)	Detailed smart home concept
Arkady (1925-1991) and Boris (1933-2012) Strugatsky	The story "Monday begins on Saturday" (1965)	Wikipedia prototype
Kir Bulychev (1934-2003)	Story "The Rusty Field Marshal" (1968) Book "One Hundred Years Ahead" (1982)	5D movies Electronic newspapers
Larry Niven (Born in 1938)	The novel "A World Out of Time" (1976)	The method of deep freezing of the body – vitrification – with the possibility of recovery after a long time
Arthur Clarke (1917-2008)	The novel "A Space Odyssey" (1968)	Tablet PC
Morgan Robertson (1861-1915)	The novel "Futility" (1898)	The novel details how the giant passenger ship Titan crashed into an iceberg and sank on the night of April. There were not enough boats on board, and most of the passengers died. After 14 years, the Titanic crashed with a coincidence of the smallest details, including the parameters of the ship and the technical characteristics of the engines

They recommend creating problem situations and ways to get out of them in the imagination. For example, survival on a desert island or at sea in a boat without oars, water and food. What will happen with the complete or partial disappearance of friction and how to use this state? How to prevent ice blockages on the river during ice drift? How to exclude or reduce hypodynamia (limitation of physical activity) at work or in the process of study?

You can practice using imagination techniques:

Agglutination. Try to combine a suitcase with a chair, a backpack with an umbrella, a briefcase with a scooter, an optical "mouse" with a person with upper limb dysfunctions, tactile information with sound, treatment with learning process (and vice versa). Please note that one object is not only "glued" to another (others), but also the object with the action of another object (for example, a vacuum cleaner + a floor washing); you should think about other options for "gluing".

Hyperbolization. Try to build a pool and / or escalator along the entire street in addition to the road. Think about what else useful you can build along the entire street. The reverse variant of the use of hyperbolization is reduction, when special services use micro cameras, micro video cameras, micro radio transmitters. What else "micro" can be done for these services? The separation by distance of the component parts is used, for example, in a catamaran (Fig. 83), where the floats (as propulsion elements) are located at the distance of several (sometimes more than 10) meters. For almost half a century, the remarkable Soviet (never realized) invention has been known, the essence of which lies in the fact that if you lay rails on a flat agricultural field at the distance of a hundred meters and install a light and durable platform with replaceable agricultural implements on them (plows, cultivators, seeders, sprinklers, mowers, threshers), then this platform, while moving, will perform all agricultural operations without tractors, combines and other equipment. Think of how this type of hyperbolization (separation of parts over a long distance) can be used on water, in the air (a good example is refueling aircraft in the air), in space (a version of a manipulator is already being developed for space needs, the "arms" length of which is a kilometer).

Figure 83. Catamaran: 1 - float; 2 - frame; 3 - sail

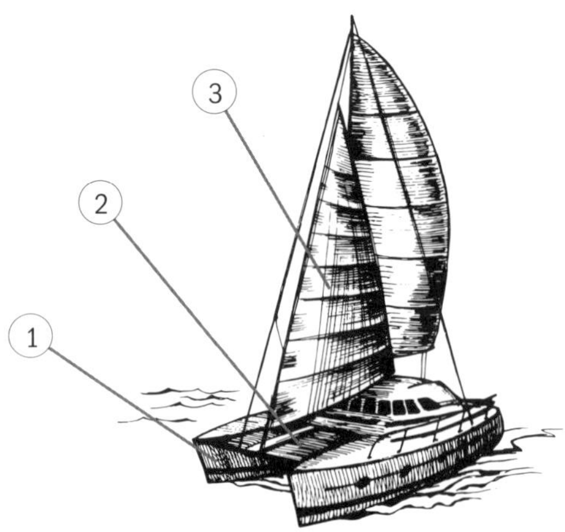

Accentuation. A simple example of the implementation of this technique is a barrier ("big hand") — a device for quickly blocking or clearing a path. It is useful to imagine what other fundamentally different barrier designs can be.

5.2 Heuristics, based on the use of productive knowledge for intensification of innovative activity

5.2.1 General information about heuristics

To intensify innovation, heuristics are used, which include the creative principles, strategies, tactics, methods, techniques, laws and patterns of TS development, based on productive knowledge.

These heuristics (many of them were developed since the time of the ancient Greeks), on the one hand, arrange the sequence of actions in the design process to create new TS, and on the other hand, sometimes directly, sometimes indirectly suggest effective solutions or indicate the most probable ways to improve the quality of TS and / or the creation of new TS generations. In addition, heuristics make it possible to overcome psychological barriers to thinking rationally that prevent the emergence of original solutions in a person's mind, or preclude the emergence of such barriers.

The prominent Soviet methodologist in the field of methodology of scientific and technical creativity G. Ya. Bush told me that after years of research, he compiled the list of about 3,000 heuristics. Unfortunately, this list was lost with the death of the scientist. However, it remains in my memory: G. Ya. Bush believed that many heuristics were overly specific (focused on narrow areas) or contain insufficiently substantiated generalizations, and the effectiveness of some heuristics was overestimated. For example, the effectiveness of brainstorming, which was widely advertised in the second half of the 20th century, based on the prohibition of criticism during the collective generation of ideas, was clearly overestimated. Regarding the brainstorming, G. Ya. Bush put it figuratively as follows: "Imagine that there is a person tied up (by the "bonds of criticism") lying on the floor. What do you think if you free this person from the fetters, can he play the violin? The answer: if earlier he could play, then he can, and if earlier he could not play, then he cannot."

In our opinion, the effectiveness of tips lists and (control) questions, which were quite common in the last century, has been overestimated (M. Tring, E. Leithwaite, A. Osborne, T. Eiloart, D. Poya, J. A. Valladares, etc.). The reader, using the Internet, can easily find these catalogues and lists (some of them have more than 200 questions in the form of advice), but, most likely, he will be disappointed, since most of them are

Heuristics can intensify innovation activity: creative principles, strategies, tactics, methods, laws and patterns of TS development based on productive knowledge.

The methodology for the formation of the ideal final result helps to overcome the psychological barriers to thinking.

haphazard and trivial (they offer advice on the level of "clearly formulate the problem").

G. Ya. Bush also say that, of course, it is useful to know all 3000 heuristics, but a person needs only to choose 10–15 of them that best correspond to the personal preferences, way of thinking, and depth of understanding in order to effectively use them in his practice. In my opinion, this list should primarily include the heuristics based on productive knowledge, presented in section 2.3.2 (table 5). Some of them have already been covered in detail in the previous sections, all others will be presented in subsequent sections.

5.2.2 The concept of the ideal final result

The methodology for the formation of the ideal final result (IFR) can contribute to overcoming the psychological barriers to thinking that prevent the formation of non-trivial solutions to improve TS — technologies, equipment, and products.

According to the Great Soviet Encyclopedia, "an ideal (*French* ideal, from *Old Greek* ἰδέα — idea, prototype) is the ideal image that determines the way of thinking and the activity of a person or a social class"[28].

From the philosophical point of view, the ideal is unattainable. For example, it is impossible to create TS to measure something with absolute precision. Even with the use of the "sinless" atomic clock developed in the USA, the error of one second will occur when measuring time over 15 billion years. However, the ideal can serve as a kind of guideline when creating new or improving any TS, which can lead to the technical solutions close to ideal[29].

In the Theory of Inventive Problem Solving (TRIZ), the following definition of IFR is adopted: "There is no TS, but the necessary function is being performed"[30]. Such a definition, on the one hand, has a certain heuristic nature, since it pushes to search for a creative solution, on the other hand, it is a psychological barrier, since it is difficult to look for something that does not exist. In my opinion, this contradiction could be resolved by such a definition of IFR: "**It seems as if** TS does not exist, but the function is being performed."

Thus, from the practical point of view, IFR is the technical solution of TS designed to fulfill a specific human need (sometimes a set of needs), in which the main useful TS characteristics, for example, performance, speed, carrying capacity, durability, the degree of efficiency, tend to maximum values, and the characteristics that determine the charge for utility (for example, energy requirement, process time, cost) tend to zero. To one degree or another, this is achieved through the use of functions, properties and relationships that are available in the analogue of TS and / or the environment, but are not involved in the implementation of a new human need[31].

In the formation of IFR, radical (in terms of final results) fantasizing[32], the use of the root "self-" in the name of TS are very heuristic, and also the assumption that the need is somehow realized without TS specially designed for it (in fact, in order to fulfill needs TS is always present, but often hidden)[33].

Thus, the following examples, formed in folk tales, fully correspond to the ideology of IFR: "self-assembled tablecloth", "flying carpet", "self-running stroller", "invisible hat", "magic sword (knife)", modern "small little men" (according to G. S. Altshuller) and others — and currently implemented in the form of various types of aircraft, helicopters (including those invisible to radars), cars, leg devices that allow you to jump and run at "superhuman speed" self-sharpening knives, nanorobots, etc.[34]

Or another illustrative example of IFR. In modern cars, when the hood is raised or lowered, a contact sensor (special TS) accordingly turns on or off a light bulb (connected by a wire to a power source) that illuminates the engine compartment. According to IFR methodology, it is possible to form the technical solution that, as if "without a sensor (special TS)" turns on and off the light bulb: you can make part of the wire from the power source to the light bulb from mercury, which, flowing when raising or lowering the hood, closes or breaks, respectively electrical circuit. In fact, this is the contact sensor, the role of which is played by the originally made wire[35].

One of the universal trends in increasing the efficiency (ideality) of TS is the development of technical solutions that maximize the utility of the vehicle and reduce utility charge. For example, the main useful characteristics of vehicles: cargo weight, transportation speed. The characteristics of the utility charge include the total weight of the vehicle (with a load), energy requirement, etc. An ideal vehicle is a vehicle in which the total weight with a load is equal to the weight of the load (that is, the weight of structures, equipment that provides movement, vehicle control, and on ships — including crew accommodation is zero). In other words, these structures and equipment should not exist[36].

The development of trucks and tankers, based on increasing their ideality, is clearly shown in Fig. 84 and 85.

The idea of towing an iceberg to deliver fresh water is highly ideal (Fig. 86).

A fairly ancient method of delivering (rafting) timber along rivers in rafts is practically ideal. The value of the ratio of utility to charge for utility in this case tends to plus infinity.

The principle of "turn harm for good" also helps shaping IFR. For example, in some hydraulic structures, the speed of water that has completed its work is excessively high, and when discharged into the river, it destroys the bottom. The solution, in which the water flow is divided into 2–3 flows, which, colliding at the confluence, mutually quench their velocities, can significantly reduce the cost of speed compensation[37].

Some tips for IFR shaping at the task setting stage are shown in section 6.2.4.

Figure 84. Improving the ideality of trucks: a — the original model of the car; b and c — subsequent models

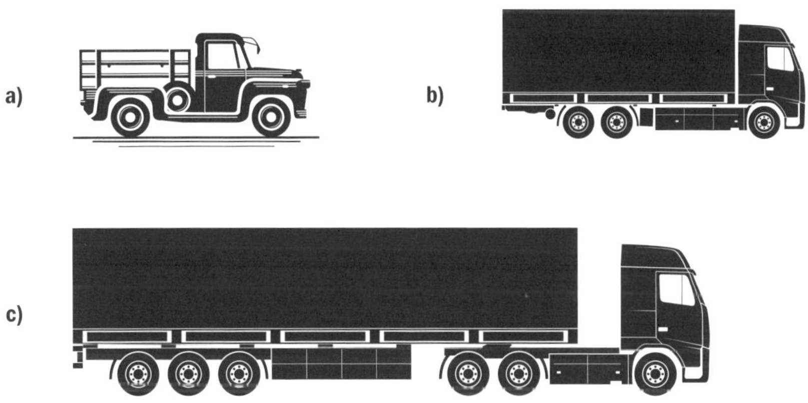

Figure 85. Improving the ideality of tankers: a — the initial tanker design; b and c — subsequent constructions

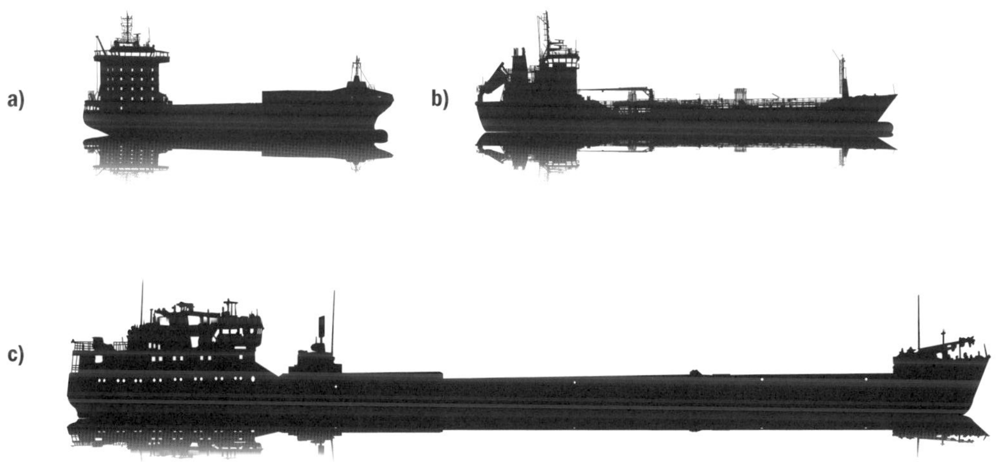

Creative activity is ineffective now, as it is based on the use of traditional methods (blind enumeration of options, inferences by analogy and association) and reproductive knowledge.

The effective model of creativity is based on the use of heuristics with productive knowledge.

Figure 86. Delivery of fresh water by towing icebergs

5.2.3 Models of creative activity

In addition to the widespread concepts of the creative process and models of scientific and technical creativity, presented in section 2.1, the reader can easily find the descriptions of 5–6 more models of creative activity (scientific and technical creativity) on the Internet upon request. However, these models seem to me either too simplified, or too complex, but all of them are not logical enough and therefore unworthy of additional attention. That is why I have developed the models of creative activity, which, on the one hand, quite imaginatively reflect the use specifics of the methods and means in scientific and technical creativity — the traditional one, based on reproductive knowledge (see section 5.1), and the heuristic one, based on productive knowledge (see section 5.2.1); on the other hand, they logically represent the interaction processes of various branches of knowledge in scientific and technical creativity.

To do this, imagine the space of all knowledge in the world in the form of the infinite set of contiguous spheres, each of which is a set of knowledge (reproductive and productive) — the solutions to one of the problems (Fig. 87).

The size of each sphere depends on the complexity of the problem, the tasks solved in it, the amount of knowledge used and the number of possible solutions. It is clear that the more complex the problem, the greater the amount of knowledge and the number of possible solutions, the larger the sphere

Figure 87. The model fragment of the imaginative representation of the space of all knowledge in the world: 1 – the spheres of knowledge-solutions set for different scientific and technical problems; 2 – the ideal final result; 3 – the areas of knowledge-solutions for different problems with competitive novelty and efficiency; 4 – the points of spheres contacts (including those from different branches of knowledge); 5 – the areas of knowledge-solutions for several problems with high novelty and efficiency due to the interaction of different knowledge-solutions, including those from different industries (shown in black); 6 – the areas of unknown knowledge

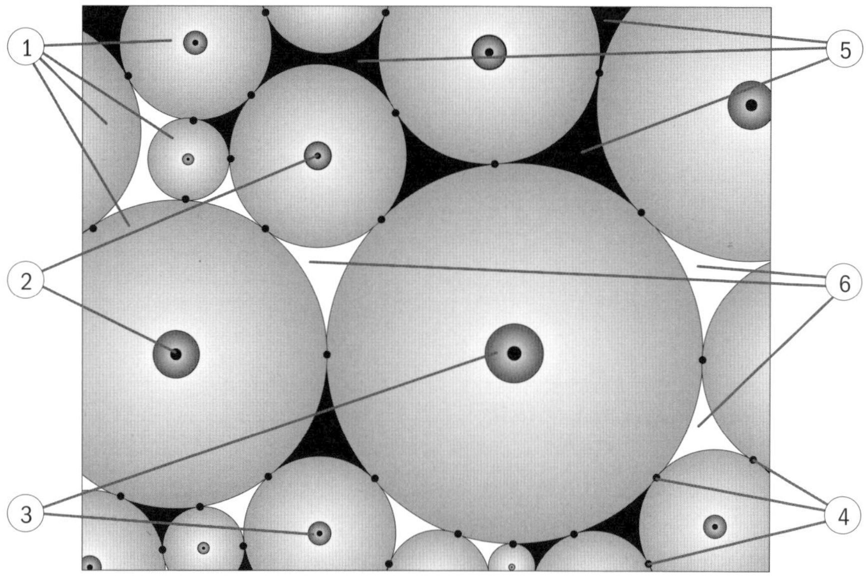

of knowledge-solutions. For example, the size of the sphere of knowledge-solutions for developing an energy converter will be much larger than for determining the number of permutations of buildings (see Fig. 61). Moreover, within each sphere (1), these knowledge-solutions are spatially distributed in accordance with the level of novelty and efficiency, which increase from the boundaries of the sphere to its center, where the ideal final result (IFR) is located (2), which, as already noted in section 5.2.2 is impossible to achieve. Each IFR is surrounded by a small sphere (3) — the area of knowledge-solutions with the greatest novelty and efficiency on one problem (first of all, exactly from this area one should choose knowledge-solutions that are most likely to be competitive). Number 4 indicates the points of spheres contacts (including those from different branches of knowledge), which, with the development of research, can form areas of knowledge-solutions (5) on several problems with high novelty and efficiency due to the interaction of various knowledge-solutions (including from different industries). Finally, the number 6 denotes the spaces between spheres of knowledge-solutions to various problems, which are the areas of unknown knowledge that have still to be discovered.

Taking into account the model — the imaginative representation of the space of all knowledge in the world — let us consider in more detail the specifics of using methods and means in scientific and technical creativity — the traditional ones, based on reproductive knowledge (Fig. 88), and the heuristic ones, based on productive knowledge (Fig. 89).

Figure 88. The model of creative activity using traditional methods and means based on reproductive knowledge: 1 — the sphere of knowledge-solutions to one of the problems; 2 — the initial ideas for models formation of TS being developed; 3 — the border of the sphere of knowledge-solutions; 4 — trials; 5 — subsequent new ideas; 6 — the area of competitive knowledge-solutions; 7 — ideal final result (IFR)

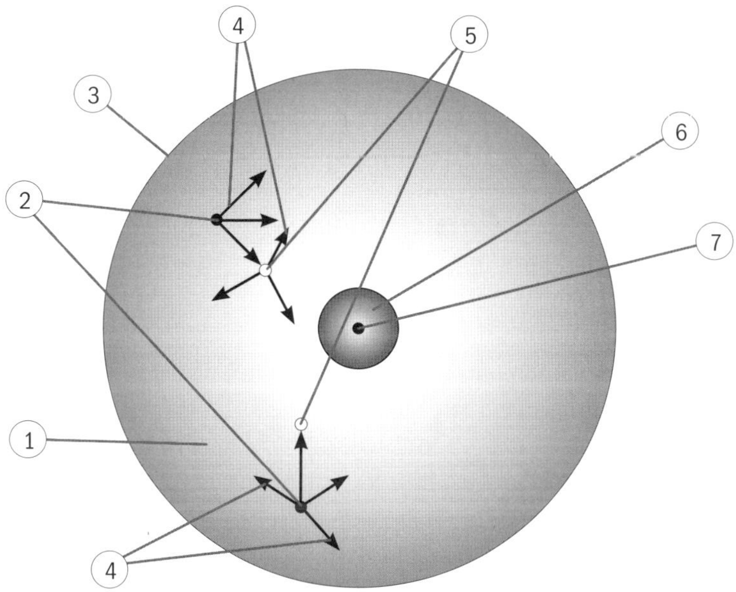

In the first case (Fig. 88), a man develops TS using traditional methods and means based on reproductive knowledge (the "blind" enumeration of options, the "trial and error" method, the inferences by analogy and the decisions caused by associations), therefore, a man, having no idea about heuristics (including IFR), will have (if he has) in the sphere (1), corresponding to the task at hand, some initial ideas (2) on the models formation of TS being developed, which are most likely to be in the zones of weak knowledge-solutions (that is, with the low level

Figure 89. The model of creative activity using heuristic methods and tools based on productive knowledge: 1 – the sphere of knowledge-solutions; 2 – heuristic tunnels; 3 – IFR; 4 – the area of competitive knowledge-solutions; 5 – initial ideas; 6 – samples; 7 – subsequent new ideas

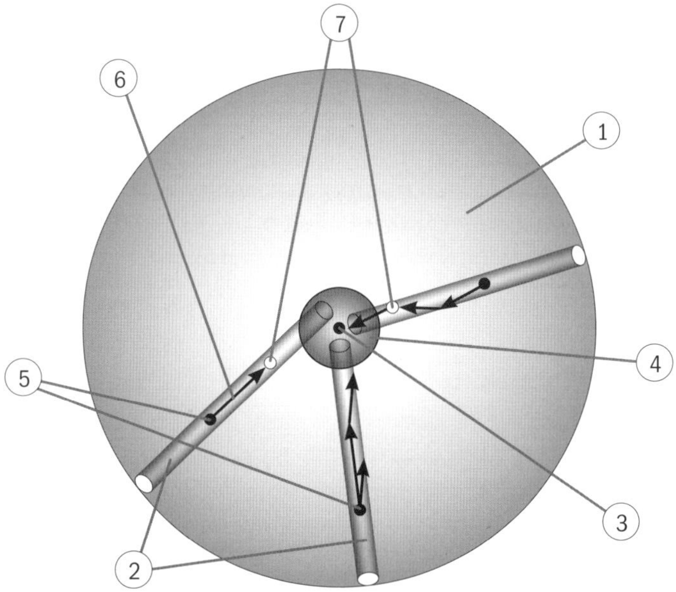

of novelty, efficiency and competitiveness), located closer to the boundaries (3) of the sphere of knowledge-solutions (1).

The formed initial ideas (2) are verified by trials (4), that is, additional research, sometimes even the creation of TS.

Sometimes, on the basis of trials (4) that test the initial ideas (2), subsequent new ideas (5) arise, which are tested with new trials. In both cases, from the trials (4), the best ones are selected, the characteristics of which most fully meet the existing needs, and the trials themselves are spatially oriented towards the area of competitive knowledge-solutions (6), as well as towards the ideal final result (7), the image of which (being important in its creativity) rarely arises in the mind due to man's lack of knowledge about the IFR and the principles of its formation. The trials with characteristics that do not come close to meeting the needs are considered erroneous, and they make new trials in another direction.

It is obvious that the use of such a creative activity model, based, in essence, on the "blind" enumeration of options, may involve a lot of erroneous tests, especially since one can identify an erroneous test not right away.

Another model of scientific and technical creativity, based on the use of heuristics (Fig. 89), is significantly (many times) more effective. A man developing a TS and having an understanding of the principles of forming an IFR first determines an IFR that corresponds to the needs, and then, based on the IFR, selects a relevant area of knowledge-solutions (1). Heuristics based on productive knowledge, reflecting the principles, laws and patterns of TS development, strategy, tactics, methods, techniques, can be figuratively represented in the form of heuristic tunnels (2), on the one hand spatially oriented to the IFR (3) and the area of competitive knowledge-solutions (4), on the other hand — limiting by the walls, soles, arches of tunnels (represented by heuristics) the implementation of trials (6) based on initial ideas (5) only within the tunnels and oriented to the center of the sphere. At the same time, due to the use of heuristics of initial ideas (5), a multitude of ideas immediately arise, and they themselves are located in tunnels much closer to the area of competitive knowledge-solutions, therefore the number of trials (6) for entering the area of competitive knowledge-solutions and generating subsequent new ideas (7) is sharply reduced. As already mentioned in Section 5.2.1, 10–15 heuristics (tunnels) are sufficient to obtain competitive solutions. The different spatial orientation of the tunnels (relative to the IFR) is explained by the different effectiveness of the heuristics for each person.

5.2.4 "Old" and "new" innovative cultures

Let us recall the classical definition of the concept of "culture" for comparing the "old" and "new" innovative cultures. According to the Great Soviet Encyclopedia, "culture (from *Lat.* Cultura — cultivation, late upbringing, development, education, reverence) is the historically determined level of society development, man's creative efforts and abilities, expressed in the types and forms of organization of life and activities of people, in their relationships, as well as in the material and spiritual values they create"[38].

This capacious definition in the "old" innovation culture is manifested in many countries in the following:

1) Insufficiently effective scientific and methodological support is used to create TS: the "trial and error" method, the inferences by analogy, as well as the decisions caused by associations, occasionally leading to "insight" (sudden enlightenment);
2) There is practically no modern software for the stages of conceptual TS design (see Fig. 45),
3) The decision-makers (DM) have points of view, which are widespread

Manifestation of the "old" innovation culture:

1. TS are created with insufficiently effective scientific and methodological support.
2. Modern software does not yet meet the needs of TS conceptual design.
3. Decision-makers often have erroneous ideas about the state and development of innovative activities in the country.

and mistaken, and misconceptions about the state and development of innovations, for example:

— the nation has a huge scientific potential (and some achievement is called with pathos — separate TS in a separate industry);

— there is a "sea" of competitive ideas in the country — only there is not enough money for their implementation;

— it is impossible to teach scientific and technical creativity — this is God's gift;

— the nation has received the sufficient number of world-class results, we just do not know how to trade them;

4) Most of scientists and specialists:

— do not know about methods, technologies, means for intensifying innovation (information about them is scattered, not systematized and difficult to access due to the fact that they were often hidden as a secret weapon);

— even if scientists do know, they do not believe in their effectiveness (in fact, this is similar to the assertion that the ability to create is a gift from God).

In the context of clause 1, let me remind you that, according to the research of the famous scientist P. Hill[39], at all (except for the regulated) stages of creating a new product (TS) and its commercialization (use), starting from the important need, it is necessary to generate at least 60 fundamentally different ideas with the optimal quantitative parameters of features, and then choose (using the known methods and means for making decisions) the best of them, which is most likely to be competitive.

In many countries, the "trial and error" method for solving a specific problem, as a rule, generates 2–3 ideas, which in practice, due to their insufficient number, often turn out to be uncompetitive.

In developed countries, the "trial and error" method yields good results due to the fact that even with an extensive approach to innovation, due to large expenditures on the acquisition of modern equipment and the involvement of a large number of highly qualified scientists and specialists, it is possible to generate a sufficiently large number of solution options, including competitive ones.

In the context of clause 2, it should be noted that in the vast world of information technology, the large and science-intensive section is rapidly developing, representing computer-aided engineering system (CAE), the annual sales of which exceed $7 billion. However, in almost all numerous domestic CAE and foreign CAE systems, at any cost (some cost is more than $2 million) and the number of modules (there are systems that include more than 300 modules), only the optimization of quantitative parameters (parametric synthesis), modeling (the result of which is, as a rule, the same parametric synthesis), graphic visualization tools, reference databases, calculation programs, decision support procedures, organizational, financial and resource design management, documentation preparation processes (in fact, all this is stage 8 of design activities; see. Fig. 45) are implemented.

The erroneous exaggeration of the scientific potential noted in paragraph 3 distracts from the problem of the necessary transition to the intensive path of innovative activity development.

In addition, decision maker's (DM) lack of understanding of the mechanisms for creating a new one has led to the widespread erroneous and dangerous opinion that there is the "sea of ideas", which means there is no particular need for new ideas, as well as in an effective instrumentarium for their formation. Firstly, there is no "sea" of ideas, and there is clearly lack of competitive ideas; secondly, any "sea" will dry up, if only to draw from it and not adequately nourish it; thirdly, there was a man who was long ahead of everyone in the thesis about the "sea". There is a unique document — a memo of the head of the US Patent Office, Charles Dewell, addressed to US President William McKinley, in which the country's chief scientific expert proposes to liquidate the organization he heads, since "everything that could be invented has already been invented." Date — 1899! In the United States, this incident is remembered as an instructive anecdote.

I comment above the fallacy of the thesis that it is impossible to teach scientific and technical creativity.

One should always learn the ability to trade, but the main thing is the competitiveness of the product on the market.

In the context of clause 4, it can be added that only a few (outstanding) scientists and specialists use effective creative methods, technologies and means (heuristics) in their professional work, and many heuristics are formed by them independently (therefore, they are outstanding). These scientists often, successfully applying their own heuristics, are unaware that some of them have long been developed and have names. But **even geniuses could increase a little their creativity potential by knowing not only their own heuristics**, but also heuristics developed by other scientists. **And for the majority of participants in innovative activity, due to the mastery of heuristics, the creative potential can be increased many times over.**

And although countries pay quite a lot of attention to the infrastructure of the innovation system (the creation of technoparks, technology transfer centers, business incubators, etc.), the improvement of legislation in the field of scientific and technical creativity, innovation, the results are still insufficient.

Thus, having an idea of the "old" innovation culture, it is possible to form the definition of "**new innovative culture" — the stable system of norms, rules, methods and means, which are mastered and actively used by society, based on productive knowledge and heuristics, ensuring effective innovation in various fields of life and at all stages — from the generation of important needs and competitive ideas to their commercialization and / or other practical use.**

For a country to be competitive, this system of norms, rules, methods and means of implementing effective innovation should be owned by the overwhelming majority of its inhabitants, a certain "critical mass" (including youth), which will make the process of mastering a new innovative culture, its use and development irreversible.

It should be noted that in practice, the way to create a competitive product (especially a science-intensive one), go-to-market and maintain it for a long time in the market is fraught with many difficulties and requires the multiple generation of competitive ideas at different stages of design activity (see Fig. 45). Let us consider this on the example of the formation and development of a small business (see Fig. 90), and, what is more, we will accept ideal conditions when an entrepreneur already has everything to start: high motivation to start his own business, unlimited material resources and financial resources, infrastructure, personnel, etc.

Fig. 90 shows that even under favorable conditions, due to uncompetitive technical solutions, a small business may not start or finish prematurely at any stage, if it does not have a new innovative culture. In practice, this is often the case. Unsuccessful attempts to start entrepreneurship lead not only to low performance of small business, but also to painful disappointments and psychological stress of its participants. As a result, losers have the sharp decrease in motivation for creative activity, and some of them join the ranks of parasites, alcoholics, drug addicts, and embark on a criminal path.

The reasons for the underdevelopment of large science-intensive business are similar.

It follows from Fig. 90 that even if a competitive product is found (created) and successfully commercialized, in order to maintain competitive advantage, you need constantly improve the product, create new types of products, diversify, and then cluster activities (see section 1.2.3). Only if these conditions are implemented the developed business with a high level of financial stability will be created.

Figuratively speaking, (according to I. A. Krylov — Russian writer) the current situation can be characterized (somewhat exaggerating) as follows: if musicians (subjects of innovative activity) do not know how to play (use only the

A new innovative culture is a sustainable system of norms, rules, methods and means, assimilated and actively used by society, based on productive knowledge and heuristics, providing effective innovation in the various spheres of life and at all stages — from the generation of important needs and competitive ideas to their commercialization and / or other practical use.

unproductive method of "trial and error", etc.), then no matter how you rearrange the chairs, even adding new musicians (these are structural transformations), no matter what scores and instruments are given, even those created by Stradivari (these are preferences, the development of the legislative framework, etc.), the orchestra will still play badly.

The way out is the creation of technologies and means of intensifying innovative activity based on productive knowledge and heuristics, as well as mass training in the mastery of these technologies and means — the formation of a new innovative culture in the country.

Figure 90. Formation of new innovation culture elements at various stages and cycles of business development

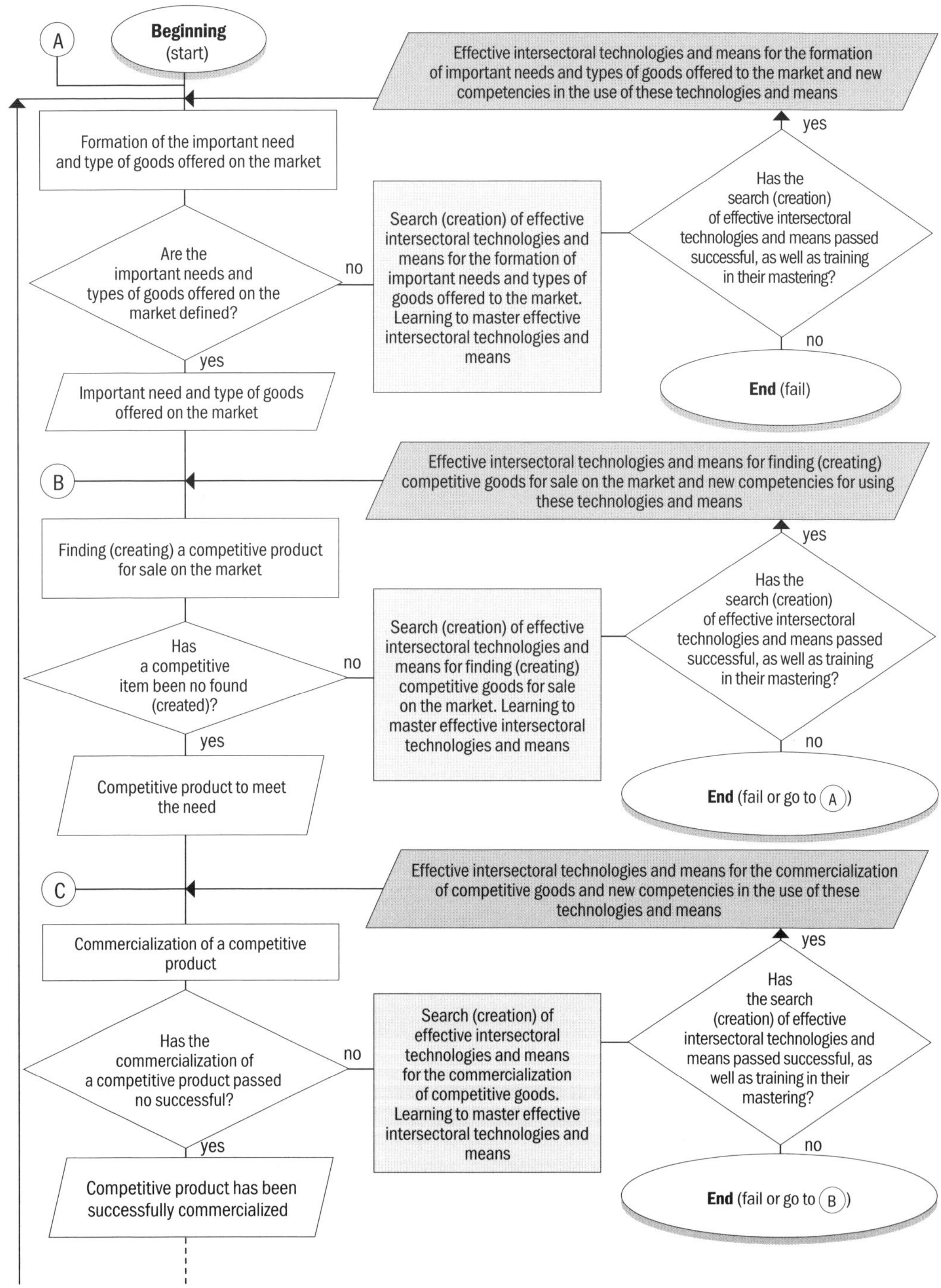

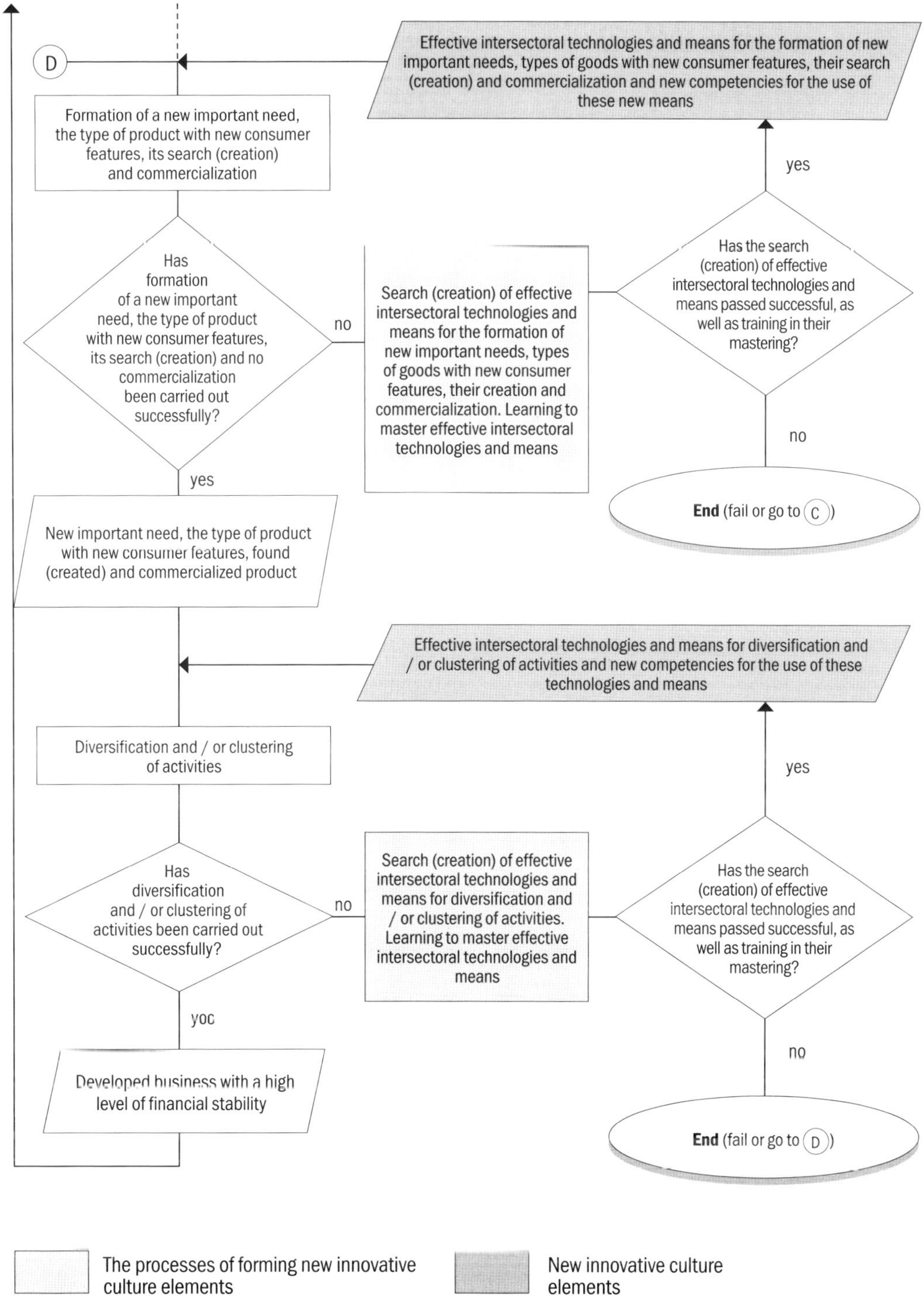
D
Effective intersectoral technologies and means for the formation of new important needs, types of goods with new consumer features, their search (creation) and commercialization and new competencies for the use of these new means
Formation of a new important need, the type of product with new consumer features, its search (creation) and commercialization
yes
Has formation of a new important need, the type of product with new consumer features, its search (creation) and no commercialization been carried out successfully?
no
Search (creation) of effective intersectoral technologies and means for the formation of new important needs, types of goods with new consumer features, their creation and commercialization. Learning to master effective intersectoral technologies and means
Has the search (creation) of effective intersectoral technologies and means passed successful, as well as training in their mastering?
no
yes
End (fail or go to C)
New important need, the type of product with new consumer features, found (created) and commercialized product
Effective intersectoral technologies and means for diversification and / or clustering of activities and new competencies for the use of these technologies and means
Diversification and / or clustering of activities
yes
Has diversification and / or clustering of activities been carried out successfully?
no
Search (creation) of effective intersectoral technologies and means for diversification and / or clustering of activities. Learning to master effective intersectoral technologies and means
Has the search (creation) of effective intersectoral technologies and means passed successful, as well as training in their mastering?
yoc
no
Developed business with a high level of financial stability
End (fail or go to D)
The processes of forming new innovative culture elements
New innovative culture elements

Chapter 6
Methods and means used at the stages of setting, solving the problem and commercialization of the result

6.1 General information about the stages of setting, solving the problem and commercialization of the result

It is convenient to consider the phased realization of human needs presented in section 1.4 systematically in three enlarged stages of scientific and technical creativity: **problem setting, problem solution, commercialization or other use of the result.**

Since the reader (I hope) has long been convinced that any human activity is the formation and implementation of designs, from the point of view of the system methodology of design activities (see Section 3.2, Fig. 45), one can imagine that **problem setting** includes[1]:

1) Analysis, synthesis and selection of general human needs;
2) Analysis, synthesis and selection of individual needs of a person;
3) Analysis, synthesis and selection of functions of TS being designed;
4) Analysis, synthesis and selection of consumer properties and parameters of TS being designed.

In fact, these four stages of problem setting are the formation of the technical task for TS development.

The solution to the problem includes (from the point of view of the system methodology of design activities, the numbering of stages is consecutive)[2]:

5) Analysis, synthesis and selection of the functional structure of TS being designed;
6) Analysis, synthesis and selection of the operating principle of TS being designed;
7) Analysis, synthesis and selection of technical decisions for TS being designed;

8) Optimization of property parameters in the technical decision of TS being designed.

If the result of solving scientific and technical creativity problem is an object of intellectual property, then they must be protected before commercialization or other use of the results (more on this in Section 6.4.2).

Then the commercialization or other use of the results of scientific and technical creativity (technical decision) may include the following works:

- The formation of a variety of technical decisions (based on the information received);
- Formation of a list of new objects for commercialization (implementation) or options for other use of the result;
- Marketing research and selection of promising technical decisions for the protection of intellectual property;
- Protection of intellectual property;
- Creation and testing of a prototype TS;
- Creation and testing of an experimental TS batch;
- Development and implementation of a business plan for TS commercialization.

The reader can easily find a lot of good guidelines, including examples of business plans for the successful commercialization of scientific and technical creativity results, on the Internet.

It should be noted that the development of solutions at the stage of commercialization or other use of scientific and technical creativity result is also carried out as a design in which, according to the system methodology of design activities (if necessary, using all design strategies and tactics), a technical decision is formed step by step at eight stages (in exceptional cases at the first seven stages of conceptual design).

At all the enlarged three stages of scientific and technical creativity for its intensification, it is advisable to use methods and means based on productive knowledge, given in section 2.3.2 (table 5). In the following sections, it will be shown that some productive knowledge is most effective when used at one stage, while others, with little adaptation, can be applied at all stages of scientific and technical creativity.

In general, for any decision to be made, it is necessary to be guided by the main principle: "The benefit from any action (design) should be many times greater than the charge for the benefit."

6.2 Problem setting

6.2.1 Once again about false needs

G.Ya. Bush divides problems that characterize false needs into three types[3]:

- Quasi-problems — expressed in the form of wishes, dreams, fantasies;
- Pseudo-problems — erroneously put forward (for example, already solved) problems;
- Chimerical problems — contradict the laws of nature (in a figurative sense, a chimera — an unreasonable, unrealizable idea).

It is shown in the systematic methodology of design activities (see Fig. 45) that with the correct formulation of the problem (formation of the technical task), great achievements (including profit) are possible, with the wrong formulation, there will definitely be big shortcomings, failures (including losses). For example, if a need for criminal activity (false need) is formed, then, although there are many known TS used by various "specialists", called window-men, pickpockets, safecrackers, robbers in the underworld, it is possible to create new, more advanced TS, but the end will always be one — a prison.

One of the most common wrong, false needs is when the goal is not the final result, but simply an activity, as lightened as possible. **A person** with such a need **is looking** for the **lost not where he has lost** (where he should be looking for), but, **relatively speaking, under a street-lamp** (where it is easier and more convenient to search).

Many experts rightly state that a well-formulated problem is already halfway solved.

Maybe someone will find the advice of American psychologists useful, who suggest using the abbreviation SMART when setting a problem, which hides five keywords, each of which unfolds into a control question:

Specific — is your goal specific?

Measurable — can you measure how close you are to it?

Achievable — is this goal achievable at all?

Relevant — do you really need to achieve this goal?

Time scheduled — is there the time limit for reaching it?[4]

Problem setting is the process of its repeated reformulation. It should be borne in mind that, as the English scientists M. Tring and E. Leithuet wrote, "The customer is always wrong."[5]

When setting a problem, some people are hindered by the erroneous criterion of common sense "it can not happen!" This most often means only that it has not happened in the past.

The most effective heuristics, which sometimes can radically change the situation even at the stage of setting the problem, are presented in the following sections.

6.2.2 Natturalhuman needs and their use when setting a problem

6.2.2.1 Continuous and periodic the most important natural needs

The influence of the 1st and 2nd stages of design activities (see Fig. 45) — "analysis, synthesis and selection of general human needs" and "analysis, synthesis and selection of person's individual needs" (at the stage of setting the problem) — on the effectiveness of design performance is very high as a whole. Let us consider in more detail the mechanisms for satisfying general and person's individual needs by creating a "new" one (see Section 2.2, Fig. 5 and 6), as well as the possibility of increasing their efficiency.

Let me remind you that if there is no TS analogue, then the technically realizable need is formulated as the antonym of the cause of the human need. For example, the need for drinking water and possible lack of a natural water source in places where people live have generated the need to create water storage tanks. At the first stage, the role of such containers was played by suitable natural objects (for example, shells, hollow bones of animals).

In the presence of analogs (prototype), the development of TS occurs when a new (in comparison with those implemented in the prototype) human need is exercised. For example, if, in the development of a new air conditioner, along with the traditional need for comfort, realized on the basis of traditional functions (to regulate the temperature and humidity of the air), we synthesize a new need to CURE (with new functions and properties), then it is possible to create a new competitive air conditioner that will supply "sea", "mountain", "forest", "steppe" air. And if a medicine is sprayed with the help of such an air conditioner, then it is possible to prevent and / or treat asthma, occupational diseases of the respiratory system (silicosis, anthracosis, siderosis, etc.)[6]. It is easy and relatively cheap to implement such an idea, and a new product (TS) will receive a solid added value due to additional consumer properties.

That is why I consider it useful to remind the reader of the brilliant statement of the modern management classic Peter F. Drucker: "It is the want that is unique, and not the means to satisfy it"[7].

Thus, the qualitative improvement of TS through the implementation of a new need, at first glance, is a very simple task. It is only necessary to select new (in comparison with those realized in the prototype) technically realizable needs from the available list of needs, and then translate them into TS being developed.

I have compiled the basic list of human needs based on the list of Russian verbs. However, this list is difficult to be observed, since it includes about 5000 items (potentially it can be several times larger), even if we exclude technically unrealizable needs from it (for example, "love", "be respected", etc.). It is even more difficult to use the list of human needs if we add the technically realizable needs to it (for example, "reduce the viscosity of oil," "increase the speed of movement," etc.), which number hundreds of thousands, if not millions[8].

That is why I have had the idea to identify in this large list the constant or the most frequent (and therefore the most important) human needs, the number of which should not (and cannot) be large (no more than 10–20 items). Such a list could become the effective tool (hint) for the qualitative improvement of TS and increasing their competitiveness through the implementation of new (primarily permanent stable) human needs. This idea has proven extremely fruitful in practice. This list, called *"The List of The Most Important Natural Needs"*, has begun to be formed and effectively used[9].

The most important human need is a constant or periodic need, but the need that most often arises at the certain periods of human life.

The analysis of various aspects of human life makes it possible to classify the following *permanent needs* (or groups of needs) as natural with great confidence:

1. Protect (prevent) from diseases and pain, discomfort (itchiness, loud sound, bright light, unpleasant odor, sweating, etc.), from the harmful effects of the environment and human actions, as well as treat.
2. Have protective and rescue facilities for critical situations.
3. Receive, store and transmit information[10].

If the first group of the most important human needs does not raise almost anyone's questions about why these needs are the most important natural (existing around the clock) human needs, then the classification of the 2nd and 3rd needs as the category of the most important human needs is far from being understood by everyone and requires clarification.

The history of the emergence of the most important natural need No. 2 is as follows.

Like millions of people in the world, on September 11, 2001, I watched in horror as, after the ramming of two skyscrapers in New York by suicide bombers in airplanes and the fire in skyscrapers, people jumped out of windows on the fiftieth floors. Of course, the first questions that arose in me: "Why do people do this?! Are there really no emergency (fire) exits, some kind of evacuation devices in these ultra-modern skyscrapers? "As it turned out, all this was. But people were cut off by fire, smoke or debris from exits, evacuation devices, and some made the decision: "A horrible end is better than endless horror!" (a similar situation is shown in Fig. 91).

Figure 91. Fire in a high-rise building

While researching the problem of fires in high-rise buildings, I have found many examples of severe consequences during fires, even on the 2nd and 3rd floors, when people were cut off from the ways and means of rescue people for various reasons, and trying to go down using improvised means (for example, on tied sheets or clothes) they were smashed to death or severely injured. It was then that the formulation of the most important (round-the-clock) human need "to have protective and life saving equipment in critical situations" arose. This need has determined the main consumer properties for such means: their size and weight should ensure their constant, easy carrying (for example, like a mobile phone, which is taken with them everywhere). I have developed and patented nine variants of such devices that meet the specified consumer properties (more about them in section 6.5).

The third group of the most important natural needs includes, according to preliminary estimates, a three-digit number of needs, most of which are periodic, but often arising.

"If one does not know to which port one is sailing, no wind is favorable".

Lucius Annaeus Seneca

"To which port one is sailing" when developing TS can be prompted by the criteria for the progressive TS development, as well as 16 groups of natural (most important) human needs.

However, for example, the needs for information (monitoring) about the state of human organs, the environment (especially for detecting dangerous deviations), of course, are constant. Also, there may be a constant need for a patient to transfer information (data of round-the-clock monitoring) about his condition to specialists. In the future, this opportunity should be realized for all people.

The most important, but *periodic needs* (many of them arise only during man's waking hours) include the following:

1. To ensure TS multifunctionality, including due to the implementation of household functions (cut, puncture, open, unscrew or screw, fasten, heat, cool, enlarge the image, clean, illuminate, etc.).
2. To ensure compatibility of functions, properties, relationships (and their parameters) in various TS, society.
3. To transform and store energy independently.
4. To change the properties and parameters of TS during their operation, transportation, storage, utilization or burial.
5. To arrange objects layout and their elements (including signs, symbols, sounds), the shape of objects, the types and characteristics of movements, the connections between objects and their elements, or replace the existing order with another.
6. To reserve (to increase TS reliability or the organizational structure, the viability of the object of living nature) technical systems, the values of their strength, deformation and other characteristics, connections, energy, time, space, information, personnel, external and internal biological organs.
7. To develop mental, creative, physical abilities, as well as to develop spiritually.
8. To entertain, amuse somebody and relax.
9. To have pleasant smells, *tactile* properties, sound (or soundlessness), color (or colorlessness).
10. To designate belonging to a nation, religion, profession, company, sport (other hobbies), team, club, etc.
11. To protect and rationally use nature.
12. To have the convenient and effective technology for the *utilization* of human waste products, etc.
13. To fulfill the need for beauty, harmony (aesthetic needs)[11].

Let's comment on some the most important natural needs. For example, the most important natural need No. 5 to "organize ..." should increase with the growth of the structural complexity of TS. Complexity without order leads to chaos.

It would seem that such periodic natural needs as the development of mental, creative, physical abilities, spiritual development (No. 7), as well as the need to have fun, entertain and relax (No. 8), are unlikely to arise in extreme situations (for example, a person is immured in a basement when a building collapsed, got lost in a cave or in a forest, got into an uninhabited island as a result of an airplane or ship crash, found himself in a war zone, etc.), because "it's not up to it." However, the facts show that there are many cases when the realization of precisely these needs in extreme situations has allowed many to avoid strong *frustrations*, sometimes even madness, and, in fact, saved people's lives. Paradoxical, but true: during the war in Iraq in 1991 (Operation Desert Storm) 600 American servicemen **received their education** on the aircraft carrier remotely.

It should be noted that, although needs No. 11 and No. 12 are very important, they have not yet become fully natural for each person and society as a whole and are included in the list rather as desirable, therefore, it is necessary to develop environmental education and upbringing (as well as corresponding regulations) constantly so that these needs can become really conscious and natural.

The same applies to need No. 13, the importance and constancy (making natural) of which are not yet realized by many. The outstanding Soviet aircraft designer O.K. Antonov said: "We know perfectly well that a beautiful plane flies well, and ugly one flies badly or even can not fly at all"[12].

There are many brilliant examples of harmony in animate and inanimate nature, art, architecture, technology (for example, the "golden ratio", or "divine proportion"), which should be known and used by analogy in the creation and development of TS[13].

6.2.2.2 Fulfillment of the most important human needs

Considering in general the extremely high importance of the most important natural needs, it is advisable to consider the possibility of their implementation when developing or improving any TS (especially those whose main function is far from the most important needs). Then if even one stable need can be realized in a technical decision, then, as a rule, the products obtained on its basis acquire new, much demanded useful properties. Sometimes the most important natural need supplants the main function of TS and becomes the main function itself (for example, a felt-tip pen for treating abrasions)[14].

The analysis of TS development and the creation of new generations TS in various industries in recent decades shows that all significant improvements of TS analogs or technical decisions that have become the basis for the creation of new generations TS are aimed at the implementation of the most important human needs. I would not say that all the developers of TS have been guided by the list of human human needs that I formed. It is possible that many of them have themselves formulated certain human needs corresponding to the above ones, and used them in their scientific and technical creativity when setting the problem. However, the circumstances, mentioned above, also indicate the verification of the high importance of all 16 groups of the most important human needs.

The examples of the implementation of the most important human needs are given in Appendix 4.

6.2.3 Criteria for the progressive development of technical systems

6.2.3.1 Essence of progressive development criterion

When setting the task of scientific and technical creativity (developing TS, creating a new generation TS or creating pioneer TS), the main design goals and the main directions for achieving these goals should be determined. However, this is not easy to do. This situation was wisely described by the ancient Roman philosopher Lucius Annaeus Seneca: "If one does not know to which port one is sailing, no wind is favorable".

Model-driven activities using heuristic tunnels (see Fig. 89), guiding them into competitive technical decisions, can help determine "to which port one is sailing".

One of such heuristic tunnels can be the criteria for the progressive development of TS, the essence of which is as follows. Analysis of the evolution of many TS over a long period of time shows that, in general, their development (increase in consumer value and utility) has proceeded either almost monotonously when the parameters of TS properties have changed, or abruptly when new useful main and auxiliary functions, as well as new properties have appeared (while maintaining positive properties or elimination of negative ones).

At the same time, the progressive change in TS properties does not occur chaotically: with a stable tendency to an increase in the parameters of some properties, others decrease or stabilize (for example, the constancy of the oscillation period of a pendulum in a clock determines its accuracy).

These progressive tendencies (increase, decrease, preservation of properties parameters constancy) are called the criteria of progressive development (CPD) of TS, and new functions, new properties (or old properties with new parameters), new functional structures, principles of operation, technical decisions of TS that provide further the preservation of these tendencies, in essence, are scientifically grounded promising directions of development (qualitative improvement) of TS. All this determines the high importance of CPD formation and use at the stages of conceptual TS design.

Let me remind you that the improvement in the parameters of one property of any TS inevitably leads to deterioration in the parameters of another property or even several properties. When designing TS, this technical contradiction makes us compromise and choose not the maximum or minimum possible values for each property[15], but lower or higher values at which other important properties will have acceptable values[16]. For example, if the parameters of the consumer property "comfort" are selected for the car, provided by the parameters of its other properties (comfortable microclimate, minimum noise level in the cabin, maximum level of damping of seat vibrations when the car is moving, adaptability of the seat to the driver, implementation of massage properties, etc.), then it is often not the maximum or minimum possible parameters of the comfort properties that are chosen, but the ones corresponding to their acceptable price. At the same time, although in practice some properties of comfort are completely abandoned (that is, the property parameter is taken to be zero), at the stage of conceptual design, the criteria for progressive development with respect to comfort should be formulated as "comfort must be increased" and strive to find the technical decision that resolves the contradiction in the best way: benefit (comfort) — charge for the benefit (cost of implementation).

In general, the tasks of TS designing are reduced to achieving one of the following goals:

- Useful functions and properties of TS are set, it is required to implement them at the lowest cost (charge for benefits);
- Allowable costs are set (charge for benefits), it is required to achieve the greatest benefit (implementation of new functions and properties and / or improvement of the parameters of existing properties).

To form the CPD of TS, in the general case, it is necessary to define two groups of properties.

1. General significant and target properties of TS, directly affecting the utility (consumer value) of TS.
2. Properties that determine the charge for utility, as well as special significant properties that do not directly affect TS utility.

For example, in relation to an electric car, the first group of properties will include the maximum speed, engine power, minimum turning radius, acceleration time to maximum speed, number of passengers, maximum permissible cargo weight, environmental and ergonomic properties, manufacturability, operation and repair, reliability, durability, aesthetic properties, and even such as, for example, the visual presentation of information in sensors and devices (such, often very important properties are called *non-metric* and are expressed only in a qualitative form).

The second group of properties will include the costs of development, manufacturing, operation, utilization of an electric car, as well as its dimensions, unladen weight (recall that its usefulness is directly determined by the number of passengers and the maximum permissible cargo weight, and not by the dimensions of the electric car), energy consumption, rotation speed electric motor, etc.

Progressive tendencies of change or stabilization of each property parameters of the first group can be identified in two ways:

— To use the actual values of changes in the parameters of a specific property in the process of progressive evolution of this TS;
— To conduct a thought experiment — varying the parameters of a specific property within wide (extreme) limits with an analysis of its impact on TS utility — and formulate identified CPD (regardless of possible deterioration of other TS properties). The latter is especially important in order to prevent the emergence (or neutralize the influence) of psychological barriers to thinking during CPD formation of TS (in contrast to the process of forming a technical contradiction in TS, the essence of which is to identify a property, the parameters of which may deteriorate with the improvement of any TS property).

In the general case, the absolute values of properties parameters of the first group must be increased.

The absolute values of properties parameters of the second group must be reduced to zero. It is easy to realize this for properties that directly express the charge for utility, for example: energy consumption, weight, dimensions (where these properties are not targeted), cost, and others, the parameters of which should be reduced. It is much more difficult to comprehend and formulate CPD for properties that do not directly express charge for TS utility, but also do not relate to properties that directly affect TS utility (that is, they do not relate to the properties of the main TS functions).

A striking example of such TS property (or its element) is the rotational motion of a technical system (or its element), the parameter of this property is the speed of the rotational motion of a technical system (or its element). The CPD of this property as applied, for example, to a refrigerator being designed, as well as to all its elements that traditionally have a rotational motion, should tend to zero. In other words, there should be no rotating elements in the refrigerator. Indeed, in a modern household refrigerator operating on the Peltier effect, nothing rotates, which means that such a refrigerator is more reliable and, as a result, more competitive.

Another example: in comparison with rotary drilling, in more modern and more efficient turbine drilling, it is not a pipe column that rotates, but only a turbodrill (Fig. 92), which makes it possible to drastically reduce the requirements for pipe strength, since there is no need to transfer rotary moment from a drilling rig to a drilling bit through the pipes column.

Often, instead of identifying and formulating the useful TS properties that need to be improved, it is much more efficient to identify and formulate properties that are opposite to the useful ones. Then the decrease in the parameters of these properties or their complete elimination can lead to the improvement in the consumer properties of TS being developed. For example, the famous English expert in the field of design methods J. K. Jones advised to go to the easily perceived or imagined (and therefore easily described) state of discomfort instead of the elusive concept of "comfort"[17].

Figure 92. Scheme of rotary (a) and turbine (b) drilling: 1 — rotary drilling rig; 2 — pipes column; 3 — drilling bit; 4 — turbine drilling rig; 5 - turbine[18]

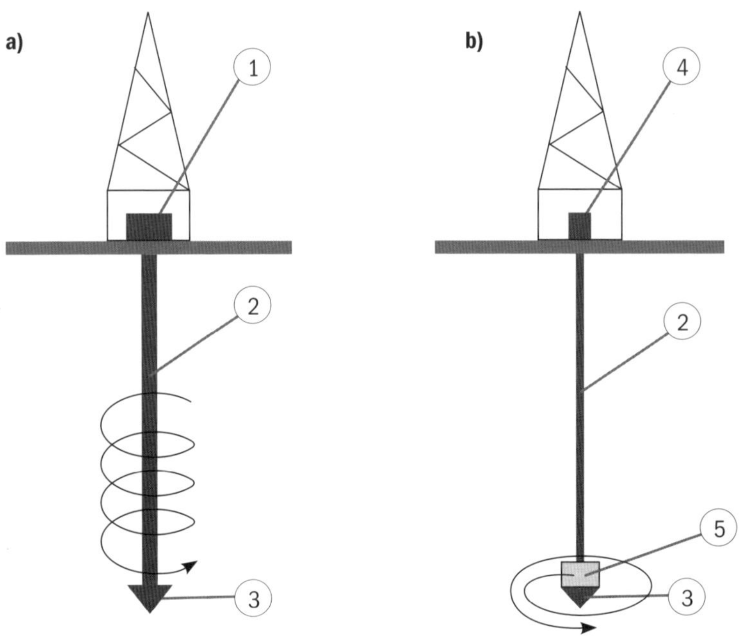

Or, for example, instead of formulating CPD of TS aimed at improving TS reliability, it is easier and more effective to formulate "CPD" leading to the increase in the unreliability of TS. Later, this technique has formed the basis of a rather effective method of "sabotage", the essence of which can be expressed by the phrase: "Towards good features and parameters through bad ones!"[19]

When forming CPD, it should be borne in mind that the total utility of several jointly functioning TS has additivity properties that depend on a number of conditions.

If two TS of the same type operate simultaneously in different (but adjacent) spaces, then the total utility is practically equal to the sum of their utility. For example, the utility of two locomotives (tractive effort) almost adds up.

With time-consistent operation of single-type TS in adjacent spaces, the joint utility increases, but usually it is less than the sum of the utilities of individual single-type TS (for example, the joint utility of multistage rocket engines)[20].

The following sections provide the methodology for the formation of CPD and examples of typical CPD for different classes of TS.

In general, the use of CPD in setting the task of scientific and technical creativity allows, together with the great Chinese thinker and philosopher Confucius, to assert: "Do not be afraid if someone runs faster than you. It is important that you move in the right direction."

6.2.3.2 Methodology for the formation of criteria for the progressive development of technical systems

When forming CPD of TS, sometimes it is necessary to take into account the tendency of the certain category of people to "demonstrative idleness" and "demonstrative consumption". Such desires (needs that took on a specific form) especially often began to arise at the end of the 18th century[21] and, unfortunately, they have continued to arise to this day. Satisfaction of such needs can be achieved when creating luxury goods: cars with interior dressed with merino leather; cases for smartphones inlaid with jewels; gold baths, taps and toilets, etc.

Taking this into account, the methodology for the formation of CPD of TS includes 11 steps.

1. For TS selected for further development (qualitative improvement), form lists of properties by groups: functional, technological, ergonomic and ecological, economic.
2. Formulate the concept of "utility" in relation to the given TS in the form of answers to the following questions:
   a) What is the main TS purpose, what social need is this TS designed to satisfy?
   b) What is the main TS property, improving (increasing, decreasing, ensuring constancy) of which increases the useful social effect of TS?

3. Highlight the target properties in the lists of properties that directly affect TS utility.
4. Select in the lists of properties special significant properties inherent in this particular TS.
5. Highlight in the lists of properties common significant properties (that is, inherent in most TS), for example: reliability, durability, dimensions, weight, energy consumption, ease of use and repair, aesthetic properties, ergonomic properties, environmental properties, manufacturability, cost, etc. It should be borne in mind that for some TS the "dimensions" and "weight" properties may be targeted.
6. Using all available relevant information sources, find the descriptions of 3–5 qualitatively different predecessors of TS under consideration that will generally represent TS class.
7. Analyze the change in target, special significant and general significant properties over time and determine the tendencies of their quantitative changes (increase, decrease, ensuring consistency), leading to the increase in the utility of the considered TS.
8. Based on the results of the properties analysis, formulate the CPD of TS under consideration, taking into account that the parameters of the properties characterizing the charge for utility, as well as properties that do not directly affect the utility of the analogue, should, as a rule, always be reduced (to zero).
9. Using all available reproductive knowledge related to this class of TS (represented by TS in question and its predecessors), as well as productive knowledge, try to formulate proposals for achieving each formulated CPD at the levels of quantitative parameters of TS, technical solutions of TS, principles of TS operation, functional TS structures, TS properties and their functions, new needs.
10. To form the additive (added) properties of the utility of TS being developed and consider the possibilities of their achievement.
11. Consider the feasibility of directions formation for the development of this TS for the realization of the need for "flaunting waste".

6.2.3.3 Examples of typical criteria for progressive development for various classes of technical systems

Table 17 shows examples of typical CPD for various classes of TS that can be used when setting the task of scientific and technical creativity.

Table 17. Examples of typical criteria for progressive development for various classes of technical systems

No.	Class of TS	Generalizing function (and its individual synonyms)	Criteria for TS progressive development			
			Functional	Technological	Environmental and ergonomic	Economic
1	Transport technologies and means (wheeled vehicles, ships, aircrafts, production lines, conveyors, pipelines, lifting mechanisms, ejection devices, including by the force of an explosion, means of movement due to the influence of various fields, etc.)	Transfer (move, drive, fly, swim, raise, lower, throw, eject, shoot, etc.)	*The speed of movement* of objects (people, cargo, objects of labor) *should be increased. The number of moving* objects (people, cargo, objects of labor) *should be increased. The range of objects movement* (people, cargo, objects of labor) *should be increased. The number of main and auxiliary functions* and the corresponding *processes and operations* that ensure the performance of the main function (transfer) *should be reduced*	*Resistance to movement* (frictional forces, aerodynamic drag, etc.) *should be reduced. The labor intensity of the processes of manufacturing* parts, assembling units of technical systems and their operation *should be reduced*	*The degree of harmful effects* of the products of associated and final activities (exhaust gases, leaks of oil products and other harmful substances, waste, worn parts, etc.) on the environment and people *should be reduced. The degree of TS adaptation* to the physiological, psychological and intellectual capabilities of a person – the user of TS *should be increased*	*The costs* of manufacturing, operation, storage, utilization and / or dumping of vehicles *should be reduced*
2	Technologies and means of disintegration (electrohydraulic, electropulse, explosive and other methods of crushing), disintegrators, crushers, mills, meat grinders, etc.	Disintegrate (grind, crush, separate, break, split, grind, cleave, etc.)	The disintegration range (for example, the number of size grades) should be increased. The degree of scattering, splintering of material during disintegration *should be reduced or* (for special purposes, for example, explosive delivery of rocks) *increased*	*The properties parameters that prevent disintegration* (adhesion force, strength, etc.) *should be reduced*, and the properties parameters that promote disintegration (porosity, fracturing, etc.) *should be increased*	*The seismic impact* of an explosion on the environment (with explosive methods of disintegration) should be reduced. *The level of emission of crushing materials* into the environment that are harmful to living nature and humans (for example, dust) *should be reduced*	*The costs* of manufacturing, operation, storage, utilization and / or dumping of disintegration means *should be reduced*

Design has two objectives: to implement specified useful TS functions and properties at the lowest cost and to implement the maximum number of new TS functions and properties and / or improve the property parameters for a given budget.

The key to success is to correctly formulate TS functions when setting a problem.

No.	Class of TS	Generalizing function (and its individual synonyms)	Criteria for TS progressive development			
			Functional	Technological	Environmental and ergonomic	Economic
3	Sorting technologies and means (screen, sieve, concentration table, filter, cyclonic collector, separator, classifier, jigging machine, flotation machine, cradle, winnower, trieur, etc.)	Sort (filter, separate, classify, grate, sift, winnow, centrifuge, float, jig, deposit, subside etc.)	*The number of functions and properties* that can potentially be used for sorting *should be increased.* *The indicator of completeness of sorting materials* by properties and parameters *should be increased*	*The property parameters* of the various materials to be sorted *should be increased or decreased* to improve the sorting process	*The degree of dangerous effect of harmful components* of sorted materials on the environment and humans *should be reduced*	*The costs* of manufacturing, operation, storage, utilization and / or dumping of sorting means *should be reduced*

6.2.3.4 Examples of directions formation for the implementation of criteria for the progressive development of technical systems at various stages of design

A simple, but, as a rule, effective way of further use of the generated CPD is that for each CPD of TS being designed, various possibilities of its implementation are considered at different stages of design activity (from the easiest one — optimization of property parameters in the technical solution of TS being designed — to the stage of analysis, synthesis and selection of functions of TS being designed, see Fig. 45). Examples of directions formation for the implementation of CPD of TS at various stages of truck design are shown in Table 18.

6.2.4 Formation of the ideal final result

At the stage of setting the task of scientific and technical creativity after the formation of CPD of TS, which determine the promising directions for improving TS, it is recommended to start forming the ideal final result (IFR) of scientific and technical creativity.

Let me remind you of the most important concepts that are used in the IFR methodology (see Section 5.2.2), with some additional explanations.

The goal of IFR formation is to overcome the psychological barriers to thinking

Table 18. Examples of directions formation for the implementation of criteria for the progressive development of technical systems at various stages of design

The criterion of progressive development	Stages of design activities					
	Optimization of property parameters for the TS being designed	Analysis, synthesis and selection of technical decisions for the TS being designed	Analysis, synthesis and selection of the operating principle of the TS being designed	Analysis, synthesis and selection of the functional structure of the TS being designed	Analysis, synthesis and selection of consumer properties and parameters of the TS being designed	Analysis, synthesis and selection of functions of the TS being designed
Cargo delivery productivity should be increased	Increase the speed of movement. Increase the volume of cargo capacity	Add a trailer (several trailers). Implement 4-way unloading. Implement a design for quick removal from a vehicle and installation of a loaded (empty) cargo container on a vehicle	Use diesel engine. Use turbojet engine. Use electric, hybrid engines	Add structural elements that provide multiplication of cargo capacities. Add a structural element that provides quick removal and installation of a loaded (empty) cargo capacity on a vehicle	New property parameters: movement speed, cargo capacity, loading/unloading time	Move cargo container from vehicle to vehicle
The degree of control over the properties of cargo during its transportation should be increased	Increase the degree of control over temperature and viscosity of cargo during its transportation	Develop a design for a cargo container that provides heating, cooling, and movement of cargo during transportation	Forced movement of cargo in a cargo container (rotation, oscillation, etc.)	Add a transformable structural element that provides cover on top of the cargo in the cargo container. Add structural elements that provide heating, cooling, and movement of the cargo	Required cargo temperature during transportation. Required cargo consistency during transportation	Heat the cargo. Cool the cargo. Mix the cargo. Generate vibrations in the cargo
…	…	…	…	…	…	…

(see Section 2.3.3), associated mainly with stereotyped thinking, *lack of flexibility* and conformism, which prevents the emergence of highly competitive, often at first glance even paradoxical, technical decisions (when the ratio of utility to costs tends to infinity), therefore, the methodologists of scientific and technical creativity at the first stage of IFR formation recommend not to think about whether the implementation of IFR is possible in principle, and not to think about the charge for utility, which consists in what costs will be required to achieve IFR or which properties will deteriorate with the improvement of other properties.

When forming IFR, one should remember about the need to resolve the constant dialectical contradiction in TS "universalization — specialization", which is the source of development. It is known from the history of technology that in the beginning, multifunctionality has developed more, but in recent decades specialization has intensified, manifested not only in the creation of dozens of TS types representing different large classes of TS (for example, ships, airplanes, cars, etc.), but also in increase in the number of varieties of relatively simple TS. For example, there are army knives, dining knives, penknives, cutting knives, for pizza, meat, cheese, mushroom, bread, fish, for curly cutting (carving), etc. One should aim for increase in the degree of integration of one TS into another TS, adjacent or being a supersystem (for example, installing a set of TS on the steering wheel of a car that perform other functions) and / or into humans, animals, plants, and even into cells.

In general, when forming IFR, it is useful to consider the possibility of using all heuristics based on productive knowledge (see Section 2.3.2, Table 5), and after the formation of IFR, one should again refer to these heuristics for their use in order to resolve contradictions in technical decisions that represent IFR.

6.2.5 Correctly formulated functions of a technical system when setting a problem as the key of its successful solution

6.2.5.1 Rules for the formulation of functions

Before reading this section, I recommend that you once again refer to section 1.4.2 "Formation of technical system's functions and properties ", where the concepts of "function" and "types of functions" are explained. Formation of the list and formulation of TS functions is, perhaps, the most difficult process in setting the task of scientific and technical creativity, to the greatest extent determining the success or failure of TS design as a whole.

I must admit that I could describe the functions of TS correctly only after much thought and repeated reformulation. But after that, almost immediately I have worked out many competitive technical decisions for the implementation of a correctly formulated function (one of such examples is described in Section 6.5).

In general, the structure of the function description coincides with the structure of technically realizable need description (see Section 1.1) and includes:

1) The description of the action (impact or reaction) of a material object or phenomenon aimed at changing the properties or state of another material object or phenomenon;
2) The description of the material object or phenomenon to which the action is directed (impact or reaction);
3) The description of the special conditions or restrictions under which the action is performed (action or reaction).

For example, "measure the temperature of the sea water at a depth of 8 km."

The rules for formulating functions are not complicated and include four main theses.

1. Action (impact or reaction) in a function should be expressed by an indefinite verb; at the same time, the verb should not indicate the way of implementation (for example, instead of "crush", "cut", "tear" it is better to "separate", and instead of "melt", "crystallize" it is better to "change the state").
2. When describing an action (impact or reaction), verbs of indirectly transitive action (for example, "improve", "exclude", "increase", "achieve", "provide", "prevent", "reduce", "deepen", etc.) should not be used.
3. When formulating an action (action or reaction), the particle "not" should not be used, that is, the function should reflect the direct action (for example, instead of "do not pass" it is better to "restrict movement").
4. When describing the object or phenomenon to which the action is directed, it is better to use the generalized description of the object or phenomenon (for example, instead of "water", "air", "soil", "material" use "substance"). Then the efficiency of the mechanism of convergence (closing-in, transfer) of knowledge increases, and, for example, technical solutions implemented in TS used to move water can be used in TS for moving air or other fluids.

Let me list the most common mistakes when formulating TS functions.

1. Instead of a function, the need is actually indicated in a declarative form (for example, to increase productivity; reduce losses, taxes, energy costs, etc.).
2. Specific TS for the implementation of the function is indicated (for example, replacing the erroneous specific function "to salt" with "to change

taste sensations", the University of Tokyo has developed a fork for hypertensive patients (see Fig. 93), which sends a weak electric current, which causes the sensation of salinity).

Figure 93. A fork for hypertensive patients developed at the University of Tokyo. The electric fork generates a salty taste, although there is no salt in the dish

3. The formulation of a function as an action is replaced by the formulation of the development (creation) of an object or phenomenon itself.

Examples of erroneous and recommended (not claiming to be the ultimate truth) formulation of TS functions are given in Table 19.

It should be noted that the functions listed in Table 19 "sort", "connect", "transfer" and others are called *generalizing* functions by me (should not be confused with the mathematical concept of "generalized function"), since they have many implemented and potentially possible varieties and have a potentially high creativity due to the high level of abstractness in the presentation of actions.

For example, at present, only a dozen fundamentally different technologies for ore washing have been implemented — increasing the concentration of a useful component in it (manual disassembly, washing, jigging, centrifugation, separation, flotation, etc.). However, the correct formulation of the function as *generalizing* "sort a substance by properties and parameters" immediately suggests a multitude of potential washing technologies based on sorting useful components and waste rock by the difference in their properties and property parameters. Under my leadership, about 1400 physical properties have

Table 19. Examples of typical mistakes when formulating TS functions[22]

Common erroneous formulation of a function	The nature of the error	Recommended formulation of a function
To reduce viscosity (oil)	The need is formulated to change one of the properties, not a function	At the high level of abstraction – to destroy (weaken) the bonds between the elements (molecules) of the substance. At acceptably low level of abstraction (which does not identify a specific technology): – heat the oil; – generate fluctuations in oil
To filter (gas, liquid), sift, classify, wash (ore), centrifuge, separate, jig (substance), disassemble (manually), deposit (substance)	Specific technology is indicated	To sort a substance by properties and their parameters
To screw, weld, magnet, rivet, fasten, glue, freeze, sew, suck, attach, tie	Specific technology is indicated	To connect
To regulate the temperature (of environment, substance)	Need is formulated	To heat or cool (environment, substance) with certain conditions (speed and frequency of temperature measurement). In this case, the function "to measure" should be implemented
To build a dam, a lintel To put a pit prop	Object-oriented approach instead of functional one	To limit the movement of substance
To pump (water) To ventilate (air)	The vulgar name of one of the technologies is indicated	To transfer the substance
To store substance (gas, liquid other)	There is no action (reaction)	To change the properties (their composition and parameters) of a substance, environment
To explore (deposit)	Need is formulated	To measure properties, parameters of a substance, environment
To vibrate	Incorrect wording	To generate oscillations
To salt (food)	Specific technology is indicated	To alter taste sensations
To measure (object)	Insufficient heuristicity of the function	To detect the presence of a property with specific parameters
To detect (object, phenomenon)		To measure properties(their composition and parameters) of a substance, environment
To drill a well, borehole	Object-oriented approach instead of functional one	To form a conductive space
To drive a working (tunnel, mine shaft, etc.)		
To dig a well, pit, ditch		

been systematized (see a fragment of the taxonomy in Fig. 17), and there are also chemical and other properties. The use of each new property (and its parameters) can become the basis for the creation of a new effective washing technology.

The high abstraction level of the generalizing function and the varieties of its implementation allows them to be interpreted, respectively, as general and individual human needs, representing the 1st and 2nd stages of the systemic methodology of design activities (see 3.2).

The common reasons analysis of the general human needs emergence that I conducted allowed me to form an interdisciplinary list of common human needs as antonyms of common reasons that covered all areas of human activity.

In this case, the emerging contradictions between TS, TS and man, TS and nature, which are individual reasons, should be considered as antonyms of the individual human needs formation in the form of varieties of the corresponding general human needs fulfillment (Table 20).

An inter-industry list of general human needs and varieties of their fulfillment, which can become the basis for the formation of corresponding individual human needs, are given in Table 20.

Table 20. Interdisciplinary list of general human needs and types of their implementation

Name of a general human needs	Varieties of generalizing function implementation
To protect	Prevent (natural disasters, man-made accidents, military threats, dangerous diseases, crime, terrorist attacks), isolate, paint, fence, screen…
To disintegrate (disconnect)	Crush, cut, bite off, dissolve, melt, pinch off, rub, tear…
To transfer	Roll, drag, throw over, jump over, fly, swim, crawl…
To form a conductive space	To drill a well, create a tunnel, a mine, laying a cable, build a pipeline, use substances and effects that facilitate the conduction of energy, substances and information (for example, capillaries)…
To make forms	To pour, sculpt, trim, sprinkle, melt, glue, freeze…
To change the aggregative state of substance	To sublime (sublimate), melt (condense), crystallize, ionize…
To integrate (connect)	To connect immovably detachably (screw on …), connect immovably permanently (weld, sew, glue, press in …), connect movably permanently (flare, calk …), bubble, carbonate (saturate the liquid with gas), mix (liquids, gases, bulk solids, gels, etc.)…
To reproduce	To print, photograph, copy, engrave, synthesize, clone…
To inform	Generate mechanical, electromagnetic and other vibrations, smell, color, sign …
To limit the movement of matter, ions	To enclose, freeze, glue, bind, sling, reduce the humidity of a substance to change its mobility, to slow down ions, use the electrophoretic effect…

Name of a general human needs	Varieties of generalizing function implementation
To sort	To disassemble, jig, filter, sift, wash, centrifuge, float, separate, divide in heavy suspensions, deposit ...
To treat	To use various types of therapy, including drugs, surgical methods, therapeutic mud, mineral waters, cellular technologies and even animals, insects, plants...
To prevent	To vaccinate, use masks, contraception methods, hygiene, sanitation, preventive physiotherapy, medical devices and equipment for the prevention of joint diseases, spinal curvatures, ENT organs, the whole body, control of human (animal) organ parameters, give up bad habits...
To develop abilities	Teach and/or learn to formulate important goals and objectives, analyze and synthesize, fantasize, reflect, model, construct, associate, develop (sense organs, memory, critical and abstract thinking, imagination, physical health, fine and gross motor skills), establish differences and similarities between objects and phenomena, generalize...
Have fun and entertain	To have fun: read, solve puzzles, play games, engage in sports, performing arts, music and other forms of art and creativity, communicate on social networks... To entertain: to participate in entertainment events as a performer, as well as a specialist ensuring the development and creation of entertainment complexes and equipment, their operation, organization and holding of entertainment events...
Fulfil aesthetic needs	To visit art galleries, museums, concerts, theatre, sports and other performances; collect works of art and other objects; decorate rooms and/or artistically transform natural objects; contemplate, listen to and/or create objects of art, architecture, literature, music, dance...
To provide multifunctionality	To separate, pierce, open, close, fix, heat, cool, generate vibrations, protect, mask, dissolve, enlarge images, cleanse, disinfect, illuminate, inform, radiate, accumulate, crush...
To change the state of TS and the environment by changing the quantitative parameters of TS properties and the environment	To heat TS and the environment surrounding the TS, cool TS and the environment surrounding the TS, generate movement of substance in TS and the environment surrounding the TS, compress or loosen substance in TS and the environment surrounding the TS, change the aerodynamic resistance of TS during movement or at rest, create a vacuum in TS and the environment surrounding the TS...
To reserve	To reserve TS, TS elements, the values of their strength, deformation and other characteristics (power, force, energy, time, space, capacity, buoyancy, lifting capacity, reliability, speed, survivability, productivity...), as well as external and internal biological organs...
To transform (energy, substance, information)	To transform: for example, chemical energy into mechanical, solar into chemical, electrical into thermal...; Modify substance: magnetize, compress...; Information: numerical, text, graphic, sound, video..., information processes: collection, processing, transmission...
To ensure compatibility	To ensure compatibility of functions, properties, functional structures, principles of operation, technical decisions of various TS, as well as in relations between states and peoples with different social systems, cultures, histories, traditions...

Name of a general human needs	Varieties of generalizing function implementation
To accelerate, slow down	To accelerate movement, process. Forms and methods of acceleration of movement, process: biological destruction, catalysis; uniform (constant) acceleration, equally slow, tangential, centripetal; collective methods of acceleration of charged particles... Slow down, process. Types of slowdown: partial, full, service, emergency... Types of process slowdown: inhibition – the process of inhibition or slowing down chemical reactions...; anabiosis – slowing down, suppression of vital activity in biological objects...; stagnation, recession – slowdown in the economy...
To change directions of movement	To reverse, turn by active and passive step-over, "plow"; change lanes: use kinematic or power steering of vehicles...
To organize	To organize the arrangement of objects and their elements (including signs, symbols, sounds), the shape of objects, types and characteristics of movements, connections between objects and their elements, or to replace the existing order with another by changing properties, parameters of properties, and combinatorics. To create (sort) a new set of objects, elements, ornaments, art objects, architecture, advertising and information from ordered objects, elements, substances, to increase the concentration of the substance in the new set...
To combine	To combine objects, their elements, technologies, their processes to obtain combinations that are new and useful...
To change taste sensations	To change taste sensations (sweet, sour, salty, bitter, alkaline, metallic, hot) not only by means of traditional seasonings (sugar, salt, pepper), but also by means of physical, chemical and other effects that change taste sensations without the use of traditional seasonings...
To generate oscillations	To generate mechanical (sound, vibrations), electromagnetic (light, radio waves, heat), damped, undamped, forced, free, self-oscillations, parametric and others...
To preserve (nature, artificial objects)	To develop, produce and use TS that ensure the elimination or reduction of emissions of carbon dioxide into the atmosphere, other harmful substances, as well as untreated discharges into water bodies and pollution of fertile soils with chemicals. Plant trees and other plants, recycle or bury organic waste. Sort waste and hand it over to collection points. To create advanced means of monitoring and protecting existing technical systems, as well as technologies, equipment and products for eliminating the consequences of dangerous man-made incidents (fires, contamination of the territory, etc.)...
To measure	To measure the properties and/or parameters of the properties of an object or phenomenon, or of the environment, for example, by means of logging or other methods of measuring physical fields...
To detect	To detect the properties and/or parameters of the properties of an object or phenomenon, usually with the help of special technologies, equipment...
To recycle	To use technical systems whose service life has expired for a new purpose, as well as production and consumption waste as material, energy and other non traditional (hidden) resources to create new TS or reduce the operating costs of existing TS...

The practice of using an inter-industry list of general needs and varieties of their fulfillment (as a basis for the formation of individual human needs) shows that the list of general human needs is practically exhaustive, and the list of individual human needs is continuously replenished.

In the context of the digitalization development in all spheres of human activity, the stages of conceptual TS design (the first 7 stages in the system methodology of design activities, determining the success or failure of the design as a whole) have practically no computer support, which could be an effective tool for conceptual TS design.

Therefore, I propose to consider the first results of computer support creation for the stages of conceptual TS design based on databases (DB), heuristics and artificial intelligence, designed to reduce the number of trials and errors multiple times and significantly reduce costs at each stage of conceptual TS design.

Considering that the entire world created and being created by man is represented by a variety of TS (see Fig. 1), general and individual human needs are associated with the search for a prototype of TS (the best analogue of TS), which should satisfy human needs. When this does not happen, there is a need to improve the TS prototype (there are very few cases when there are no TS analogues in the world to satisfy human needs, so such cases are not considered yet).

Thus, **at the first stage** of the system methodology of design activities (the first stage of conceptual design), the DB of general human needs (GN) can be formed on the basis of an inter-industry list of general human needs (Table 20) with synonyms of general human needs included in it.

At the second stage, individual human needs (IN) can be formed on the basis of a DB of fulfillment varieties of general human needs aimed at resolving contradictions identified in the prototype of TS, as well as on the basis of the most important human needs selected from the list (DB).

At the third stage, it is necessary to specify the individual human needs in the function (F) of the TS being improved, and also to add to these functions new others, formed on the basis of the DB of "controlling" and "good" verbs; **at the fourth stage** — to specify the list of functions by properties (P) of the TS being improved, and also to add new others, formed on the basis of the DB of the ontology of physical properties, taxonomies of chemical, biological properties and the DB of "good" adjectives (including "smart" and "clean"). **At the fifth stage**, for each function in the TS being improved, form elements of the functional structure (FS) corresponding to these functions, and also reserve the most important general significant TS properties (using the DB of general significant TS properties). **At the sixth stage**, find or synthesize a new operating principle (OP) of the TS being improved, using an automated system for searching or synthesizing OP. **At the seventh stage**, based on the data generated at each stage of conceptual design, form a consolidated technical decision (TD) for the TS being improved. **At the eighth stage**, if necessary, using CAD, optimizes the parameters of properties in the technical decision (TD) of the TS being improved.

At the same time, at all stages of conceptual design it is planned to use neural networks for computer support of heuristics.

The possibilities of using generalizing functions and individual human needs in education are described in Chapter 7.

6.2.6 Laws and patterns of technical systems development and their use when setting a problem

6.2.6.1 General information about the laws of technical systems development

The creation and development of technical systems is a complex process influenced by many factors, in particular: the current needs of man and society, the objective level of TS state and technical progress in general, the presence and level of competition. However, since the second half of the 19th century, the idea that the process of creating and improving TS is subject to the influence of some objective laws and patterns has begun to be heard more and more often.

Let us first of all consider the laws of TS development and possibilities of using them at the stage of setting the problem of scientific and technical creativity.

In the Philosophical Encyclopedia[23], the law (Greek νόμος, Latin lex) is defined as the tendency of change, movement, development inherent in the nature of the phenomena of the real world, which determines the general stages and forms of formation and self-organization process of specific developing systems of natural phenomena, society and the spiritual culture of mankind.

Methodologists of scientific and technical creativity, especially G. S. Altshuller[24] and his followers[25], have done great deal of positive work to popularize the opinion about the origin and development of TS based on laws and patterns and formulated many laws of TS development. However, most of them "are short of" the high status of the law, and some are based on false premises (examples of such "laws" will be given in Chapter 8 "G. S. Altshuller, of course, is great, but...").

The generally accepted laws of TS development (which will be discussed in more detail below) are the laws of stage development, progressive evolution, and the correspondence between TS functions and structure.

The processes of creating and improving TS are influenced by objective laws and patterns.

The generally accepted laws of TS development are the laws of stage development, progressive evolution and the correspondence between TS functions and structure.

6.2.6.2 The law of technical systems stage development

In my opinion, the law of TS stage development is the most significant in scientific, technical and other creativity, and the consequences from it are useful at all stages of creativity (setting, solving a problem, commercializing or other use of the results).

This law, formulated by S.S. Tovmasyan in 1972[26], says: the emergence of TS, and then its qualitative development occurs step by step during the transfer from man to TS in the strict sequence of the following fundamental functions of labor (**FFL**):

T — Technological FFL directly related to the transformation of labor subject;

E — Energy FFL associated with the provision of energy for the processing of labor subject;

O — FFL of operation, associated with the operation of labor subject process transformation and providing energy process;

P — FFL for planning the quantity and quality of products (or actions).

For example, if it was necessary to carry stones and in the absence of TS, at first man (M) performed all FFL himself: he lifted and carried a stone (technological FFL), using his muscular strength (energy FFL), he controlled the process (FFL of operation) and planned where to get it and where to put the next stone (FFL of planning).

Symbolically, this state can be represented as follows:

M = T + E + O + P;

TS = 0.

To facilitate and intensify the work of carrying loads, man created primitive TS — a drag, then a wheelbarrow (Fig. 94). At the same time, a new technology arose due to the fact that the technological FFL was transferred to TS, but the rest of the FFL remained for the man.

Figure 94. TS transferred to technological FFL: a) implementation option — drag; b) implementation option — a wheelbarrow transported by a man

a) b)

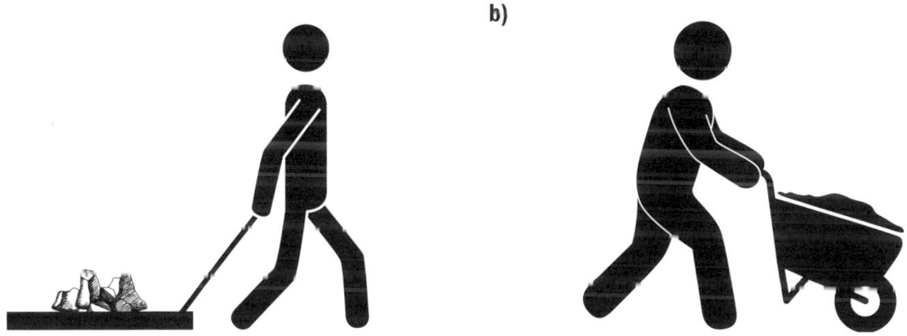

Thus, the first stage of the law implementation can be symbolically represented as:

M = E + O + P;

TS = T.

At the second stage, man put a draft animal to the cart, which then he replaced with an engine (Fig. 95), transferring the energy FFL to TS. Then, symbolically, this stage of the law implementation looks like this:

M = O + P;

TS = T + E.

Figure 95. Technological and energy FFL is transferred to TS: a) implementation option — a cart with a draft animal; b) implementation option — a car

a)

b)

At the third stage of law of development of TS implementation, along with the technological and energy FFL, FFL of operation (O) is transferred (Fig. 96), and the planning FFL (P) is performed by man:

M = P;

TS = T + E + O.

At the fourth stage of law of development of TS, along with the technological (T), energy (E) and FFL of operation (O), the planning FFL (P) is transferred:

M=0

TS=T+E+O+P

To show the universality of the law of the staged TS development, we will consider, from the point of view of this law, the stages of the emergence and development of various TS (transport, education technology and library technology) in Table 21.

Analysis of Table 21 shows that it is during next FFL implementation in TS that qualitatively new consumer properties appear.

It is very simple to determine FFL, which can be a reference point and even a hint for the formation of the most promising direction of TS development (Fig. 97). The

Figure 96. FFL of operation, along with the technological and energy FFL, is transferred to TS: implementation option – vehicle control systems in emergency situations (ABS, SRS, etc.)

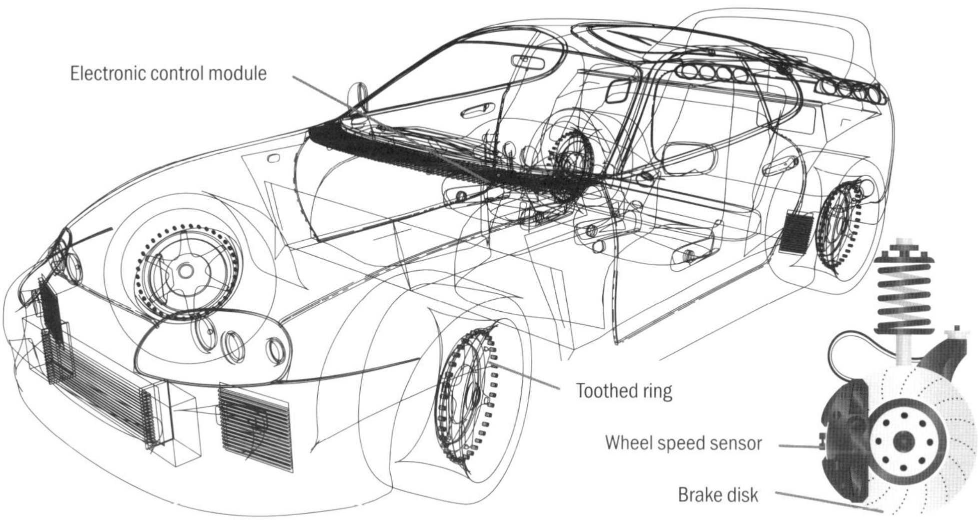

figure shows that it is necessary to define which FFL have already been implemented in the existing TS analogue (prototype), and then, knowing the strict sequence of FFL (T → E → O → P), the implementation of which leads to the creation of qualitatively new TS, choose the next one (after implemented in TS) FFL.

If all FFL are implemented in TS (as, for example, in a car), then you should return to the technological FFL and consider the possibility of its implementation with new properties (for example, the car must move not only on land, but also on water). To ensure such technological FFL, as a rule, subsequent FFL (energy, control, planning) must be implemented in a new way.

After another implementation of all FFL, all stages should be repeated again, based on the fact that technological FFL should ensure the movement of the car not only on land, water, but also through the air (perhaps such TS will no longer be called a car). Further the stages that provide going to the space, other galaxies At the same time, the implementation of each new technological FFL will lead to the emergence of a new TS generation.

It should be noted that the law of staged TS development can be effectively used in the formation of a well-founded scientific and industrial policy. However, nescience or ignorance of it leads to the planning (and implementation) of R&D (research and development work) based on insufficiently effective improvement of TS only by optimizing the parameters of TS properties.

Figure 97. Algorithm for promising directions formation for TS development based on the law of staged development

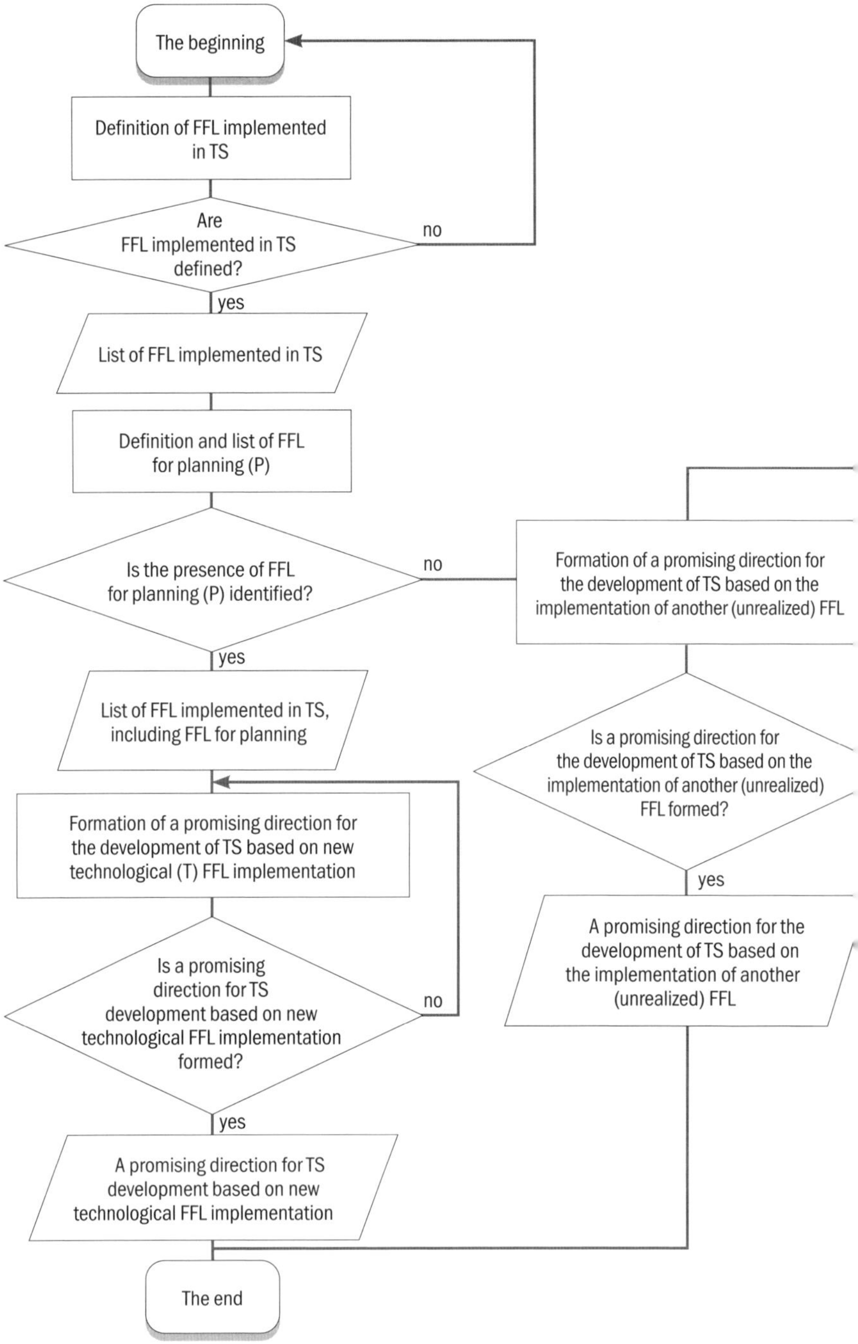

6.2.6.3 The law of progressive evolution of technical systems[27]

The operation of the law is based on logic, which determines the following most probable ways of hierarchical exhaustion of TS capabilities, implemented in the form of several subordinate cycles (Fig. 98).

Table 21. Stages of creation and development of transport, education and library systems in the context of the staged development law[28]

Industry or TS name	Stages of origin and development of TS based on the division of FFL between M and TS			
	Stage 1 M=E+O+P; TS=T	Stage 2 M=O+P; TS=T+E	Stage 3 M=P; TS=T+E+O	Stage 4 M=0; TS= T+E+O+P
	TS origin	TS development		
Transport	Drag, wheelbarrow, trolley transported by man	Cart with draft animals; self-carriage; first car	A car with anti-lock braking, traction control and other systems	Taxis without a driver are already available at Heathrow (London) and Narita (Tokyo) airports
Education	Traditional educational technologies (Humboldt model of the university)	Computerization of learning. Multimedia educational resources. Development of distance education	Formation of educational standards. Development of national and international certification of training courses. Creation of an automated control system of an educational institution. International e-learning standards (IMS, ADL). Metadata for the integration of educational resources. Cooperation in the creation of training courses based on the use of standardized educational objects. The Bologna Process. Nostrification of educational documents. Knowledge codification	Automated monitoring of the needs and results of educational and scientific activities. Automated planning of research priorities and training for various sectors of the economy and industry
Library	Traditional library technologies	Mechanization of library processes. Computerization of library activities	MARC formats. Interlibrary protocols and interfaces. The use of cloud computing technologies. Multi-platform and multi-format automated library and information systems (ILS). Integration of ILS into IT environments for supporting various types of activities	Automated monitoring of readers' interests and acquisition planning. Implementation of functions of purposeful implicit new knowledge extraction from various sources in ILS. Library without librarians

A. With unchanged general and individual human needs (respectively, GN and IN), functions (F), consumer properties (CP), functional structure (FS), principle of action (PA) and technical decision (TD) in TS, its properties parameters (Pp) are improved (cycle A)[29].

B. After the exhaustion of cycle A possibilities (improvement of TS by changing the parameters of its properties), transition to more rational TD takes place, after which the development proceeds along cycle A again. In this case, the A — B cycle is repeated until approaching the best TD for the given PA and FS.

C. After exhausting the possibilities for perfecting TS by improving TD, transition to new, more efficient PA can occur, after which the development of TS again proceeds along the B — A cycle, while the C — B — A cycle is repeated until approaching the best PA for many known, at present time, physical, chemical, biological and other effects and phenomena[30].

D. After exhausting the possibilities for improving PA, it is advisable to move to more efficient FS.

E. After exhausting the possibilities for improving FS (with the implementation of cycles D, C, B, A on the basis of the new FS), it is advisable to revise the consumer properties (CP) of TS (changing the CP parameters, excluding and / or forming new CP).

F. After exhausting the possibilities of the cycle E — D — C — B — A, it is advisable to form a new list of functions (F) performed by TS.

G. After the F — E — D — C — B — A cycle is exhausted, it is advisable to revise the individual needs (IN).

H. After the exhaustion of the possibilities of the cycle G — F — E — D — C — B — A, it is advisable to replace the general human need (GN)[31].

The method of using the law of progressive TS evolution for positioning the considered TS in the hierarchy of its capabilities exhaustion and determining the next stages in the development of this TS consists in the following sequence.

1. To identify publications (articles, patent applications, patents, etc.) related to the considered TS over the past 3–5 years (for pioneering TS, it is advisable to identify all publications since its inception), using all available information materials.
2. To group publications by TS novelty levels (see Section 2.2) and determine their frequency (quantity per unit of time) for each level.
3. If, as a result of the analysis, it turns out that the frequency of publications devoted to considered TS and, for example, to the first level of novelty (changes in the parameters of TS properties; see Table 4, Section 2.2), is high, then this TS has not yet exhausted the possibilities of cycle

A (Fig. 98) and it should be further developed by optimizing the property parameters.

4. If the frequency of publications on the first level of novelty fades out or has stopped, then TS has exhausted its possibilities of development due to optimization of property parameters (cycle A is exhausted) and you should go to cycle B (change TD) without changing GN, IN, F, CP, FS, PO (Fig. 98).
5. If, as a result of the analysis of publications frequency on TS under consideration, it turns out that most publications are devoted to the second level of novelty (TD) and the frequency is consistently high and even increasing, then cycle B has not yet been exhausted and TS should be continued to develop through the development of new TD and if necessary, optimize the parameters of its properties (cycle A).
6. If the frequency of publications on the second level of novelty fades out or has ceased, then TS has exhausted its possibilities of development due to cycles B and A and you should go to cycle C, and then to B and, if necessary, to A without changing GN and IN, F, CP, FS (see Fig. 98). Following this logic, depending on the situation, it is necessary to stay on the current cycle or go to the higher one (up to cycle H).

After the implementation of another higher cycle, descend successively along the lower cycles, including cycle A, if necessary.

The attentive reader will determine that the law of progressive TS evolution fully corresponds to the basic tactics of iterative approximations (see Section 3.2.3.2) in the system methodology of design activities, and will be right, since the law actually justifies this tactic.

6.2.6.4 The law of correspondence between functions and the functional structure of a technical system

The essence of the law lies in the fact that any element of the functional structure of TS must correspond to each useful function (main, auxiliary) that provides the consumer properties of TS, and the entire functional structure that unites all elements of the functional structure must correspond to the main function.

This law, in essence, refutes the formulation of the ideal final result proposed by G. S. Altshuller, the developer of TRIZ, and supported by his followers: "There is no technical system (which means there is no functional structure of TS. — V.P.), but the function is being performed" (see Section 5.2.2). I repeat, with the correct formulation of IFR, IT SEEMS AS IF there is no technical system and functional structure (but they certainly exist as well as a functional

Figure 98. The most probable ways of hierarchical exhaustion of TS capabilities

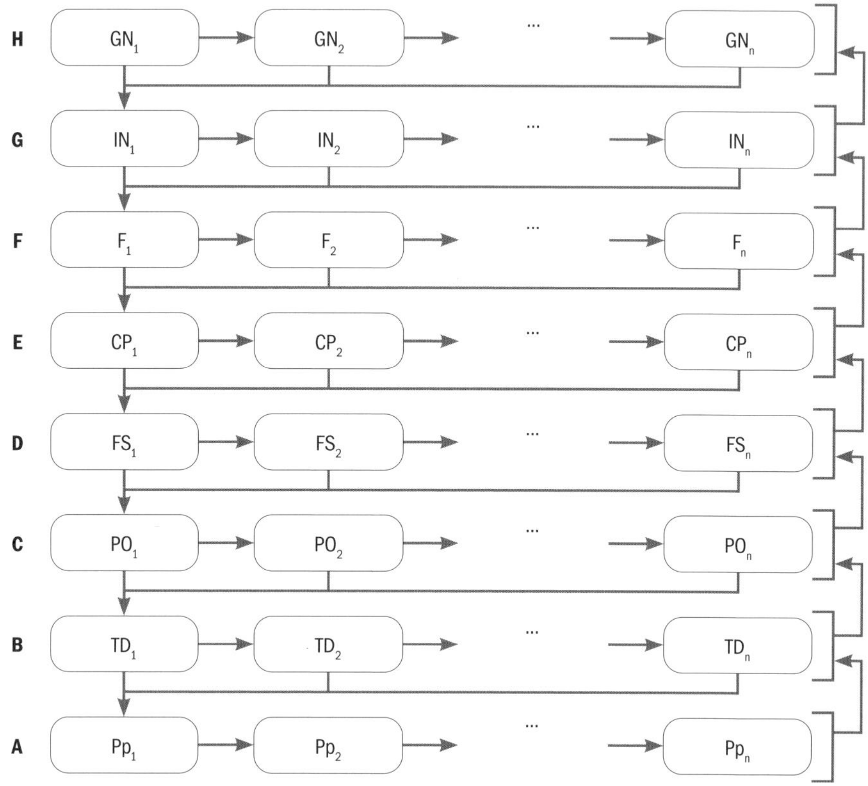

structure corresponding to it, that can be implemented in a latent form), so the function is fulfilled.

Inconsistency between function and functional structure can arise in three cases, which are presented in table 22.

6.2.7 Patterns in the development of technical systems

6.2.7.1 General information about patterns in the development of technical systems

In recent decades, scientists and specialists have been increasingly engaged in identifying and formulating patterns (trends) in TS development.

The patterns of TS development (qualitative improvement) presented in Table 23 are the most common in publications (although in general they have not yet been sufficiently developed).

Table 22. Examples of possible inconsistencies between function and functional structure and ways to prevent inconsistencies

The reason for inconsistency emergence	Examples of inconsistency manifestation	Ways to prevent inconsistency
An element is excluded from TS functional structure	The inconsistency arisen between functions and functional structure impedes the performance of useful functions. Unless special measures are taken, it is impossible to travel safely, for example, in a car if a wheel is lost or when the engine is stopped in a single-engine aircraft	When designing TS, structural elements or reserves (of power, strength, durability, energy, capacity, plasticity, buoyancy, pressure, speed, force, load capacity, reliability, survivability, time, light, color, viscosity, calorific value, productivity, space, material, functionality, dimensions, weight, permeability, stability, information, density, hardness, duration, and other capabilities) must be incorporated into the design to ensure safe way out in an emergency
The external conditions of TS operation or functioning change	A regular ballpoint pen may not write at low temperatures	When designing TS, the change (use of new elements) of TS functional structure should be provided, which ensures the performance of new functions, the need for which has arisen due to changed external conditions
Human need changes	The emergence of new human needs (for example, housing or vehicle comfort increasing) makes it necessary to implement new functions that are not available in the previous TS functional structure	When designing TS, it should be possible to build up (maybe modular) elements of the functional structure, eliminating the discrepancy between new needs, new functions and the previous functional structure of TS

Table 23. Common patterns of TS development list[32]

No.	The name of TS development pattern
1	The pattern of TS development with a change in the properties of the surfaces of TS elements
2	The pattern of TS development with switching to new types of converted energy or field
3	The pattern of TS development with switching to new types of mechanical movement of objects in TS
4	The pattern of TS development with a change in degrees number of objects motion freedom in TS
5	The pattern of TS development with a change in the geometric characteristics of the mechanical contact of TS elements
6	The pattern of TS development with a change in the configuration of elements of TS functional structure and the connections between them in space
7	The pattern of TS development with a change in the relative position of a pair of elements in TS
8	The pattern of TS development with a change in the geometric characteristics that determine the shape of TS elements
9	The pattern of TS development with switching to new types of TS elements symmetry
10	The pattern of TS development with a change in the characteristics of the empty space of TS elements
11	The pattern of TS development with a change in the internal structure of TS elements
12	The pattern of TS development with switching to new connection types of TS elements

When developing new TS or improving its analogue, use the patterns of TS development (trends in changes in the characteristics of the functioning or structure of TS, leading to improvement in consumer properties):

— when changing the surface properties of TS elements;

— when switching to new types of converted energy or field;

— when switching to new types of mechanical movement of objects in TS;

— when changing the degrees number of objects movement freedom in TS.

The basis of the modern understanding of these patterns was laid by the works of G. S. Altshuller, Yu. S. Meleschenko, E. P. Balashov, A. F. Kamenev, A. I. Polovinkin, S. S. Tovmasyan, N. A. Shpakovsky. Among foreign studies, devoted to the patterns of TS development, the works of D. Mann, G. Linde and P. Hill are the most often mentioned.

Based on TS studies, the pattern of TS development can be simplified in the form of an oriented graph, the nodes of which are the stages of TS improvement due to increasing the number of useful properties and / or improving their parameters, and the arcs represent transitions (and time intervals) between the

Figure 99. Simplified representation of the pattern of TS development and its use in scientific and technical creativity: TS1n, TS2n, TSN — new technical solutions obtained on the basis of the stages of TS development, the sequence of which is due to the increase in the number of useful TS properties and / or improvement of TS properties due to increasing their parameters; TS1m, TS2m, TSM — new technical solutions obtained on the basis of stages, the sequence of which is due to an increase in the number of useful TS properties and / or improving TS properties by reducing their parameters

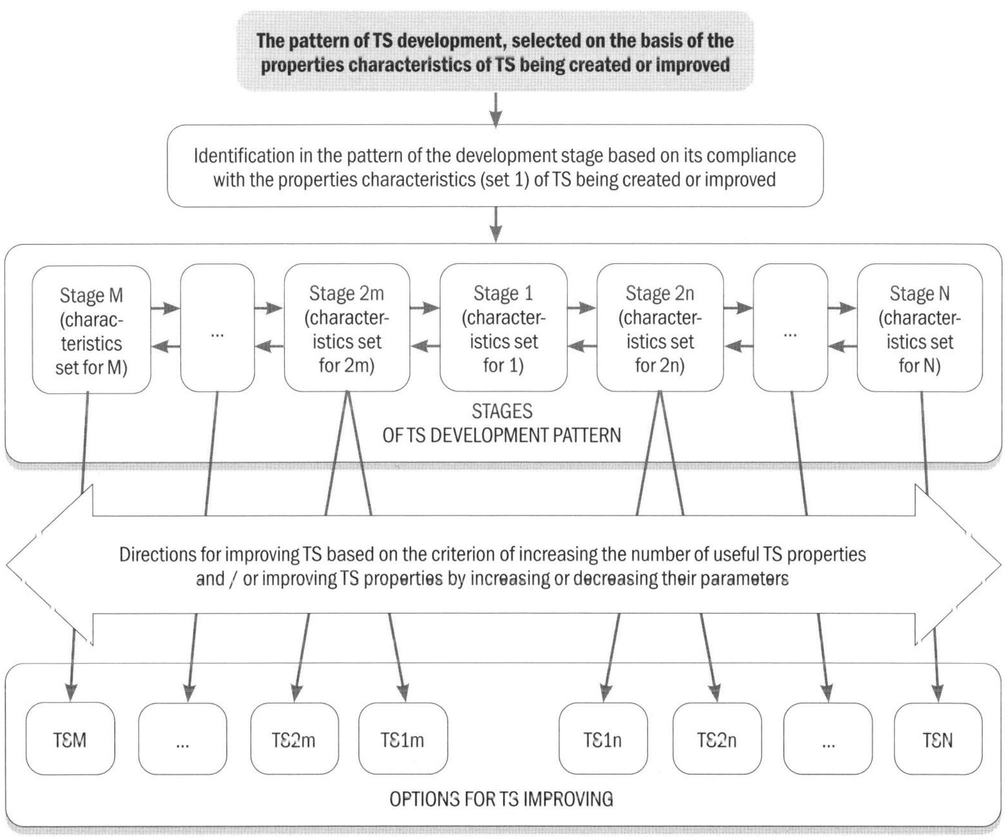

stages. In this case, the use of the pattern consists (Fig. 99) in determining the correspondence of the characteristics of the improved TS to the characteristics of any stage (or, in other words, determining the position of improved TS on the graph), and the heuristic conclusion is that, with a high probability, the most promising options for improving TS (TS1n, TS2n, TSN or TS1m, TS2m, TSM with new useful properties or improved property parameters) can be determined (instead of chaotic search by "trial and error" method) depending on the need based on the characteristics of the stages following the current stage of development of TS being improved on the graph, or on the basis of the characteristics of the stages preceding the current stage of development of TS being improved.

However, even a superficial analysis of TS development patterns, described in the literature and indicated in Table 23, has showed that the stages of TS development in some of these patterns are not substantiated and even contradict the laws of nature, and in addition, the emergence of new TS necessitates constant correction of all known patterns and seeding of their list with new ones.

For example, several decades ago, the TS development pattern was described when the geometric characteristics of the mechanical contact of TS elements (TS with the environment, including with objects of animate and inanimate nature or with another TS), indicated under No. 5 in Table 23, were changed. Along with this, the quantity of these characteristics, their meanings and changes

Figure 100. Increase in the cross-country ability and / or speed of vehicles depending on the geometric characteristics of TS contact with the environment (at the point, along the line, along the area) or with another TS (volumetric contact), where G is the force of gravity, P is the driving force, T – friction force

Contact at the point	Contact along the line	Contact on the surface	Contact on volume
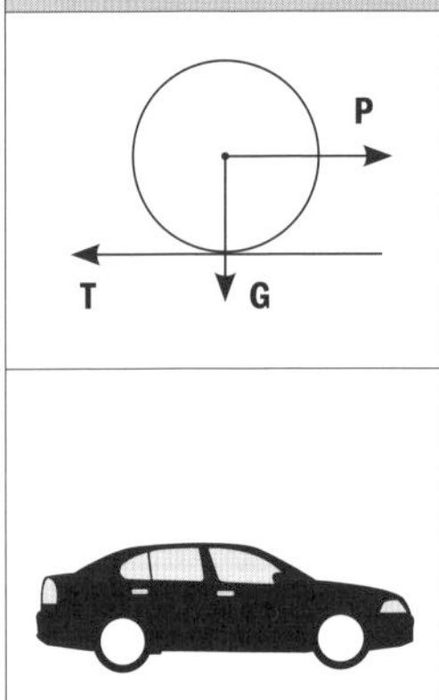	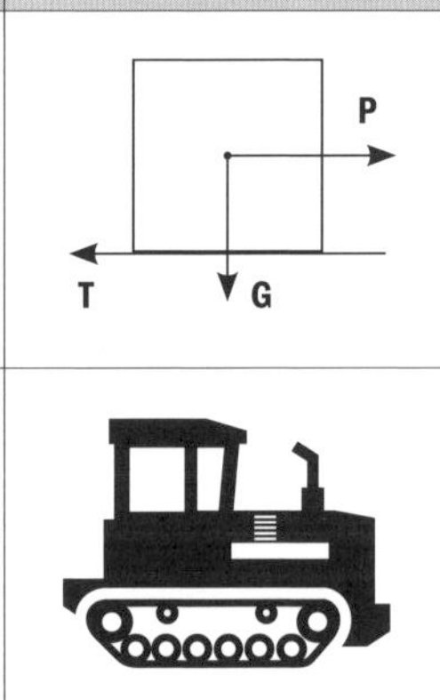	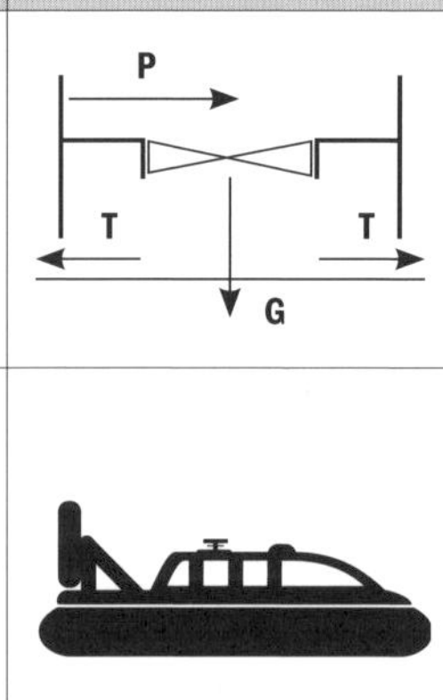	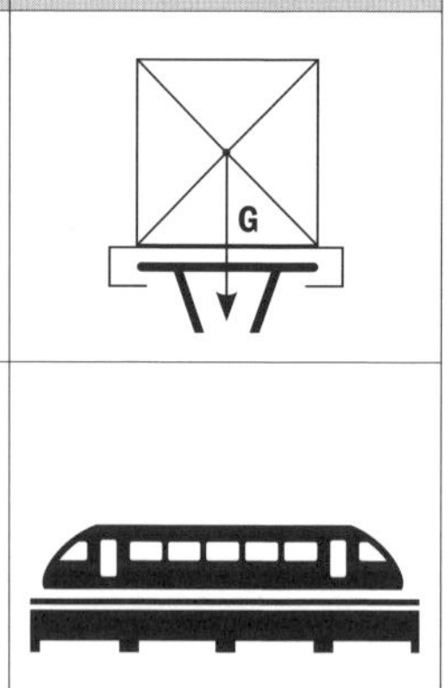
Car	Caterpillar	Hovercraft	Magnetic levitation train

tendencies (considered as "point — line — surface — volume") were considered quite logical and comprehensive. Fig. 100 shows examples to illustrate this pattern.

In the potentially highly effective technical decision "Wheeled propulsion device" (author G. I. Denisenko), which allows to expand the possibilities of wheeled transport when driving over rough terrain or on the surface of a liquid medium due to the controlled deviation of the wheel rim sections, the transition from the contact of the wheel with the surface of the medium has been implemented "at the point" to the contact "with many points" (that is, an intermediate value in the trend of changes" point — line ") (Fig. 101). In addition, it would be logical to assume the existence of new stages in TS development, which are combinations of the previously described stages, on the condition that they satisfy the logic of sequential increase in the number of elementary contact points (for example, "combination of points and lines", etc.).

Figure 101. Wheeled propulsion device capable of overcoming any off-road: in the wheeled version on asphalt (a), when driving on water (b), rough terrain (c): 1 – rim section; 2 – pneumatic tires; 3 – hub; 4 – section deflection device[33]

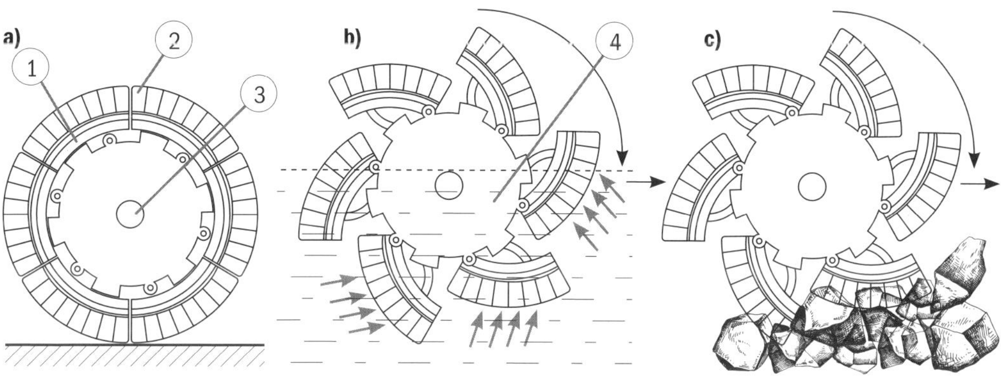

The creation of protective TS in a car is the confirmation of the universality of the pattern of TS development, depending on the geometric characteristics of a person's contact with TS (Fig. 102).

However, the logically justified expedience of creating protective TS based on the contact of a person and TS in volume in modern cars is realized only partially due to the multiplication of airbags, the number of which is close to 20. At the same time, the psychological barrier to thinking, which makes it possible to move from the multiplication of "contacts on the surface" to "contact by volume", has not been overcome. One of the reasons lies in the fact that the majority of specialists (paradoxical as it may sound) do not know (this was confirmed by our survey) about the existence of this and other patterns of TS development.

Figure 102. Increase in the effectiveness of protective devices in a car depending on the geometric characteristics of the contact between a man and TS

Contact at the point	Contact along the line	Contact on the surface	Contact on volume
			?
Headrest	Seat belt	Airbag	

All this has prompted me, together with colleagues, to carry out large-scale research, the results of which are presented in the following sections. These results include:

— Practically, for the first time formulated **definition of TS development pattern — revealed stable tendencies of successive changes in the characteristics of the functioning or structure of TS, leading to improvement in the consumer properties of TS;**

— Revised descriptions of 12 patterns given in Table 23 (see Sections 6.2.7.2–6.2.7.13);

— Description of the new pattern of TS evolution (see Section 6.2.7.14);

— Generated generalized descriptions of TS development patterns, justified by the logic of possible changes in TS characteristics and confirmed by experimental data on TS development in various industries (see Section 6.2.7.16);

— Methods of using TS development patterns in scientific and technical creativity (see Section 6.2.7.17);

— Justification of the possible number of TS development patterns (see Section 6.2.7.18).

When describing TS development patterns, various possibilities of improving the consumer properties of TS were considered, including due to increasing the parameters of TS properties and / or the environment (for example, with the increase in the number of points of contact of the vehicle with the environment, its passing ability increases), as well as due to the decrease in the properties parameters of TS and / or medium (for example, with the decrease in the number of points of contact of driving tools (mall hammer, mallet, hammer, pick) with the medium, the concentration of energy during impact increases). At the same time, in the first case, the mental movement along the stages of the pattern in search of the best technical solution on the graph occurs from left to right (see Fig. 99) from the stage corresponding to TS being improved, and in the second case — from right to left.

6.2.7.2 The pattern of TS development with a change in the properties of the surfaces of TS elements

There is the pattern of TS development with a change in the properties of the surfaces of TS elements, which consists in the sequential implementation of the next stages of development in the course of TS improvement (Fig. 103).

1. Smooth surface.

Figure 103. An example of sequential implementation of development stages associated with changing the surface properties of TS elements

Stage 1	Stage 2	Stage 3
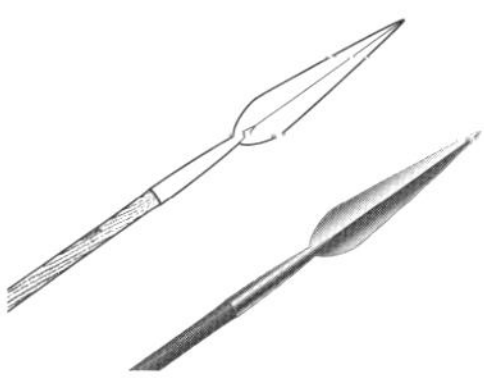	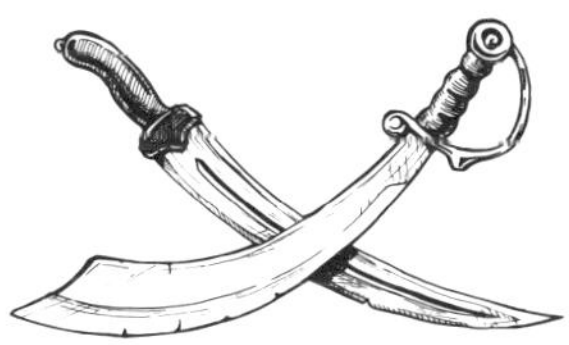	
Smooth shaft	Handle with lugs for fixing in the hand	Winding for handle deforming to the shape of the fingers and soaking up sweat

2. A surface with no particular form convex irregularities, located chaotically, or a rough surface.
3. A surface with convex irregularities of a certain (given) shape, located chaotically, or irregularities of no particular shape, arranged in an orderly manner.
4. A surface with irregularities of a certain (given) shape, arranged in an orderly manner.
5. Surface with indentations or through holes of irregular or regular shape, located randomly or orderly along the surface.
6. Surface with contact properties determined by the chemical composition and molecular (or nano-) structure of the substance on the surface (antifriction, adhesion, adsorption, reflective, etc.) or the properties of the subsurface material (magnetization, high absorbency, elasticity, etc.).
7. Surface with the ability to change properties.

8. Surface with the ability to replace properties and control their parameters.
9. A surface consisting of sections with the ability to replace properties and control their parameters[34].

The logic of the sequence formation of development stages: the increase in the depth of location and distribution density of the sources of contact properties of TS element surface with the simultaneous increase in a variety of surface contact properties achieved at the stage of development.

It should be noted that in Fig. 103–115, illustrating the stages of patterns, not all stages are presented (in particular, because some possible stages have not yet been implemented in TS).

6.2.7.3 The pattern of TS development with switching to new types of converted energy or field

There is a pattern in TS development during the transition to new types of energy converted into TS, which consists in the sequential implementation of the next stages of development in the course of improving TS (Fig. 104).

Figure 104. The example of the sequential implementation of development stages associated with the transition to new types of energy converted into TS

Stage 1	Stage 2	Stage 3

Rowing boat	Sailing vessel	Steam powered vessel

Stage 4	Stage 5
	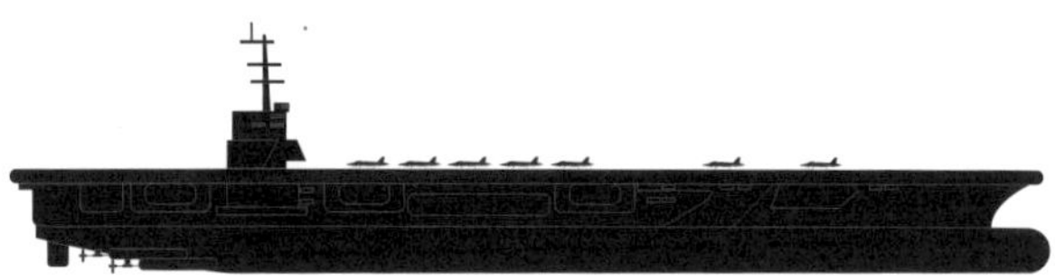
Diesel vessel	Nuclear powered vessel

1. Transformation of muscular power energy.
2. Transformation of mechanical energy (potential, including the transformation of the gradient of the gravitational field, and kinetic one).
3. Transformation of thermal energy.
4. Transformation of electromagnetic field energy or chemical energy.
5. Transformation of nuclear energy or the energy of elementary particles stream.

The logic of the formation of the sequence of development stages: the increase in the stream density of the energy converted into TS.

6.2.7.4 The pattern of TS development with switching to new types of mechanical movement

There is a pattern in TS development during the transition to new types of mechanical movement of objects in TS (including elements of TS, objects of animate and inanimate nature or other TS), which consists in the sequential implementation of the next stages of development in the course of improving TS (Fig. 105).

1. Uniform translation.
2. Uniformly accelerated (uniformly slowed down) translational motion.
3. Irregular translational motion.
4. Uniformly rotary motion.
5. Uniformly accelerated (uniformly slowed down) rotational motion.
6. Irregular rotary motion.
7. Uniform complex motion.
8. Uniformly accelerated (uniformly slowed down) complex motion.
9. Irregular complex motion.

The logic of the sequence of development stages formation: the increase in the irregularity of motion and at the same time, the complication of motion trajectory.

6.2.7.5 The pattern of TS development with a change in degrees number of elements motion freedom in TS

There is a pattern in TS development with a change in degrees number of mechanical motion freedom of TS elements, which consists in the sequential implementation of the next stages of development in the course of improving TS (Fig. 106).

1. Lack of mobility.
2. One degree of freedom.

Figure 105. The example of the sequential implementation of development stages associated with the transition to new types of mechanical movement of objects

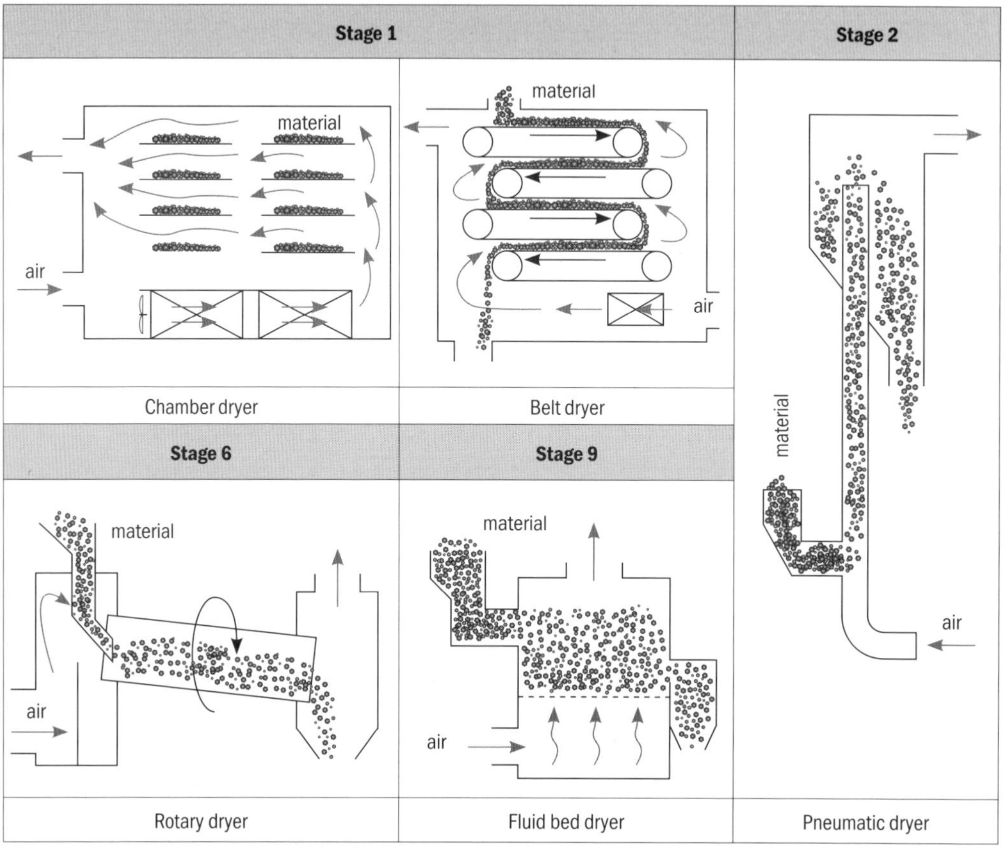

Figure 106. The example of the sequential implementation of development stages associated with a change in degrees number of mechanical motion freedom of TS elements

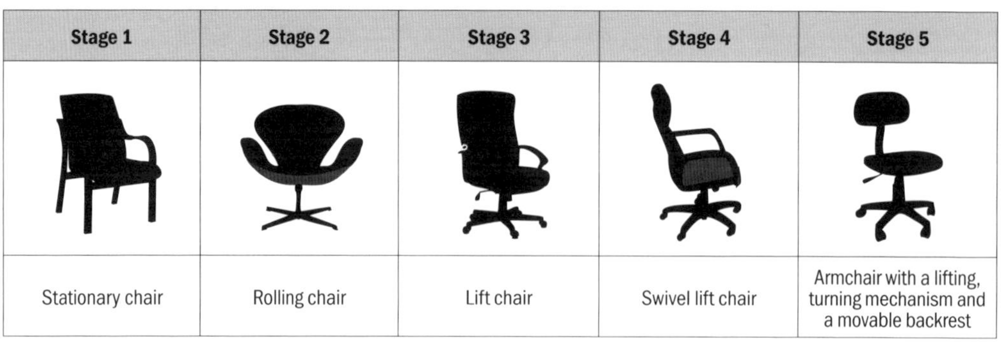

3. Two degrees of freedom.
4. Three degrees of freedom.
5. Four degrees of freedom.
6. Five degrees of freedom.
7. Six degrees of freedom

The logic of the sequence of development stages formation: the increase in degrees number of mechanical motion freedom of TS elements.

6.2.7.6 The pattern of TS development with a change in the geometric characteristics of objects mechanical contact

There is a pattern in TS development when the geometric characteristics of objects mechanical contact (TS with the environment, including with objects of animate and inanimate nature, or with another TS) change, which consists in the sequential implementation of the next stages of development in the course of improving TS (Fig. 107).

Figure 107. The example of the sequential implementation of development stages associated with a change in the geometric characteristics of objects mechanical contact

Stage 1	Stage 3	Stage 4	Stage 6	Stage 7
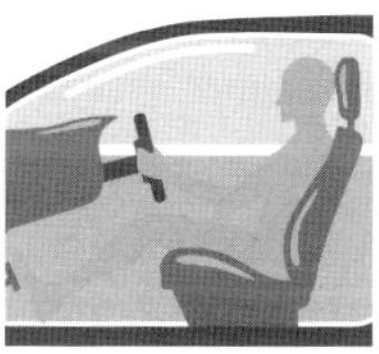	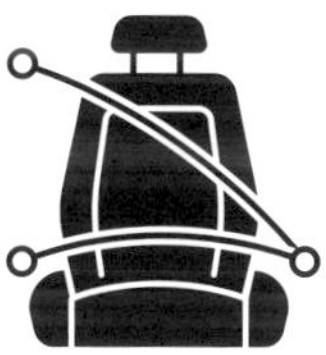	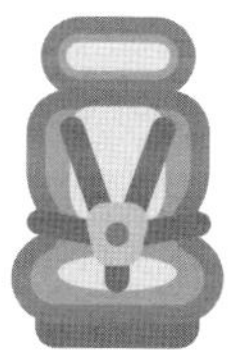		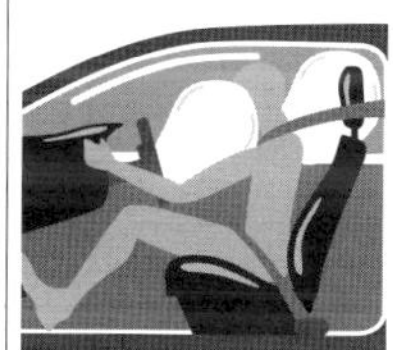
Headrest	Seat belt	Racing car seat belt	Airbag	Multiple airbags

1. Point.
2. Many points.
3. Line.
4. Lots of lines.
5. Combination of points and lines.
6. Surface.
7. Lots of surfaces.
8. Combination of points, lines and surfaces.

TS development patters:

- with a change of the geometric characteristics of objects mechanical contact;
- with a change of elements configuration of TS functional structure and the connections between them in space;
- with a change of the relative position of a pair of elements in TS;
- with a change of the geometric characteristics that determine the shape of the object;
- with a switch to new types of object symmetry;
- with a change of the characteristics of objects empty space;
- with a change of the internal objects structure;
- with a switch to new types of connection of TS structural elements;
- with a change of spatial distribution of the object's substance properties and their anisotropy.

9. Volume.
10. Lots of volumes.
11. Combination of points, lines, surfaces and volumes.
12. Lack of contact[35].

The logic of the sequence of development stages formation: expansion of the area of objects mechanical contact, starting from a point, sequentially in three mutually perpendicular directions, and at the same time the increase in the length, area and volume of the contact area, respectively.

6.2.7.7 The pattern of TS development with a change in the configuration of the elements of TS functional structure and the connections between them in space

There is a pattern in TS development when the configuration of the elements of TS functional structure and the connections between them in space change, which consists in the sequential implementation of the next stages of development in the course of TS improvement (Fig. 108).

Figure 108. The example of the sequential implementation of development stages associated with a change in the configuration of the elements of TS functional structure and the connections between them in space

Stage 1	Stage 2	Stage 3	Stage 5
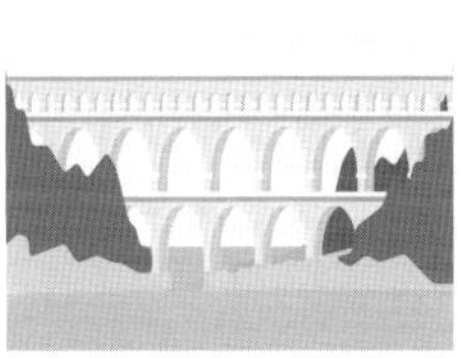	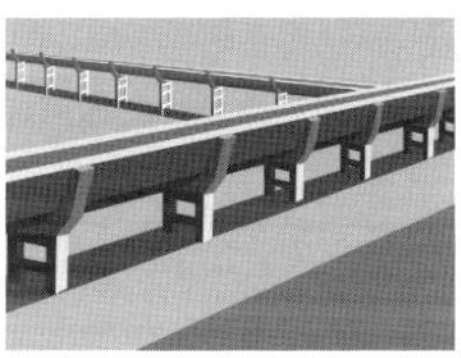	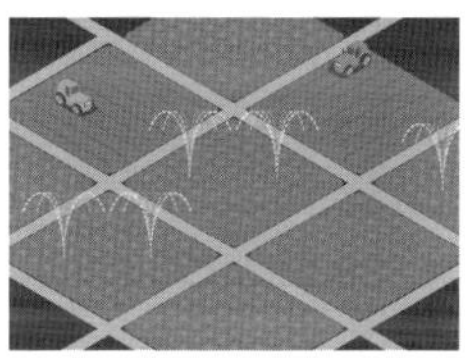	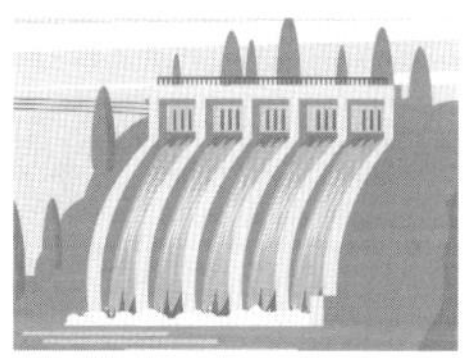
Aqueduct	Irrigation canal system	Irrigation system in a desert area	Dam with spillway regulating sluices

1. Linear or ring structure.
2. Branched (hierarchical) structure or structure of the "star" type.
3. Multiply connected (including network, cellular) structure.
4. Hybrid (combined) structure.
5. Time-changing structure.

6.2.7.8 The pattern of TS development with a change in the relative position of a pair of elements in TS

There is a pattern in the development of TS when the relative position of a pair of TS elements changes, which consists in the sequential implementation of the next stages of development in the course of TS improvement (Fig. 109).

Figure 109. The example of the sequential implementation of development stages associated with a change in the relative position of a pair of elements in TS

Stage 1	Stage 4	Stage 5
		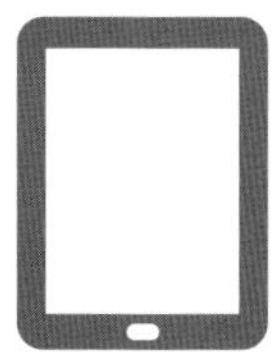
Personal computer	Notebook	Tablet PC

1. TS element does not touch another TS element.
2. TS element is in contact with another TS element.
3. TS element completely covers (envelopes) the other TS element.
4. TS element partially penetrates (enters) another TS element.
5. TS element is located inside another TS element, but is in contact with the external environment.
6. TS element is located inside another TS element and does not come into contact with the external environment.
7. TS element is split and divided into parts within another TS element.
8. TS element is dissolved inside another TS element.

The logic of the sequence of development stages formation: convergence of a pair of elements in TS — from the absence of mechanical contact between them to merging into a single element.

6.2.7.9 The pattern of TS development with a change in the geometric characteristics that determine the shape of TS elements

There is a pattern in TS development with a change in geometric characteristics determining the area of space occupied by TS elements, which consists in the sequential implementation of the next stages of development in the course of TS improvement (Fig. 110).

Figure 110. The example of the sequential implementation of development stages associated with a change in geometric characteristics determining the shape of objects

Stage 1	Stage 3	Stage 4	Stage 6	Stage 7	Stage 9
	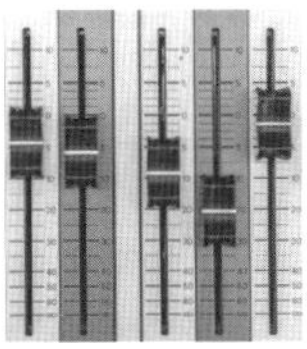				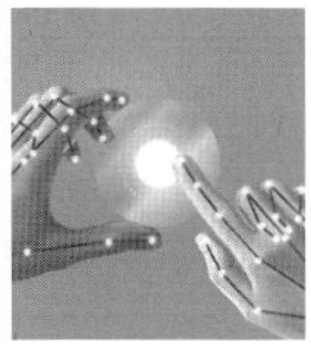
Button	Slider	Mouse wheel	Touch screen	Track ball	Gesture recognition system

1. Point.
2. Many points.
3. Straight line.
4. Curved line.
5. Many straight and / or curved lines.
6. Flat surface.
7. Curved surface.
8. Many flat and / or curved surfaces.
9. Volume.
10. Lots of volumes.

The logic of the sequence of development stages formation: the complication of the geometric shape of the space area occupied by TS elements, with the expansion of this area, starting from the conditional point, sequentially in three mutually perpendicular directions.

6.2.7.10 The pattern of TS development with switching to new types of TS elements symmetry

There is a pattern in TS development during the transition to new types of TS elements symmetry, which consists in the sequential implementation of the next stages of development in the course of improving TS (Fig. 111).

1. Asymmetry.
2. Bisymmetry.

3. Axial (axial, n-th order) symmetry.
4. Central (spherical) symmetry[36].

The logic of the sequence of development stages formation: the simplification of a geometric object (from surface to line to a point), relative to which TS elements are symmetrically located.

Figure 111. The example of sequential implementation of development stages during the transition to new types of TS elements symmetry

Stage 1	Stage 2	Stage 3	Stage 4
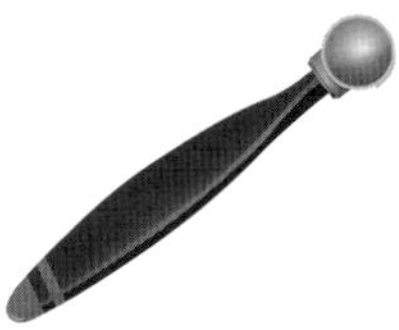			
Blade	Two-bladed propeller	Multibladed propeller	Tilt-rotor propeller

6.2.7.11 The pattern of TS development with a change in the characteristics of the empty space in TS

There is a pattern in TS development with characteristics change of the empty space in TS, which consists in the sequential implementation of the next stages of development in the course of improving TS (Fig. 112).

Figure 112. The example of the sequential implementation of development stages with a change of empty space characteristics in TS

Stage 1	Stage 4	Stage 8	Stage 9

Full brick	Embossed brick	Hollow brick	Foamed concrete block

1. Solid TS element without empty space.
2. Empty space without direct contact with TS element.
3. Empty space touches TS element.
4. Empty space is partially wedged into the TS element.
5. Isolated empty space inside TS element.
6. A lot of isolated empty spaces inside TS element.
7. Through space inside TS element.
8. A lot of through empty spaces inside TS element.
9. Porous structure of the object.
10. Capillary structure of TS element.
11. Microcapillary or microporous structure of TS element[37].

The logic of the sequence of development stages formation: the sequential increase in the number and volumes of empty spaces, a variety of their distribution (ordered and / or disordered) in the TS element, with the complication of the forms (from correct to incorrect) of these empty spaces.

6.2.7.12 The pattern of TS development with dividing TS elements into parts

There is a pattern in TS development with dividing TS elements into parts, which consists in the sequential implementation of the next stages of development in the course of improving TS (Fig. 113).

Figure 113. The example of the sequential implementation of development stages with dividing TS elements into parts

Stage 1	Stage 2	Stage 3	Stage 4	Stage 5	Stage 6
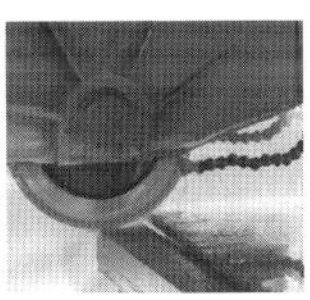				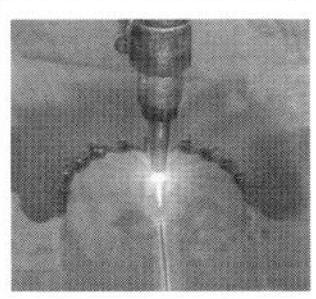	
Material processing with one-piece solid tool	Processing of material with multiple solids a solid tool, "divided into parts"	Sandblasting: impact on the material with a dispersed structure	Water jet slitting: material processing with a stream of molecules	Plasma cutting: the impact of ions flow	Laser cutting: the material is processed by a stream of light radiation (photons)

1. Single one-piece TS element.
2. TS element, divided into parts.
3. TS element, separated to a dispersed structure.

4. TS element, separated to molecules.
5. TS element, separated to atoms or ions.
6. TS element, subdivided to elementary particles.

The logic of the sequence of development stages formation: the decrease in size and at the same time the increase in the number of individual parts of TS element, jointly performing the principle, main, or auxiliary function of a given TS element.

6.2.7.13 The pattern of TS development with switching to new connection types of TS elements

There is a pattern in TS development during the transition to new connection types of TS elements, which consists in the sequential implementation of the next stages of development in the course of improving TS (Fig. 114).

Figure 114. The example of the sequential implementation of development stages during the transition to new connection types of TS elements

Stage 1	Stage 2	Stage 3	Stage 5	Stage 6
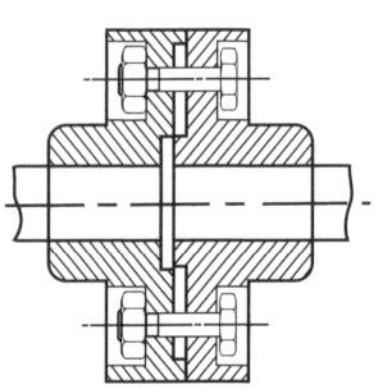	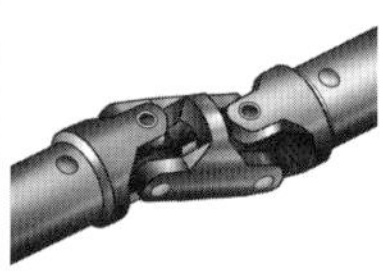	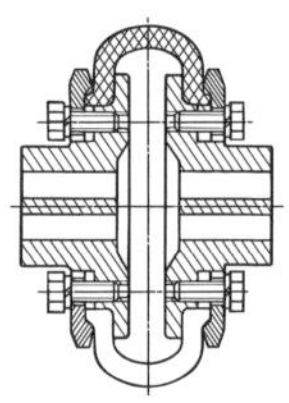		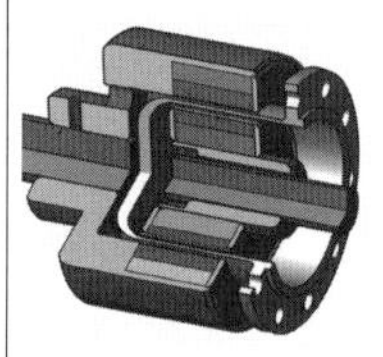
Fixed coupling	Articulated coupling	Gum coupling	Hydrocouple	Magnetic coupling

1. Rigid fixed connection.
2. Rigid movable connection (articulated).
3. Elastic connection.
4. Plastic connection.
5. Connection with liquid and / or gaseous medium.
6. Connection by field.
7. No connection[38].

The logic of the sequence of development stages formation: the increase in the mobility of the connection, including due to changing the element properties (for example, increasing the elasticity and / or fluidity) that performs the functions of the connector.

6.2.7.14 The pattern of TS development with a change in the spatial distribution of TS element's substance properties and their anisotropy

I have been the first who have formulated a new pattern in TS development with a change in the spatial distribution of TS element's substance properties and their anisotropy, which consists in the sequential implementation of the next stages of TS development in the course of its improvement (Fig. 115).

Figure 115. The example of the sequential implementation of development stages with a change of the spatial distribution of TS element's substance properties and their anisotropy

Stage 1	Stage 2	Stage 3	Stage 4	Stage 7
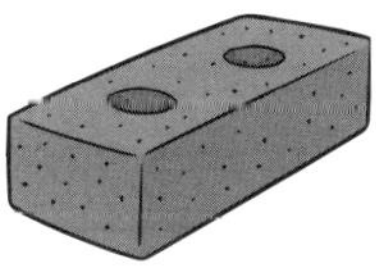	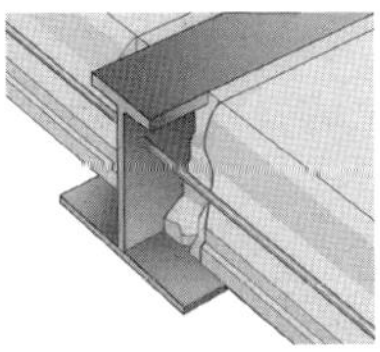		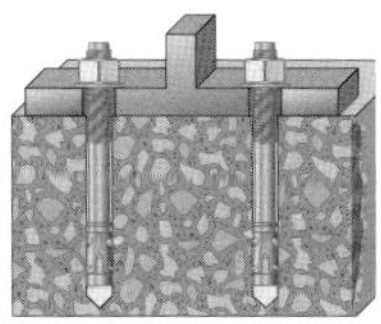	
Solid concrete	Hardening of flooring with rolled steel	Iron concrete	Roof bolting	Steel fiber concrete

1. Substance properties in TS element are evenly distributed.
2. Various properties of the substance in TS element are evenly distributed within the regions with clear boundaries.
3. The regions of TS element substance have various properties with clear boundaries between the regions, evenly distributed in TS element.
4. The regions of TS element substance have various properties with clear boundaries between the regions and are unevenly distributed in TS element.
5. The regions of TS element substance have various properties with "blurred" boundaries between regions with different properties.
6. The regions of TS element substance have various properties with "blurred" boundaries between the regions, which are evenly distributed in TS element.
7. The regions of TS element substance have various properties with "blurred" boundaries between the regions, which are unevenly distributed in TS element.
8. The regions of TS element substance have various properties, distributed in regions unevenly, without defined boundaries between regions or with the time-varying distribution of regions and / or boundaries.

The logic of the sequence of development stages formation: the increase in the number and uneven distribution of substance regions with various properties within TS element, as well as the elimination of clear boundaries between regions with various properties.

6.2.7.15 On the high heuristicity of generalized descriptions of TS development patterns and the approach to the formation of such descriptions

Analysis of TS development patterns, described in the form of the sequences of TS development stages in Sections 6.2.7.2–6.2.7.14, and the experience of their use have led me to the conclusion that, although in general they effectively perform the role of generators of specific, associatively colored new technical solutions that have new competitive properties in comparison with analogs, it is possible to significantly increase the possibility of using patterns if they are presented in the form of generalized descriptions, which are the result of mental actions to combine all species concepts (stages of development) included in the pattern in one generic concept, reflecting the essence of changes in the development of TS, — the generalized description of TS development pattern.

The generalized description of a pattern contributes to deeper understanding of its essence, which makes it possible to supplement it with new probable stages (intermediate and subsequent). At the same time, the sequence of new stages in a pattern is determined by the logic of TS development and new TS emergence in different industries, confirming this logic.

For example, the hypothesis of the existence of TS development pattern with a change in the geometric characteristics of the mechanical contact of objects (No. 5 in Table 23) has been formed on the basis of the sequence of species concepts (stages of TS development) "point — line — surface — volume" previously presented in the works of experts. However, the general logic of the change is to increase the number of elementary points of contact — from one to infinity. In this case, a line can be represented as a set of points, a surface — as a set of lines, a volumetric body — as a set of surfaces. And then it is easy to come to the conclusion that the described sequence probably misses the intermediate stages of development that logically complement it: "plurality of points", "plurality of lines", "plurality of surfaces", "plurality of volumes", which is confirmed by appropriate technical decisions, for example, wheel propulsor (see Fig. 101). In addition, it is logical to assume the existence of new stages in the development of TS, which are combinations of the previously described stages, provided that they satisfy the formulated logic of the sequential increase in the number of elementary points of contact (for example, "the combination of points and lines"), a set of points forming planes and volumes geometrically regular or irregular shapes with a regular or irregular spatial arrangement.

Based on this logic, the sequence common for all patterns has been formed, leading to the increase in the number of useful TS properties and / or improvement in the properties parameters: "the reasonable set of elements of a geometrically regular shape and with a regular arrangement in space → the reasonable set of TS elements of a geometrically regular shape and with an irregular arrangement in space or geometrically irregular shape with a regular or irregular spatial arrangement."

For example, powerful (more than 20 meters) underground horizontal or flat bed deposits of minerals are mined by a room-and-pillar development system, where ore is removed from chambers located between regular, most often round (sometimes more than 20 meters in diameter) support pillars supporting the worked-out space (fig. 116).

Figure 116. Room-and-pillar system of underground mining of a thick horizontal or flat bed seam mineral deposit: 1 – support ore regular pillars of geometrically regular shape; 2 – chamber; 3 – roof bolting; 4 – isolines of the minoral content[39]

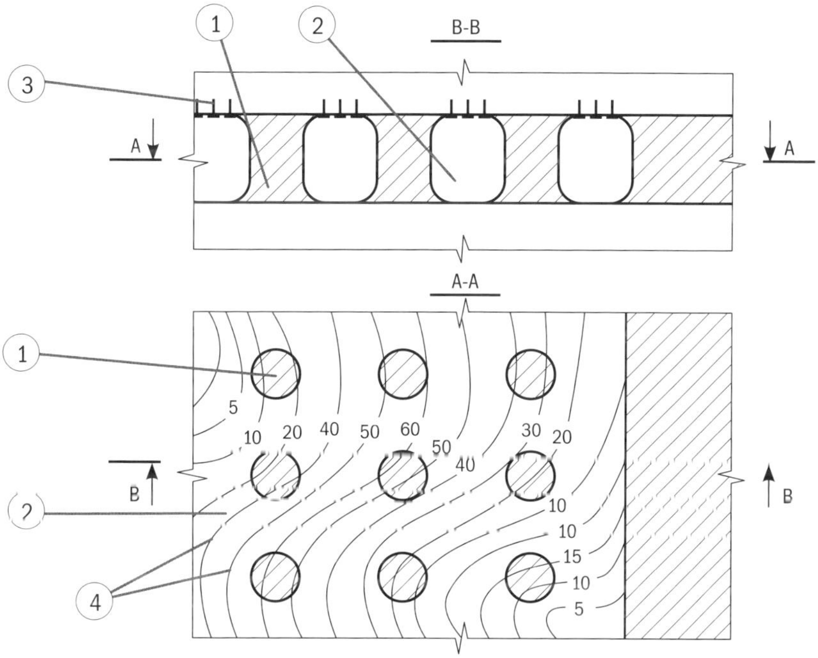

One of the main useful properties of this system is the maintenance of the working space (chambers between the supporting pillars), that is, the prevention

The efficiency of using TS development patterns in scientific and technical creativity can be significantly increased if they are presented in the form of generalized descriptions.

The generalized description is the result of the mental unification of all species concepts (stages of development), included in the pattern, into a single generic concept reflecting the essence of changes during the development of TS.

of its collapse. The big disadvantage of this system is that the mineral in the support pillars, as a rule, remains forever in the bowels. Losses of mineral resources exceed 40% of its total volume contained in the mined seam. With the regular (that is, at equal distances) arrangement of support pillars of the correct geometric shape, some of them may end up in zones of increased mineral content, thereby increasing its losses.

Soviet (and then Russian) scientist V. I. Borshch-Kompaniets[40] worked out the technology for the development of thick seam deposits with support pillars of irregular geometric shape, of various sizes, located irregularly, but in zones of low mineral content, which sharply reduced production losses while maintaining the function of support pillars to support the working space (Fig. 117). Metaphorically (for metaphors see Section 6.2.12) such a technical decision can be called ordered disorder.

Figure 117. Improved room-and-pillar system of underground development of a thick horizontal or flat bed seam mineral deposit: 1 – support ore irregular pillars of irregular shape; 2 – chambers; 3 – support ore irregular pillars of regular shape; 4 – roof bolting; 5 – isolines of the mineral content[41]

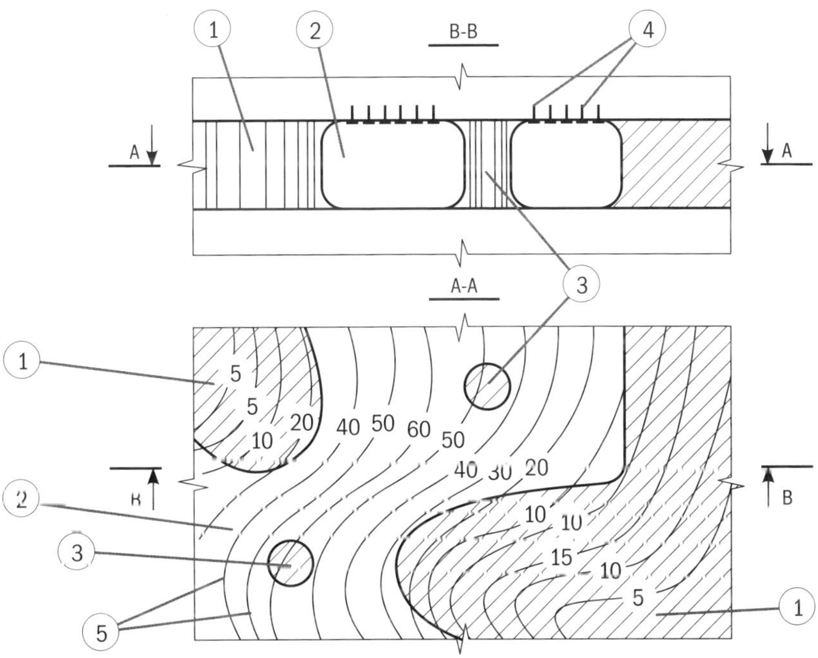

Or another example, when, for aesthetic reasons, as well as to preserve the environment, they refuse to arrange regular balconies in order to save a growing tree nearby (Fig. 118), or arrange balconies so that the tree passes through them.

Figure 118. House built around a tree

The generalized descriptions of TS development patterns, formed taking into account above mentioned approaches, are given in the next section.

6.2.7.16 Generalized descriptions of TS development patterns and forecasting new stages of TS development

The following generalized descriptions of the identified and substantiated patterns of TS development have been developed, which are presented in a text form, as well as in chains of pictograms.

1. **TS development pattern with a change in the surfaces properties of TS elements**, which consists in the fact that the improvement of TS consumer properties occurs when new surface properties are implemented in TS: with a change of external surface properties (first, quantitative, then qualitative characteristics — for example, from the increase in the smoothness of the surface to the use of natural bulges or depressions and then the creation of artificial ones, first geometrically ordered, and then of disordered sizes and shapes); then, with a change of the internal properties of the surface material (first with fixed parameters, then with controlled parameters)[42].

The sequence of the pictograms illustrating this generalized description is shown in Fig. 119 in a simplified form, where 1 is a smooth surface; 2 — a surface

Figure 119. Generalized description of TS development pattern with a change of the surface properties of TS elements (in the form of pictograms)

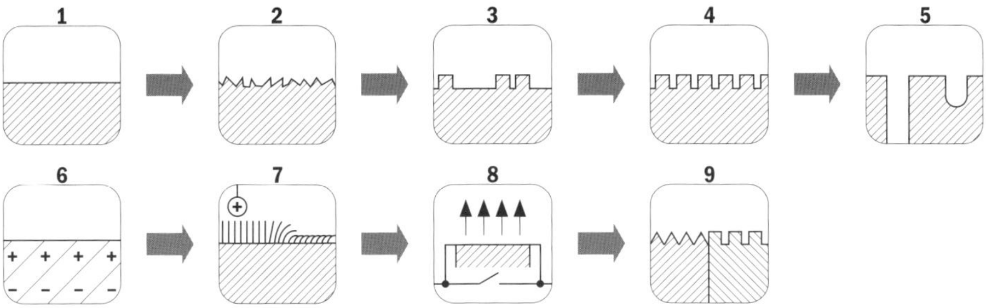

with convex irregularities of arbitrary shape, located chaotically, or a rough surface; 3 — a surface with convex irregularities of a certain (given) shape, located chaotically, or irregularities of an arbitrary shape, arranged in an orderly manner; 4 — a surface with irregularities of a certain (given) shape, arranged in an orderly manner; 5 — a surface with depressions or through holes of irregular or regular shape, located randomly or orderly along the surface; 6 — a surface with contact properties determined by the chemical composition and molecular (or nano-) structure of the surface substance; 7 — a surface with the ability to replace properties; 8 — a surface with the ability to replace properties and control their parameters; 9 — a surface consisting of areas with the ability to replace properties and control their parameters[43].

2. **TS development pattern with a transition to new types of energy converted in TS**, which consists in the fact that the improvement of TS consumer properties occurs with the use of energy converters with the higher flow density of the converted energy — from TS without energy or field converters to TS with energy converters (sequentially): biological, mechanical, thermodynamic, electromagnetic and chemical, flow of elementary particles or nuclear.

Figure 120. Generalized description of TS development pattern with a transition to new types of energy converted in TS (in the form of pictograms)

The sequence of the pictograms illustrating this generalized description is shown in Fig. 120 in a simplified form, where 1 is the transformation of muscular strength energy; 2 — the transformation of mechanical energy (potential, including the energy of the gravitational field, and kinetic); 3 — the conversion of thermal energy; 4 — the energy transformation of the electromagnetic field or chemical energy; 5 — the transformation of nuclear energy or the energy of elementary particles stream[44].

3. **TS development pattern with a transition to new types of mechanical motion of TS elements**, which consists in the fact that the improvement of TS consumer properties occurs with a change of the nature of TS elements motion (sequentially or simultaneously) in the directions of increasing the unevenness of movement (from uniform to uniformly variable and further to uneven), complication of motion trajectory (from translational rectilinear to translational curvilinear, then to rotational, complex).

Figure 121. Generalized description of TS development pattern with a transition to new types of mechanical motion of TS elements (in the form of pictograms)

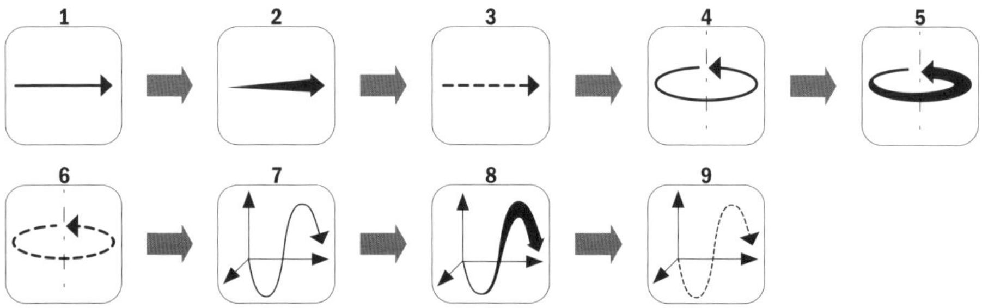

The sequence of the pictograms illustrating this generalized description in a simplified form is shown in Fig. 121, where 1 — uniform horizontal translational motion; 2 — uniformly accelerated (equally slowed down) horizontal translational motion; 3 — uneven horizontal translational motion; 4 — uniform rotary motion; 5 — uniformly accelerated (equally slowed down) rotational motion; 6 — uneven rotational motion; 7 — even complex motion; 8 — uniformly accelerated (equally slowed down) complex motion; 9 — uneven complex motion[45].

4. **TS development pattern with a change in degrees number of TS elements motion freedom**, which consists in the fact that the improvement of TS consumer properties occurs[46] with the increase in degrees number of freedom of TS elements mechanical motion.

The sequence of the pictograms illustrating this generalized description in a simplified form is shown in Fig. 122, where 1 — lack of mobility; 2 — one degree of freedom; 3 — two degrees of freedom; 4 — three degrees of freedom; 5 — four degrees of freedom; 6 — five degrees of freedom; 7 — six degrees of freedom (the number of degrees of freedom is shown by numbers inside the pictograms).

Figure 122. Generalized description of TS development pattern with a change of degrees number of freedom of TS elements motion (in the form of pictograms)

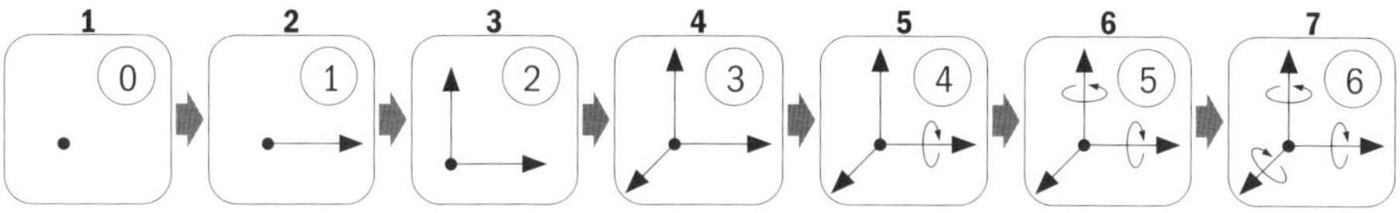

5. **TS development pattern with a change in the geometric characteristics of mechanical objects contact**, which consists in the fact that the improvement of TS consumer properties occurs with the increase in the number of contact points, and this number tends from one point to infinity (continuous contact), and the form of contact becomes more complicated from uniformly or irregularly positioned points, geometrically correct, and then irregular solid or dashed lines to flat or curved surfaces and volumetric contact areas[47].

Figure 123. Generalized description of TS development pattern with a change of the geometric characteristics of mechanical objects contact (in the form of pictograms)

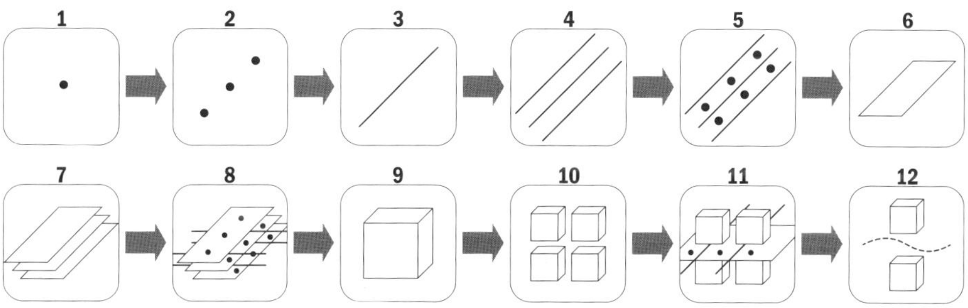

The sequence of the pictograms illustrating this generalized description in a simplified form is shown in Fig. 123, where 1 is a point; 2 — a set of points; 3 — line; 4 — many lines; 5 — combination of points and lines; 6 — surface; 7 — many surfaces; 8 — combination of points, lines and surfaces; 9 — volume; 10 — a set of volumes[48], 11 — combination of points, lines, surfaces and volumes; 12 — no contact.

6. **TS development pattern with a change of the configuration of TS elements functional structure and connections between them in space**, which consists in the fact that the improvement of TS consumer properties occurs with the sequential complication of the configuration of TS functional structure elements and connections between them in space, expressed in increase in the number of connections between TS elements and reasonable decrease in the regularity of these connections arrangement.

Figure 124. Generalized description of TS development pattern with a change of the configuration of the elements of TS functional structure and the connections between them in space (in the form of pictograms)

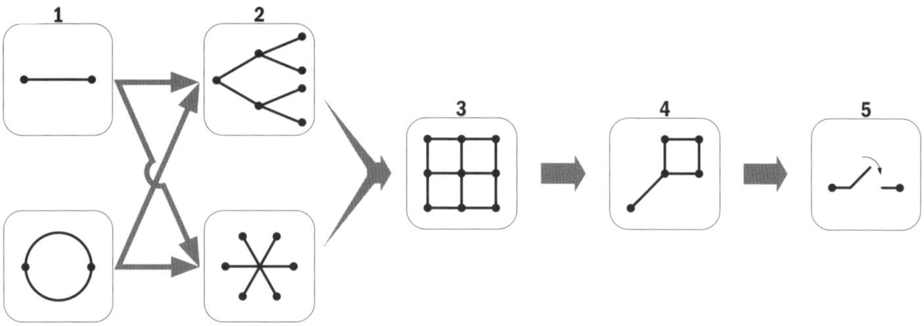

The sequence of the pictograms illustrating this generalized description in a simplified form is shown in Fig. 124, where 1 is a linear or ring structure; 2 — the branched (hierarchical) structure or structure of the "star" type; 3 — multiply connected (including network, cellular) structure; 4 — hybrid (combined) structure; 5 — time-varying structure.

7. **TS development pattern with a change of the relative position of a pair of elements in TS**, which consists in the fact that the improvement of TS consumer properties occurs with the sequential convergence of a pair of elements in TS — from the absence of contact between these elements to the increase in the contact surface area between them and the mutual penetration of elements and then to the subsequent merging into a single element.

Figure 125. Generalized description of TS development pattern with a change of the relative position of a pair of elements in TS (in the form of pictograms)

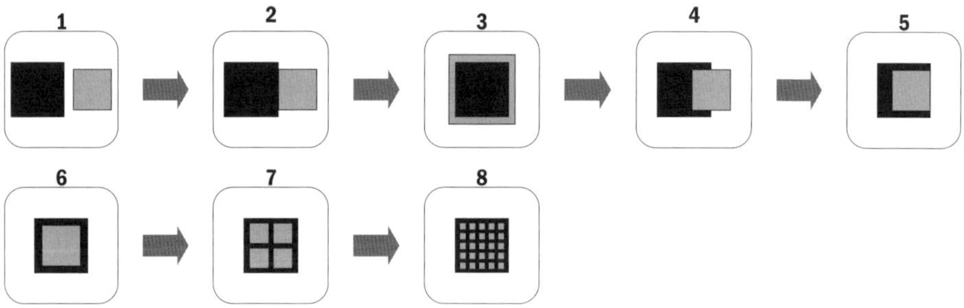

The sequence of the pictograms illustrating this generalized description in a simplified form is shown in Fig. 125, where 1 — TS element does not touch another TS element; 2 — TS element is in contact with another TS element; 3 — TS element completely covers (envelops) another TS element; 4 — TS element partially penetrates (enters) another TS element; 5 — TS element is located inside another TS element, but it is in contact with the external environment; 6 — TS element is located inside another TS element and does not come into contact with the external environment; 7 — TS element is divided and distributed in parts within another TS element; 8 — TS element is dissolved inside another TS element.

8. **TS development pattern with a change of the geometric characteristics that determine the shape of TS elements**, which consists in the fact that improvement in TS consumer properties occurs with a change of the geometric properties of integral or composite TS elements that are part of TS, directed towards the increase in the number of points in space occupied by TS element, first in one direction (from a point to a set of points and lines), then in a two-dimensional region of space (from a point to a set of points and lines in a plane and further to a plane), then in the three-dimensional region of space (from a point to a set of points, lines and surfaces, further to volume and a set of volumes).

Figure 126. Generalized description of TS development pattern with a change of geometric characteristics that determine the shape of objects (in the form of pictograms)

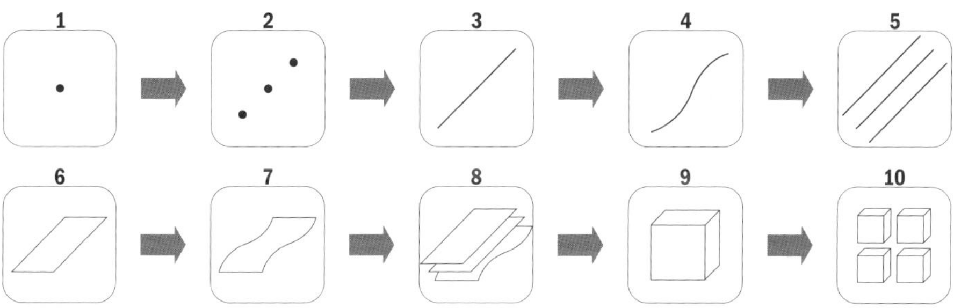

The sequence of the pictograms illustrating this generalized description in a simplified form is shown in Fig. 126, where 1 is a point; 2 — a set of points; 3 — straight line; 4 — curved line; 5 — a lot of straight and / or curved lines; 6 — flat surface; 7 — curved surface; 8 — a lot of flat and / or curved surfaces; 9 — volume; 10 — many volumes[49].

9. **TS development pattern with a transition to new types of symmetry of TS elements**, which consists in the fact that the improvement of TS consumer properties occurs during the transition to new types of symmetry of integral or composite TS elements, starting with asymmetry in the direction of simplifying an imaginary geometric object, relative to which the considered elements TS are symmetrically arranged. Simplification direction:

from the surface (bilateral symmetry) to the line (axial symmetry) and further to the point (spherical symmetry).

Figure 127. Generalized description of TS development pattern with a transition to new types of symmetry of TS elements (in the form of pictograms)

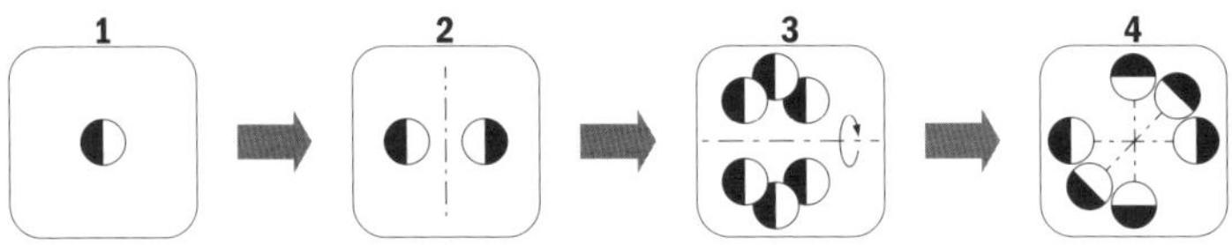

The sequence of the pictograms illustrating this generalized description in a simplified form is shown in Fig. 127, where 1 is asymmetry; 2 — bilateral symmetry; 3 — axial (axial, n-th order) symmetry; 4 — central (spherical) symmetry[50].

10. **TS development pattern with a change of characteristics of empty space in TS**, which consists in the fact that improvement of TS consumer properties occurs with deepening of empty space area — from the surface of TS element under consideration to its center, with decrease in size and simultaneous increase in the number of areas of empty space inside the considered TS element, as well as with increase in the number of through areas of empty space inside the considered TS element.

Figure 128. Generalized description of TS development pattern with a change of the empty space characteristics in TS (in the form of pictograms)

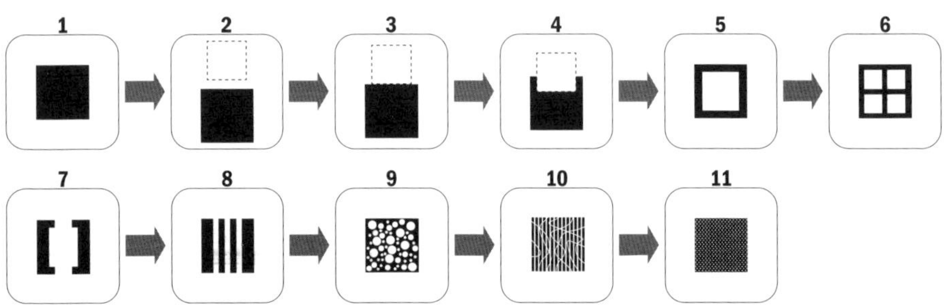

The sequence of the pictograms illustrating this generalized description in a simplified form is shown in Fig. 128, where 1 is a solid TS element; 2 — the area of empty space without direct contact with TS element; 3 — the area of empty space is in contact with the element of TS; 4 — the area of empty space is partially wedged into TS element; 5 — empty space inside TS element; 6 — fragmented (divided) empty space inside TS element; 7 — the through area of empty space inside TS element; 8 — several through areas of empty space inside TS element; 9 — porous structure of TS element;

10 — capillary structure of TS element; 11 — microcapillary or microporous structure of TS element[51].

11. **TS development pattern with a division of TS elements into parts**, which is that the improvement of TS consumer properties occurs with decrease in size (to elementary particles) and simultaneous increase in the number of separate parts of TS element, jointly performing the principal, main or auxiliary function of this TS element.

Figure 129. Generalized description of TS development pattern with a division of TS elements into parts (in the form of pictograms)

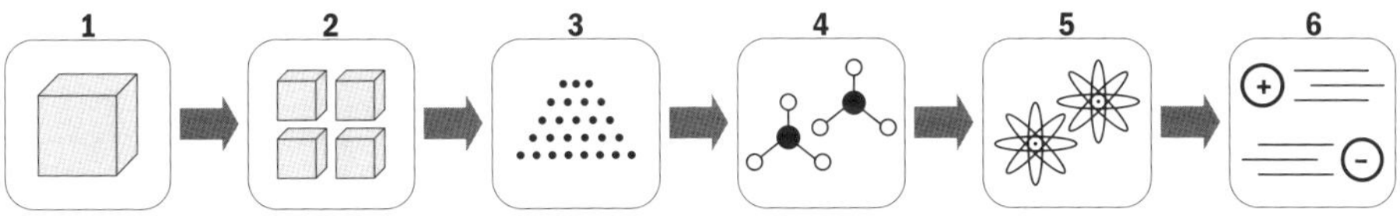

The sequence of the pictograms illustrating this generalized description in a simplified form is shown in Fig. 129, where 1 is a single integral TS element; 2 — TS element, divided into parts; 3 — TS element, divided to a dispersed structure; 4 — TS element, separated to molecules; 5 — TS element, separated to atoms or ions; 6 — TS element, divided to elementary particles[52].

12. **TS development pattern with a transition to new types of TS elements connection**, which consists in the fact that the improvement of TS consumer properties occurs in the direction of increasing the mobility of the connection (subject to the increase in the efficiency of the connection functions) and in the direction of increasing the elasticity and fluidity of TS element performing the functions of connector, and then — in the transition to the absence of mechanical connection.

Figure 130. Generalized description of TS development pattern with a transition to new types of TS elements connection (in the form of pictograms)

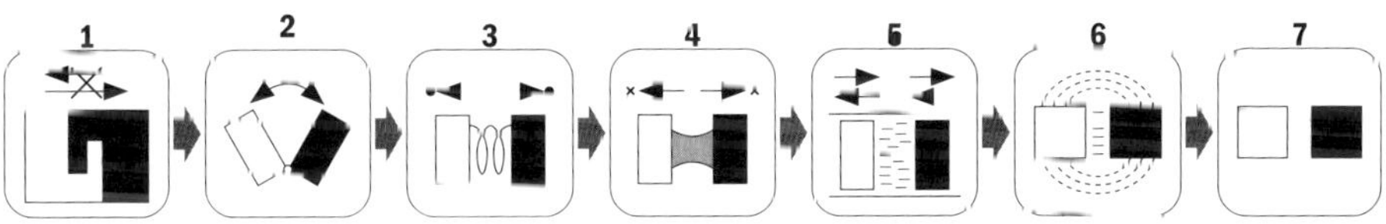

The sequence of the pictograms illustrating this generalized description in a simplified form is shown in Fig. 130, where 1 is rigid fixed connection; 2 — rigid movable connection (articulated); 3 — elastic connection; 4 — plastic connection; 5 — connection by liquid and / or gaseous medium; 6 — connection by field; 7 — no connection[53].

It can be identified so many patterns of TS development how many TS consumer properties exist, that is, tens of thousands.

"Good" verbs allow you to formulate TS functions that increase its consumer value. Good adjectives help describe the properties of TS that enhance its value. Accordingly, "bad" verbs and "bad" adjectives describe the functions and properties of TS that impair its consumer value.

The number of "good" verbs (TS functions) and adjectives (TS properties) should be increased, and the number of "bad" verbs should be decreased.

13. **TS development pattern with a change of the spatial distribution of the substance properties of the TS element and their anisotropy**, which consists in the fact that TS consumer properties improve with a change in the spatial distribution of TS element substance properties and their anisotropy in the direction of increasing the number and non-uniformity of the distribution of TS element substance regions with various properties and reducing the clarity of the boundaries of areas with various properties.

Figure 131. Generalized description of TS development pattern with a change of the spatial distribution of the object's material properties and their anisotropy (in the form of pictograms)

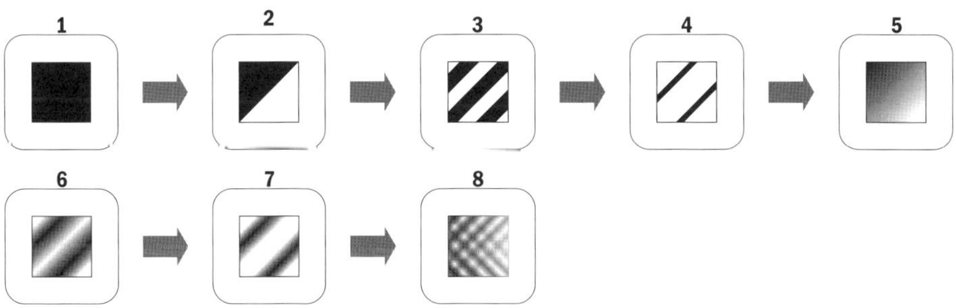

The sequence of the pictograms illustrating this generalized description in a simplified form is shown in Fig. 131, where 1 is a set of material properties evenly distributed in the object; 2 — the areas of the substance of TS element with various sets of properties with clear boundaries between areas and uniformly distributed properties within the boundaries; 3 — the areas of TS element substance with various sets of properties with clear boundaries between areas and uniform distribution (alternation) of areas with various properties; 4 — the areas of the substance of TS element with different sets of properties with clear boundaries between the areas and uneven distribution (alternation) of areas with various properties; 5 — the areas of substance of element TS with various sets of properties with "blurred" boundaries between areas with various properties; 6 — the areas of TS element substance with various sets of properties with "blurred" boundaries between areas and uniform alternation of areas with various properties; 7 — the areas of substance of TS element with various sets of properties with "blurred" boundaries between areas and uneven alternation of areas with various properties; 8 — the areas of the substance of TS element with various sets of properties, distributed unevenly, without definite boundaries between them, or with time-varying distribution of areas and / or boundaries between them[54].

It should be noted that, depending on the characteristics of the stages of all development patterns, TS can be used either in the direct or in the reverse sequence of the stages indicated in the generalized descriptions of the patterns.

6.2.7.17 The technique of using TS development patterns

To effectively identify promising directions for improving TS on the basis of their development patterns, the technique has been developed and tested that allows predicting their further development during TS analysis and forming new options for technical decisions that are likely to have better consumer properties in comparison with prototypes.

The essence of the developed technique for identifying promising directions for improving TS based on the patterns of TS development is to perform the following steps, the sequence of which is schematically shown in Fig. 99 and 132.

Figure 132. The sequence of steps to identify promising areas for improving TS

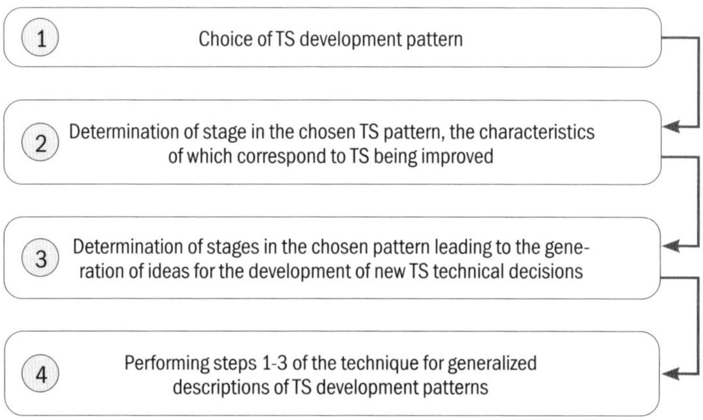

Step 1. Based on the analysis of the characteristics of TS being improved (the contradictions present in TS), it is necessary to select TS development pattern, which allows resolving these contradictions.

Step 2. In the chosen TS development pattern, it is necessary to determine the stage, the characteristics of which correspond to TS being improved.

Step 3. Based on the selected stage in the pattern, it is necessary to determine the following or preceding stages, leading to the generation of ideas for the development of new TS technical decisions, potentially having new competitive properties and / or improved properties parameters.

If the analysis of the next or previous stages has not lead to the generation of ideas for the development of new TS technical decisions, potentially having new competitive properties and / or improved properties parameters, then you should go to step 4.

Step 4. The same actions, as in steps 1–3, should be performed with the generalized descriptions of TS development patterns (only you should operate not with stages, but with trends in generalized descriptions).

Sometimes, at step 3, there is no need to determine, on the basis of the selected stage in TS development pattern, the next or previous stage (or trends) leading to the generation of ideas for the development of new competitive technical TS decisions, but you need only to achieve, under special conditions, the properties already available in TS being improved. For example, in the pattern of TS development during the transition to new types of energy converted to TS (see Fig. 104) or in the generalized description of this pattern (see Fig. 120), the characteristics of a laptop (TS being improved), including the need for electrical energy, correspond to the characteristics of stage 4 (the gradient of electromagnetic field transformation or the gradient of chemical energy concentration). However, in many parts of India or African countries, there is no industrial generation of electricity (which could be used directly or to recharge batteries), therefore, to obtain electrical energy, you can use the steps in this pattern, preceding stage 4 (that is, transformations, for example, of biological (muscular) energy into mechanical, and then into electrical) (see Fig. cl.4.31). Variants of using water or wind energy instead of biological (muscular) energy are not excluded. The following combination is also possible: thermal energy → mechanical energy → electrical energy.

6.2.7.18 Forecasting new patterns of TS development

This book describes 13 well-grounded patterns of TS development. However, how many of these patterns can be identified? The answer is in determining TS development pattern: TS development pattern is the identified stable tendencies of successive changes in the characteristics of the functioning or structure of TS, leading to improvement in the consumer properties of TS. Thus, theoretically, it can be identified so many patterns of TS development how many TS consumer properties exist, that is, tens of thousands. The task of identifying and justifying the description of TS development patterns is of great scientific and practical importance.

6.2.7.19 Environmental laws

There are more than 50 environmental laws formulated by various specialists, the names of which I consider it appropriate to give in Appendix 8, although even the names (and even more so interpretations) of some of them seem to be controversial. Almost all environmental laws[55] are not fully comprehended from the point of view of the possibilities of their use in scientific and technical creativity. Perhaps the first acquaintance of the reader with these laws will be more fruitful than mine, and will serve as the impetus for the emergence of effective ideas for the creation of TS aimed at preserving and developing the ecosphere.

6.2.8 "Controlling", "good" and "bad" verbs for the formation of TS functions

Let me remind you that the most important stage of scientific and technical creativity and, accordingly, the design stage, which is repeated many times at the stages of setting, solving a problem, commercializing or using the results of creativity in other way, is the formation of TS functions.

Since each function is based on a verb of an indefinite form, many researchers, based on the analysis of TS functions, have tried to identify and form the list of verbs that would determine the effectiveness of TS and even the direction of its qualitative improvement. Such verbs have been called "strong", "most influential", "controlling".

As an example, here I give the list of "controlling verbs" formed by Don Koberg and Gene Bagnal[56]:

Multiply	Compress	Thicken
Divide	Complement	Stretch
Exclude	Hide	Displace
Suppress	Freeze	Throw away
Invert	Soften	Protect
Separate	Inflate	Isolate
Regroup	Bypass	Integrate
Unite	Take away	Symbolize
Distort	Delete	Abstract
Rotate	Simplify	Analyze
Align	Repeat	

Not all of these "controlling verbs" (e.g., "abstract", "symbolize", "simplify") can be used in the direct formulation of functions, but they are useful for choosing the right course of action.

Developing the generally fruitful idea of "controlling verbs", I have come to the idea, on the one hand, about the advisability of expanding the list of "controlling verbs" based on their usefulness for the formation of TS functions leading to increase in TS consumer value, on the other hand, about the need of formation of a verbs list (TS functions) leading to decrease in TS consumer value.

For simplicity, I have called the first group of verbs "good" and the second "bad."

At the stages of setting, solving a problem, commercializing or using the result in other way — wherever functions are formed for choosing existing TS, creating or improving TS, — functions should be formed on the basis of "good" verbs that prevent or reduce the negative consequences of TS functions manifestation characterized by "bad" verbs.

The "good" verbs can include:

Accumulate
Amortize
Vary
Vibrate
Excite
Churn
Build in
Burn out
Crystallize
Freeze
Evaporate

Corrode
Carbonate
Harmonize
Generate
Damp
Drain
Bend
Radiate
Grind
Ionize
Heal

Modify
Overturn
Transform
Spread out
Dissolve
Compress
Glue
Transform
Shield
Streamline
Mask

Considering, for example, that there are more than 27 thousand verbs in the Russian language, the list of "good" verbs can be supplemented. The obviously "bad" verbs — the basis of harmful functions in TS — include, for example, the following: stink, rot, wear out, smear, smoke, pollute, clog. Most likely, "bad" ones will be the verbs with the prefix "pere-"(over) (overplay, oversalt, overheat), characterizing something excessive, as well as with the prefix "nedo-" (under-), characterizing something unfinished, insufficient.

It should be noted that many "good" verbs can be "bad" if they form the basis of harmful or neutral functions in TS, and vice versa — "bad" verbs can be the basis of useful functions in special means (for example, when using pungent gases or coloring liquids to disperse an aggressive crowd)[57].

6.2.9 "Good" and "bad" adjectives to form TS properties

I have made a similar division into "good" and "bad" with adjectives (for example, there are more than 20 thousand of them in Russian language), which describe TS consumer properties.

"Good" adjectives describe properties that provide increase (sometimes abruptly) in the consumer value created by TS in comparison with its counterparts (prototype).

"Bad" adjectives describe negative properties that arise during operation, storage, transportation, as well as during utilization or burial of TS after the end of its life or obsolescence. These negative properties reduce TS consumer value.

As in the case of verbs, when choosing existing TS, creating or improving TS, the consumer properties of TS should be formed on the basis of "good" adjectives, which prevent or reduce the negative consequences of the manifestation of TS properties characterized by "bad" adjectives.

The examples of "good" adjectives formed with the prefixes "bes-" (without, less, free), "bez-" (without, less, free, non), "anti-" (anti), "sverkh-" (super), the antonyms of adjectives formed using the stem "sam-" and the connecting vowel "o" (such adjectives there are about 200), words based on "auto" (the first part of compound words meaning "self"), as well as the most important human needs are shown in Fig. 133–136.

Figure 133. Examples of "good" adjectives based on the prefixes "bes-" (without less, free, non) (a) and"bez-" (without less, free, non) (b)

a) **Prefix "bes-" (without, less, free, non)**

transliteration	translation
beskalorijnyj	calorie free
beskarbyuratornyj	carburetorless
beskarkasnyj	frameless
beskolesnyj	wheelless
beskondensatornyj	condenserless
beskonechnyj	endless
beskontaktnyj	contactless
beskrajnij	endless
beskrovnyj	bloodless
besplamennyj	flameless
besporovyj	imporous
besposhlinnyj	duty free
bespredmetnyj	pointless
besprovodnoj	cordless
bessvyaznyj	incoherent

b) **Prefix "bez-" (without, less, free, non)**

transliteration	translation
bezavarijnyj	accident-free
bezalkogol'nyj	non-alcoholic
bezbalansirnyj	balanceless
bezbar'ernyj	barrier-free
bezbiletnyj	ticketless
bezboleznennyj	painless
bezbumazhnyj	paperless
bezvizovyj	visa-free
bezvodnyj	waterless
bezvozvratnyj	non-refundable
bezvrednyj	harmless
bezvremennyj	timeless
bezgranichnyj	limitless
bezdificitnyj	deficit-free
bezdrozhzhevoj	yeast-free

transliteration	translation
besslednyj	traceless
bessolevoj	saltless
bessrochnyj	indefinite
besstochnyj	drainless
besstupenchatyj	stepless
besstykovoj	weldless
besschetnyj	countless
bestravmatichnyj	non traumatic
bestranshevyj	trenchless
besturbovoj	turbineless
besformennyj	shapeless
beskhrebetnyj	spineless
bescvetnyj	colorless
(prozraznyj)	(transparent)
beschuvstvennyj	insensible
(nechustvitel'nyj	(insensible)
besshovnyj	seamless
	etc.

transliteration	translation
bezdymnyj	smokeless
bezlopostnoj	leafless
bezlyudnyj	deserted
bezmotornyj	motorless
bezognevoj	fireless
bezokonnyj	windowless
bezopornyj	unsupported
bezoskolochnyj	shatterproof
besosnovnoj	baseless fail
bezothodnyj	waste-free
bezuglevodnyj	carbon-free
bezuglerodnyj	carbon-free
bezudarnyj	shockless
bezyinventarnyj	non-inventorial
bezynduktivnyj	non-inductive
bezynercionnyj	non-inertial
bezyskrovyj	non-spark
	etc.

Figure 134. Examples of "good" adjectives based on the prefixes "anti-" (anti) (a) and "sverh" (super-) (b)

a) **prefixes "anti-" (anti)**

transliteration	translation
antialkologol'nyj	anti-alcoholic
antibakterial'nyj	antibacterial
antiblikovyj	anti-glare
antiblokirovochnyj	anti-blocking
antibronhokonstriktornyj	anti-bronchoconstrictor
antivibracionnyj	anti-vibration
antivirusnyj	anti-viral
antigennyj	antigen
antidotnyj	antidote

b) **prefixes "sverh" (super-)**

transliteration	translation
cverhabsorbiruyushchij	superabsorbent
sverhvysotnyj	superhigh
cverhvyazkij	superviscous
sverhglubokij	superdeep
sverhdal'nobojnyj	ultra-long-range
sverhzvukovoj	supersonic
sverhletuchij	supervolatile
sverhlipuchij	super-sticky
sverhluchistyj	super-radiant

transliteration	translation
antikislotnyj	anti-acid
antikoagulyantnyj	anti-coagulant
antikolebatel'nyj	anti-vibration
antikorrozijnyj	anti-corrosion
antimagnitnyj	anti-magnetic
antimonopol'nyj	anti-monopoly
antiobledenitel'nyj	anti-icing
antiokislyayushchij	antioxidant
antiprigarnyj	anti-stick
antirezonansnyj	anti-resonance
antiselevoj	anti-mud
antisepticheskij	antiseptic
antiskleroticheskij	anti-sclerotic
antisolevoj	anti-salt
antispasticheskij	anti-spastic
antistaticheskij	anti-static
antistressovyj	anti-stress
antitaksicheskij	anti-toxic
antiudarnyj	anti-shock
antifrikcionnyj	anti-friction
antishchelochnoj	anti-alkaline
	etc.

transliteration	translation
sverhmelkij	super-small
sverhmoshchnyj	super-powerful
sverhplavuchij	super-buoyant
sverhplastichnyj	super-plastic
sverhplotnyj	super-dense
svekhporistyj	super-porous
sverhprovodyashchij	super-conducting
sverhprochnyj	super-strong
sverhredkij	super-rare
sverhsil'nyj	super-strong
sverhskol'zskij	super-slippery
sverhskorostnoj	super-fast
sverhsypuchij	super-free-flowing
sverhtverdyj	super-hard
sverhtekuchij	super-fluid
sverhtoksichnyj	super-toxic
sverhtonkij	super-thin
sverhtyaguchij	super-viscous
sverhtyazhelyj	super-heavy
sverhustojchivyj	super-stable
sverhkhrupkij	superfragile
sverhchastotnyj	superfrequency
sverhchistyj	superpure
sverhchuvstvitel'nyj	supersensitive
sverh"emkostnoj	supercapacity
sverh"yarkij	superbright
	etc.

Many "good" adjectives are formed with the help of foreign language prefixes "re-" (for example, recuperative), "ultra-" (ultra-short), "infra-" (infrasonic), "pseudo-" (pseudo-boiling), etc.

Many "good" adjectives can not only serve as a hint for the creation and improvement of TS, but also serve as the basis for the formation of new industries. For example, on the basis of the adjective "carbon-free" (see Fig. 133), carbon-free energy develops, which, according to scientists, is designed to stop global warming.

The examples of "bad" adjectives that are the basis of TS properties, the

Figure 135. Examples of "good" adjectives using the antonyms of adjectives with the prefix "ne-" (not) (a) and adjectives formed using the stem "sam-" (self, auto) with a connecting vowel "o" (b)

a) **prefixes "anti-" (anti)**

transliteration	translation
vodonepronicaemyj	waterproof
gazonepronicaemyj	gas-tight
neavtoklavnyj	non-autoclave
neb'yushchijsya	shatterproof
nevidimyj	invisible
nevpityvayushchij	non-absorbent
negnushchijsya	rigid
nezamerzayushchij	non-freezing
nezatverdevayushchij	non-hardening
neizvlekaemyj	non-removable
neiznashivayushchijsya	non-wearing
nekroshashchijsya	non-crumbling
nelinyayushchij	non-fading
nemnushchijsya	non-wrinkling
neobratimyj	irreversible
nepod"emnyj	non-lifting
nepotoplyaemyj	unsinkable
nerazryvnyj (nerazryvnoj)	non-breaking (non- bursting)
nerastvoryayushchijsya	non-dissolving
nervushchijsya	tearproof
neskoroportyashchijsya	non-perishable
neslipayushchijsya	non-sticky
neubivaemyj	non-kill
neulovimyj	elusive
neelektrizuemyj	non-electrifying
puleneprobivaemyj	bullet-proof
pylenepronicaemyj	dust proof
	etc.

b) **prefixes "sverh" (super-)**

transliteration	translation
samovarnyj	self-boiled
samovzvodnyj (samozaryadnyj)	self-cocking (self-loading)
samovosplamenyayushchijsya	self-igniting
samovosstanavlivayushchijsya	self-healing
samodvizhushchijsya	self-propelled
samozazhivayushchij	self-healing
samozamedlyayushchijsya	self-decelerating
samozatachivayushchijsya	self-sharpening
samoischezayushchij	self-extinguishing
samoletnyj	aircraft
samonagrevayushchij	self-heating
samonastraivayushchijsya	self-adjusting
samoohlazhdayushchij	self-cooling
samoochishchayushchijsya	self-cleaning
samorazvivayushchijsya	self-developing
samorazmorazhivayushchijsya	self-defrosting
samoraskryvayushchijsya	self-opening
samoreguliruyushchijsya	self-adjusting
samosval'nyj	dump
samosglazhivayushchij	self-smoothing
samostoyatel'nyj	independent
samoubirayushchijsya	self-cleaning
samounichtozhayushchijsya (samorazrushayushchijsya)	self-defeating (self-destructive)
samouplotnyayushchijsya	self-compacting
samoupravlyayushchijsya	self-steering
samouskoryayushchijsya	self-accelerating
	etc.

If you are setting a task, it is very creative to use:

- — the adjectives "smart" and "clean";
- — synonyms of verbs and adjectives;
- — metaphors.

Figure 136. Examples of "good" adjectives based on "avto" – the first part of compound words meaning "self", "auto" (a), and the most important human needs (b)

a) **Stem "avto-" (self, auto)**

transliteration	translation
avtoblokirovochnyj	auto-blocking
avtogennyj	autogenous
avtodoil'nyj	auto-milker
avtoionnyj	auto-ion
avtoklavnyj	autoclave
avtomaticheskij	automatic
avtomodel'nyj	auto-model
avtonomnyj	autonomous
avtopilotnyj	autopilot
avtopogruzochnyj	auto-loading
avtopoil'nyj	auto-drinker
avtoritarnyj	authoritarian
avtorotacionnyj	auto-rotary
avtosmazochnyj	auto-lubricant
avtostradnyj	auto-stradny
avtoscepochnyj	auto-coupling
avtotransportnyj	motor transport
avtohromnyj	autochromic
avtohtonnyj	autochthonous
	etc.

b) **The most important human needs**

transliteration	translation
akkumuliruyushchaya	accumulating
baktericidnaya	bactericidal
garmonichnaya	harmonious
generiruyushchaya	generating
gibridnaya	hybrid
zashchitnaya	protective
informiruyushchaya	informing
krasivaya	beautiful
lechebnaya	therapeutic
mnogofunkcional'naya	multifunctional
normaliziruyushchaya	normalizing
oboznachayushchaya	designating
ohranyayushchaya	protecting
ochishchayushchaya	purifying
preobrazuyushchaya	transforming
profilaktiruyushchaya	prophylactic
razvivayushchaya	developing
razvlekayushchaya	entertaining
rezerviruyushchaya	reserve
sovmestimaya	compatible
uporyadochivayushchaya	ordering
uspokaivayushchaya	soothing
utiliziruemaya	recyclable
energosberegayushchaya	energy saving
	etc.

occurrence of which should be prevented: excessive, harmful, insufficient, dirty, fragile, expensive, noisy, weak, smoky, smelly, poisonous, explosive, etc. As you can see, many "bad" adjectives are the opposite of good. Let me remind you that often properties based on "bad" adjectives are implemented in useful

TS functions (for example, in noise, gas grenades, smoke candles and other special means).

Returning to the "good" adjectives, I note that the use of properties based on the "good" adjectives "smart" and "clean" is extremely effective in choosing existing TS, creating and improving TS. Moreover, I believe that the first person to suggest developing TS with properties based on the adjective "smart" is a genius! Therefore, due to the extremely high creativity of the adjectives "smart" and "clean", a separate paragraph is devoted to them. There are great prospects for improving TS and / or creating new TS generations for the adjective "open".

6.2.10 High creativity of using adjectives "smart" and "clean" when setting a task

It is a very difficult task to generate (invente) new important human needs, functions and consumer properties of TS for the qualitative improvement of TS, create TS of new generations, and moreover to create pioneering TS. And in order to make these processes effective, at the end of the last century, the idea to create TS was born that would become smart, (almost) like man, that is, capable of complex reasoning and rational conclusions, including the ability to perceive, capture, recognize, generalize, fix in memory, make the decision.

This idea turned out to be very fruitful. For example, it is not easy, without knowing heuristics, to form competitive functions and properties of TS (in particular, a residential house) to fulfill the human need to build a house and make life comfortable, safe, entertaining (with useful and interesting leisure time), and economical in it. But one has only to assume that even if the house is "smart", which means that it is capable of fulfilling and even anticipating human desires, it is easy to form the functions and properties of such a house, in which many, not fully realized dreams, can become the reality. Let's consider some popular solutions based on the highly creative adjective "smart".

Smart "home" recognizes "the owner and other residents (for example, by fingerprints, retina, smell, voice), opens the lock, turns on the necessary light, provides the desired temperature in each room, and if necessary, turns on the music center, TV on desired channel (fig. 137). At a given time, it cleans the premises himself. If the owners go on vacation or on a long business trip, then the "smart" house imitates their presence (for example, closes the curtains in the evening, opens them in the morning). Moreover, the owner can monitor each room over the Internet in real time. If someone illegally enters the "smart" house, then the house calls the police and sends a pre-recorded message, as well as sends an SMS to the owner, turns on the light and takes pictures of the unwelcome guests. The taken photographs can be viewed immediately via the Internet. In the event of a fire, the smart home turns off gas and electricity, blocks ventilation to prevent the spread of fire and smoke, turns on the smoke exhaust system, and notifies the owners and firefighters of the fire. In the event of a leak, the anti-leakage system turns off the water and takes other measures to neutralize the consequences of the leak.

Figure 137. A variant of the layout of the "smart" home

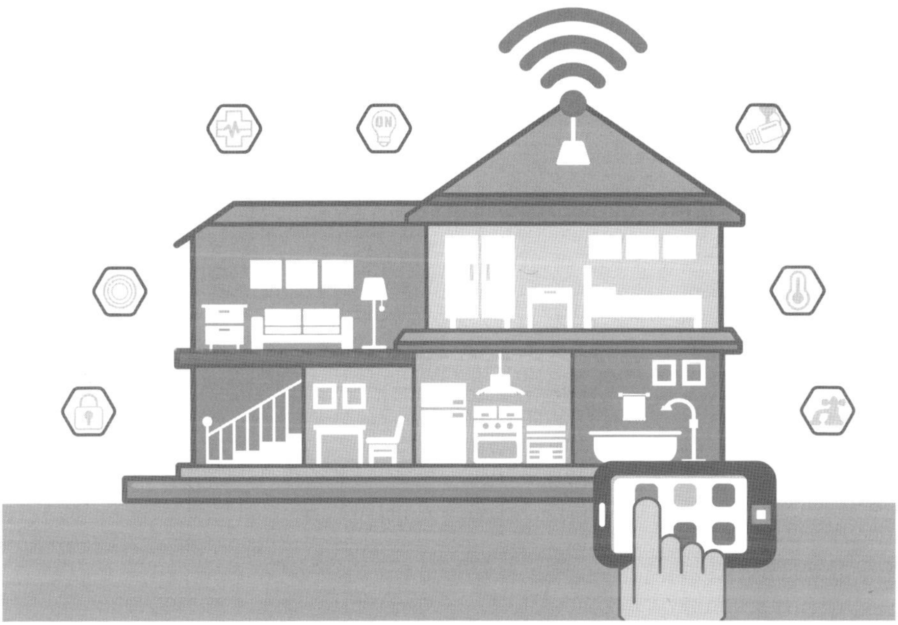

The "smart" kitchen bakes bread and prepares breakfast itself. A "smart" refrigerator can inform the owner about the expiration date of the products, tracks which products are running out, informs the owner, sends the order to the store and checks if everything is delivered.

"Smart" underwear can check itself, for example, the patient's condition and make the necessary injections on time. Clothing for athletes controls the frequency of respiration, pulse, pressure of man during physical exercise and informs man about their current data (Fig. 138). Clothes made of "smart" fabrics warm in cold weather and cool in heat.

In the near future, a "smart" mobile phone will be able to monitor man's condition constantly and send data (for example, cardiograms) to a doctor in case of dangerous deviations. If man needs new connections, then a smartphone, "feeling" man nearby with interests close to its owner in the field of sports, culture, education, science, similar interests and views on politics, religion, can give the owner a sign so that he can identify and speak with him (if he objects, then his phone can "beep" about it).

Figure 138. Prototype of "smart" clothing for athletes

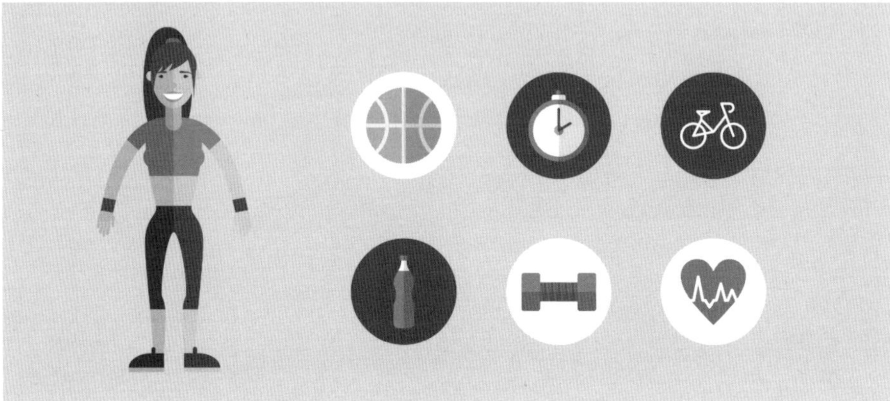

Figure 139. "Smart" pill with a sensor that monitors the regularity of medication intake

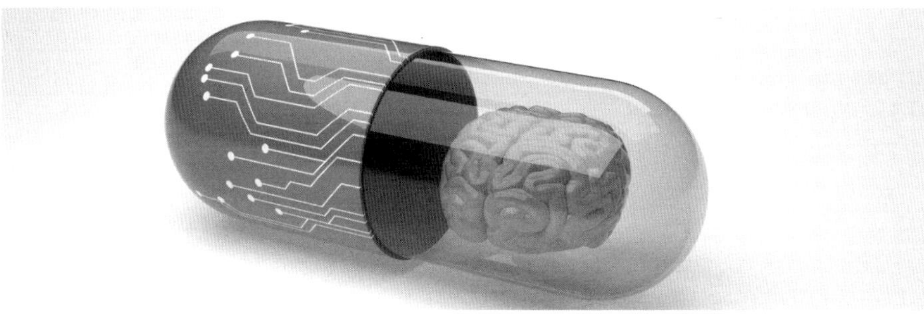

It is the word "smart" that has prompted the creation of an amazing individual vehicle — a gyro scooter (Fig. 140), which is called "smart slippers", since it is controlled only by the pressure of the foot.

The adjective "smart" is so creative that any reasonable person can quite easily suggest how to use it to create new TS (even a new industry) and qualitatively improve any known TS. I recommend looking at the surrounding TS, remembering other TS with the thought "how to make them smart", and new functions and properties that are attractive, at least for you, can be certainly formed. For example, a city, energy, agriculture, medicine, trade, logistics can be infinitely "smart".

Great opportunities for the implementation of the idea of "smart" are in nature.

Figure 140. "Smart slippers" - gyro scooter

It is believed that monkeys, elephants, dolphins, rats, crows, dogs, and cats have the greatest intelligence. American, English and other biologists have made an amazing discovery: plants can hear, see, feel touch, smell, communicate with each other and even read our minds![58] For example, taste sensors are found in the roots of plants. They help finding nutrients underground. An interesting observation has been made by scientists regarding the fact that the plant's sense of taste is also used in self-defense. For example, when a tomato bush tastes the saliva of a caterpillar, which begins to chew on the leaves, the plant releases a special substance that binds the caterpillar's digestive enzyme. Or, for example, it turned out that many plants have their own musical predilections. Primroses, phloxes and tobacco prefer Wagner's operas, mimosas and hyacinths love Tchaikovsky, and cyclamens love jazz music[59]. Genetic engineering can develop these and many other smart qualities in animals, plants and other organisms for the realization of human intelligent needs.

High creativity of the adjective "pure" consisits in the high potential generation of many directions based on the word "pure," to create new TS generations of various classes. For example, these include "pure metals", "clean water", "clean air", "pure breed", etc. The creative possibilities of the adjective "clean" are extremly expanded by the adjectives associated with it, for example, "ecological" (food, fuel, production, appliances, furniture, dishes, cosmetics, etc.).

6.2.11 Creativity of using synonyms of verbs and adjectives when setting a task

When formulating TS functions and properties (taking into account the rules for formulating TS functions specified in section 6.2.5.1, recommendations on the use of "controlling", "good" and "bad" verbs given in section 6.2.8, recommendations on the use of "good" and "bad" adjectives, as well as the adjectives "smart" and "clean" specified in sections 6.2.9 and 6.2.10), it is advisable to consider synonyms of verbs and adjectives taken as the basis of the formulated functions and properties of TS.

Synonyms in the Russian language, although according to the classical definition they are "different in sound, but have the same or similar meaning", often include words the meaning proximity of which is maximal heuristic. For this reason, the synonyms of verbs and adjectives can serve as a hint for the development (use) of original TS. For example, it is interesting that the verb "**to transfer**" has 115 synonyms, among which are "to lower", "to roll off", "to throw". You can pick up 81 synonyms for the verb "**to sort**", including "to decompose", "to spray"; to the verb "**to separate**" — 153, among which "to reject", "to isolate", "to split off", "to rip off", "to comb out".

The adjective "new" has 67 synonyms, including "fresh", "zero"; to the adjective "smart" one can find 112 synonyms, including "thin", "prophetic"; to the adjective "pure" — 236, including "naked", "intense", "crystal"; to the adjective "protective" — 32, including "defensive", "barrage", "shielding", as well as the adjectives with clarifying meanings of the word "protective": "warm-", "light-", "dust-", "snow — ", "ice", "steam", "radio", "chemical", "electrical protection".

6.2.12 Metaphors and their use when setting a task

Metaphor — a kind of trope (a word or expression used in a figurative sense in order to enhance the completeness and imagery of an object or phenomenon description), which is based on associations by similarity or contrast (see Section 5.1.5) or analogy (see 5.1.4). In other words, a metaphor is a hidden comparison (precisely a hidden one, which, unlike an explicit comparison, does not contain the conjunctions "like", "if", "as if", "exactly"; it does not contain the words "similar to", "it reminds")[60].

High imagery of the description of the objects or phenomena being developed contributes to their more complete representation, which is very important at the stage of setting a problem. It is an extremely difficult task to find and, more over, to formulate a good metaphor. Aristotle rightly considered the ability for metaphorical thinking (the perception of similarities in different things) as a sign of genius.

Considering that the most important mechanism for creating a new one is based on the convergence (closing-in, transfer) of knowledge and is especially effective in the convergence of knowledge from traditionally non-overlapping branches of knowledge, we consider various types of metaphors, including those

Table 24. Types of metaphors

Types of metaphors	Examples of metaphors
Riddles	The fur coat is new – but it has a hole on the hem (the key – ice hole). No hands, no hatchet, but a hut is built (the key-nest)
Proverbs and sayings	The husband is the needle, the wife is the thread. Strike while the iron is hot
Names	In Central Asia, the name Aizhan means "merry moon"; the Lithuanian name Audra means "storm"; in ancient Russia, in order to scare away evil spirits, boys were called Nekras(not handsome), Nezvan (not invited), Nezhdan(not expected); Alexey – "protector" (Greek name); Olga – "sacred" (Scandinavian name)
Metaphors with positive and negative characteristics of man and his activities	Words with a positive metaphorical meaning: sun, rose, star; words with negative metaphorical characteristics (for some reason, they are recalled more often than positive ones): carcass, skin, bear, den, yoke, pigsty, donkey, poorhouse, poker, stump, porridge, husk, pawn, aphid, fever, slave, jackal
Metaphor in literature	"Pointed fir-tree eyelashes over the blue eyes of the lakes" (words from the song "Long will Karelia dream"), "The camomiles hid, the buttercups drooped when I froze from the bitter words" (words from the song "Why do you girls love beautiful ones")
Metaphor in religion	An excerpt from St. John Chrysostom: "Let us cleave to Him (the Lord) and cleave with deeds ... He (the Apostle John the Theologian) inspires us with unity with Him by many comparisons. Look: He is the head, we are the body, and how much distance can there be between the head and the body? He is the foundation, we are the building; He is the vine, we are the branches; He is the groom, we are the brides; He is the shepherd, we are the sheep; He is the way, we are walking; we are the temple, He is the dweller; He is the firstborn, we are brothers; He is the heir, we are co-heirs; He is life, we are living; He is the resurrection, we are being resurrected; He is the light, we are being enlightened"
Metaphors in scientific and technical creativity	Worm gear transmission. The sphere of many knowledge-solutions for one of the scientific and technical problems (see Fig. 87 "Model fragment of the figurative representation of all knowledge space in the world"). Heuristic tunnel (see Fig. 89 "Model of creative activity using heuristic methods and tools based on productive knowledge"). "The city is the unity of the dissimilar"

that, at first glance, have no relation to scientific and technical creativity (table 24).

The metaphor of Aristotle is truly brilliant in its consistency, completeness and imagery of object description:"The city is the unity of the dissimilar." Judge for yourself, have the development institutions in the city of Moscow provided everything for the improvement of this huge megalopolis?

Let us consider from the point of view of this metaphor what variants of UNITY (systems with unified management) can be in the city, as well as the variants of differences of UNLIKE people (table 25).

Variants of UNITY in a megalopolis

1. Systems of external and internal security.
2. The health care system.
3. Systems of upbringing, education, culture and recreation.
4. Energy and transport systems.

Table 25. Differences between people, which are the basis of their OTHERNESS, and the necessary measures to meet the needs of UNLIKE people in the development of a megalopolis

Differences between people, which are the basis of their OTHERNESS	The necessary measures to meet the needs of UNLIKE people in the development of a megalopolis
Confession	It is necessary to have places of worship for people of different faiths (churches, mosques, synagogues, Buddhist khuruls and datsans (temples), etc.)
Age	There should be a sufficient number of kindergartens, schools, playgrounds and libraries, a well-developed clubs movement, areas of communication and activities for adolescents. The age-related problems of mature and especially elderly people should be taken into account, for whom the implementation of everyday needs at work, at home and even on vacation is difficult
Gender differences (genetic, anatomical, behavioral)	Gender differences are evident in almost all areas. For example, different toys are provided for boys and girls, it is believed that they have to play different games, because boys use larger space for games, unlike girls. This is why gender differences must be taken into account in all systems that reflect UNITY
Anthropological differences	For example, clothing should be designed to accommodate variations in the physical constitution of people of different races
People's height and weight	It should take into account big differences in height and weight of people, such as the creation of furniture, clothes, vehicles
The level of health and physical capabilities of people	When creating urban infrastructure (especially housing and communal services, transport, education, culture, recreation, public catering), it is necessary to take into account the large differences in the level of health of people, especially people with disabilities (motor, visual impairment, hearing, speech impairments, mental and psychiatric disorders, etc.)
National culture, clothing, cuisine	It is advisable to create zones in the city that reflect different national cultures; opening of shops selling national clothes, food products, as well as food outlets based on national cuisines; classes in national languages, art, sports, organization of national mass media, etc.
Purposes of transfer in transport	It is necessary to create special types of transport for schoolchildren, tourists injured in accidents and seriously ill patients
Differences in attitude towards animals	Lovers and especially owners of pets need to create conditions for their maintenance and communication with pets, their training, preventive measures and treatment
The difference in schoolchildren's creative and other abilities, goals of education, their professional orientation	Schoolchildren with different creative abilities should be taught according to different programs, which should also take into account different learning goals: practical activity, obtaining secondary vocational education or entering a university, and then postgraduate study, etc.), professional orientation (for example, there should be different programs for children gravitating towards the humanities or technical sciences)
Income and wealth	There should be created, including cheap ones, but high-quality housing, clothing, furniture, household appliances, food, provided (also cheap, but high-quality) social, educational, medical and other services

5. Systems of housing and communal services.
6. Trading system.
7. Systems of other sectors of the economy (including banking, public catering, etc.).

8. Moral and volitional unity of people in massive critical situations.
9. Communication and information systems.
10. Control system.

Newcomers have special problems associated with changes in the usual conditions, isolation from loved ones.

Recommendations for the formulation of metaphors in scientific and technical creativity include the following reasoning.

The metaphor helps to comprehend, explain the state of TS at all stages of its life cycle (see Section 1.4.7) or the essence of the phenomenon if it:

— Transparent in its meaning;
— Easier than TS or the phenomenon that the metaphor helps to explain.

The phrase TS "sleep mode" can be consibered as an example, which figuratively explains many features of TS in the corresponding state, as opposed to simply switched off TS. Or, for example, the mechanism of convergence (closing-in, transfer) of knowledge becomes clearer from the model of the figurative representation of the world all knowledge-solutions space for various scientific and technical problems (Fig. 87), and the advantages of using heuristic methods and means that dramatically reduce the number of trial and error in scientific and technical creativity, are clearly visible in the model of creative activity, representing these heuristics in the form of tunnels that orient a man in the right direction (Fig. 89).

The most striking property of another object or phenomenon is often used (for example, "chameleon glasses") if you would figurative understand the metaphor of an object or phenomenon. The metaphor, in contrast to the comparison based on the use of conjunctions "as", "as if", "exactly", should be shorter than the comparison, because it does not have these unions, as well as the words "similar", "it reminds".

When formulating a metaphor, along with the goal of the most complete TS description or the essence of the phenomenon, mnemonic capabilities should be used (mnemonics is the art of memorization). For example, if mushrooms are metaphorically called natural vacuum cleaners (having likened the toxins in the soil to dirt and dust), then, most likely, a man will remember for a long time that it is the mushrooms that actively suck out toxins from the soil

It is necessary to develop metaphorical thinking persistently, which, I remind, is a sign of genius.

6.2.13 Focal Object Method

The method of focal objects (MFO) was proposed by E. Kunze in 1923, the professor at the University of Berlin, and in the 50s it was improved by C. Weiting in the USA.

Use the method of focal objects to overcome the barriers to thinking, to be creatively liberated to formulate a problem or form new functions and properties of TS.

When setting a task, as well as commercializing and / or using the results of scientific and technical creativity to form a variety of human needs, functions, properties, functional structures, principles of action, technical decisions of TS, use the method ("ascent-descent").

MFO is distinguished by its simplicity, invariance to the subject area and is very effective for overcoming psychological barriers to thinking, creative liberation in the formation of new functions and properties for improving TS analogue. Often in this case, a hint is provided by inferences caused by random associations connected with objects of animate and inanimate nature, other TS (see Section 5.1.5). However, the effectiveness of a single, random association is low. First, figuratively speaking, you can sit for a very long time, like I. Newton under a tree, and wait for an overripe apple to fall on your head. Secondly, even if the association does arise, it does not guarantee a creative result. Taking this opportunity, let me remind you: nothing guarantees a creative result, but the probability of getting it, of course, increases with the number of associations. Therefore, the main focus of MFO is the activation of associative thinking.

The essence of MFO consists in the following. They randomly choose from a variety of information sources (encyclopedic dictionaries, other books and newspapers) any words: names of TS, objects of animate and inanimate nature, including those representing parts of speech (for example, in Russian language there are 12 of them — adjectives, pronouns, numbers, conjunctions, etc.), and then, as it were, focus (hence the word "focal" comes in the name of the method) the features of these words on the analogue of TS being improved and synthesize a new technical decision on their basis.

As a rule, at the first stage, it is enough to use 4–5 words. If there is TS among them, it is necessary to formulate its main functions and properties. If objects are from living or inanimate nature, then their basic properties are used, if they are verbs or adjectives, they must be used directly: verbs — to formulate functions, adjectives — as properties.

For example, if the analogue being improved is a chair, and a randomly chosen word is "a lamp", its functions are "to generate light radiation", "to heat" and others, and the properties are glass, transparent, colored and others, then it is easy, by focusing these features of the lamp on the chair, find new solutions: heated chair, illuminated, transparent, ultraviolet, etc.

Sometimes a numeral can be very useful. For example, if the analogue being improved is a weapon, then a randomly chosen numeral can suggest a new solution (multi-barreled mortar and machine gun, multi-door cabinet, three-barreled gun).

It can be effective to combine a analogue being perfected with a randomly selected object (a noun with a noun), for example: a pen-clock, a boat-tent, a crutch-flash-light, and even beach slippers-treasure hunter (which, if a man is walking on the beach, signal people who have fallen metal (gold, platinum, silver, etc.) jewelry into the sand (and lost).

A randomly chosen pronoun can also be productive. Perhaps the pronoun "our" (instead of the usual "my")[61] has been a clue to the creation of a tandem bicycle.

Some scientists — specialists in the field of scientific and technical creativity — suggest using the ready-made list of nouns with striking properties from traditionally non-intersecting areas of human activity to facilitate the choice of nouns and increase its efficiency. In the literature on scientific and technical creativity, you can find several options for such lists. I have also compiled the original list of 100 nouns in my own way that are recommended for use in MFO. This list is given in Appendix 5.

To expand the list of randomly selected words, G. Ya. Bush recommended forming semantic associations on their basis (see Section 5.1.5), figuratively calling them "associations' garlands." If you need to focus on finding new properties, I recommend going straight to the database or adjectives dictionary.

The MFO scheme, including the use of the "associations' garland" and the taxonomy of adjectives, is shown in Fig. 141.

Figure 141. Schematic representation of the focal object method

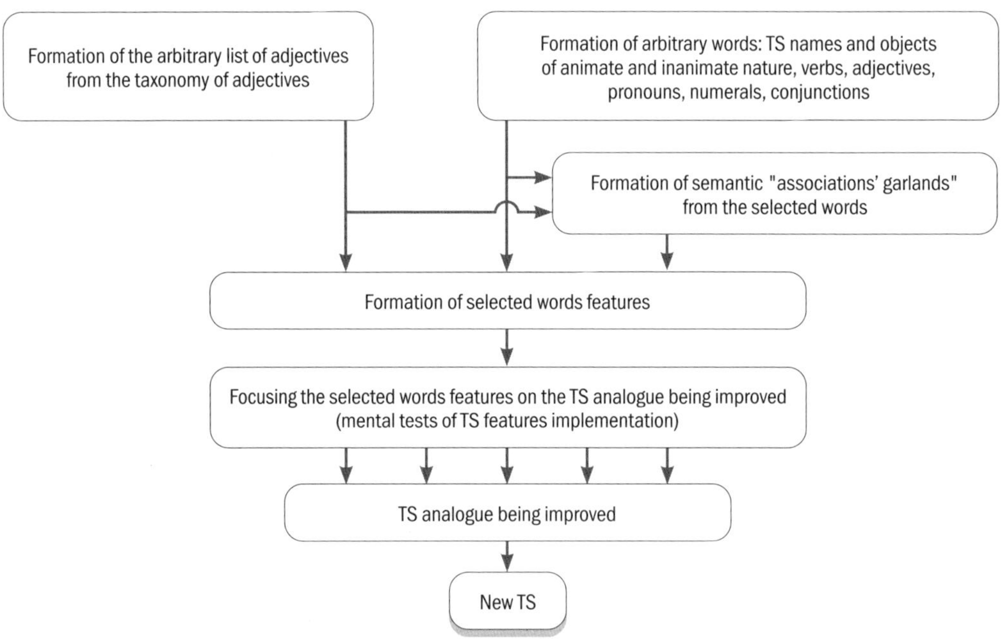

6.2.14 The "ascent-descent" method and its use when setting a task

The method of "ascent-descent" developed by me is the effective method of forming the necessary variety of human needs, functions, properties, functional structures, principles of operation, technical decisions (TD) of TS, objects, phenomena, as well as relations between TS and their elements, leading to the implementation of TS design strategies (see Section 3.2. 2). It is based on the sequential use of forms of inference: induction (the transition from the particular to the general) and deduction (the transition from the general to the particular and the singular). In induction and deduction, the mental operations of abstraction and concretization are used, respectively. In this case, the unit (taxon) of the particular can be descriptions (characteristics) of a known object, phenomenon, need, function, property and, in general, any unit of knowledge about an object (object, phenomenon), process, relation. Common to any taxon an existing or potentially possible taxonomy (classifier, thesaurus, ontology, etc.) is in which taxa are included on the basis of particular stable characteristics.

The "ascent-descent" method can be effectively used at all stages of scientific and technical creativity (design) — formulation, solution of the problem and commercialization or other use of the results, including with implementation of all strategies and design tactics instead of or with limited use of ineffective traditional methods and means based on reproductive knowledge (see Section 5.1).

In general, the "ascent-descent" method includes four stages, schematically shown in Fig. 142. Initial taxa 1 at the stage of setting the problem (forming the technical specification, task (TT) for TS development) can alternately be the characteristics of general, particular, technically realizable human needs, functions, consumer properties (see stages 1–4, Fig. 45), and at the stages of solving objectives or commercialization of results — particular characteristics of TS analogues, selected for improvement and described at stages 5–8 of their development (see Fig. 45), as well as the scope of analogues use.

The "ascent-descent" method can also be used to form a variety of general human needs (Fig. 143). For example, a particular characteristic (the name of an individual need) "the need for protection from natural disasters and the reduction of the severity of their consequences" (taxon 1, Fig. 143, stage 1) is selected.

On the basis of induction (the transition from a particular characteristic to a general one), we conclude that the general characteristic of taxon 1 is "protection", and taxonomy is "general needs for the protection of people" (stage 2). At the same time, we seem to be making the "ascent" to taxonomy.

At stage 3, using deduction ("descent"), we select taxa from information sources (or form potentially possible).

Figure 142. Use of the "ascent-descent" method to form the required variety of taxa – human needs, functions, properties, functional structures, principles of action, technical decisions of TS, objects, phenomena, as well as relations between TS and their elements, where ▢ – selected new taxa 2 and 4 are marked with a shade

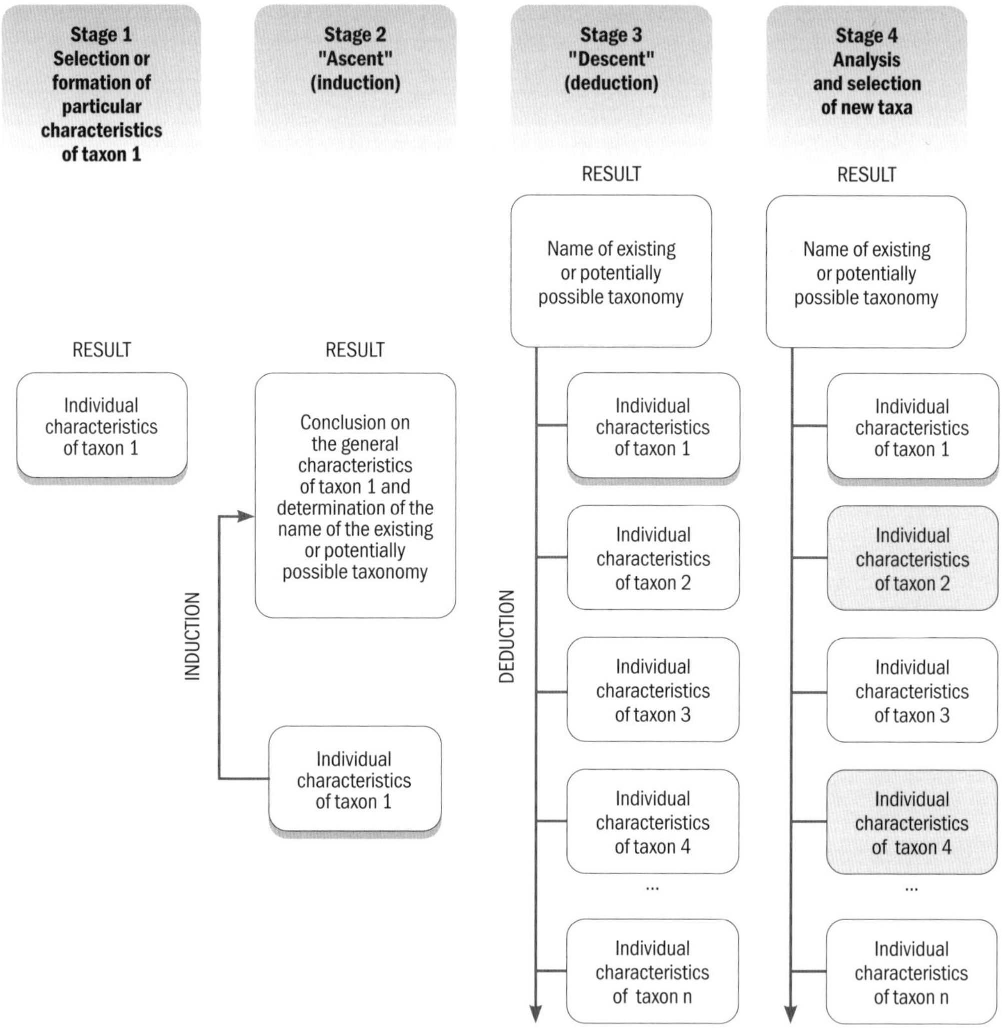

At stage 4, we select new individual needs for the protection of people from the existing (or formed) taxonomy.

The selected new individual needs can be considered as general, and after deduction (descent), a new taxonomy can be formed (including using the list of the most important human needs).

Figure 143. Example of the use of the "ascent-descent" method for the formation of a variety of individual needs in the protection of people based on a given individual need, where ☐ – the selected new needs are marked with a shade

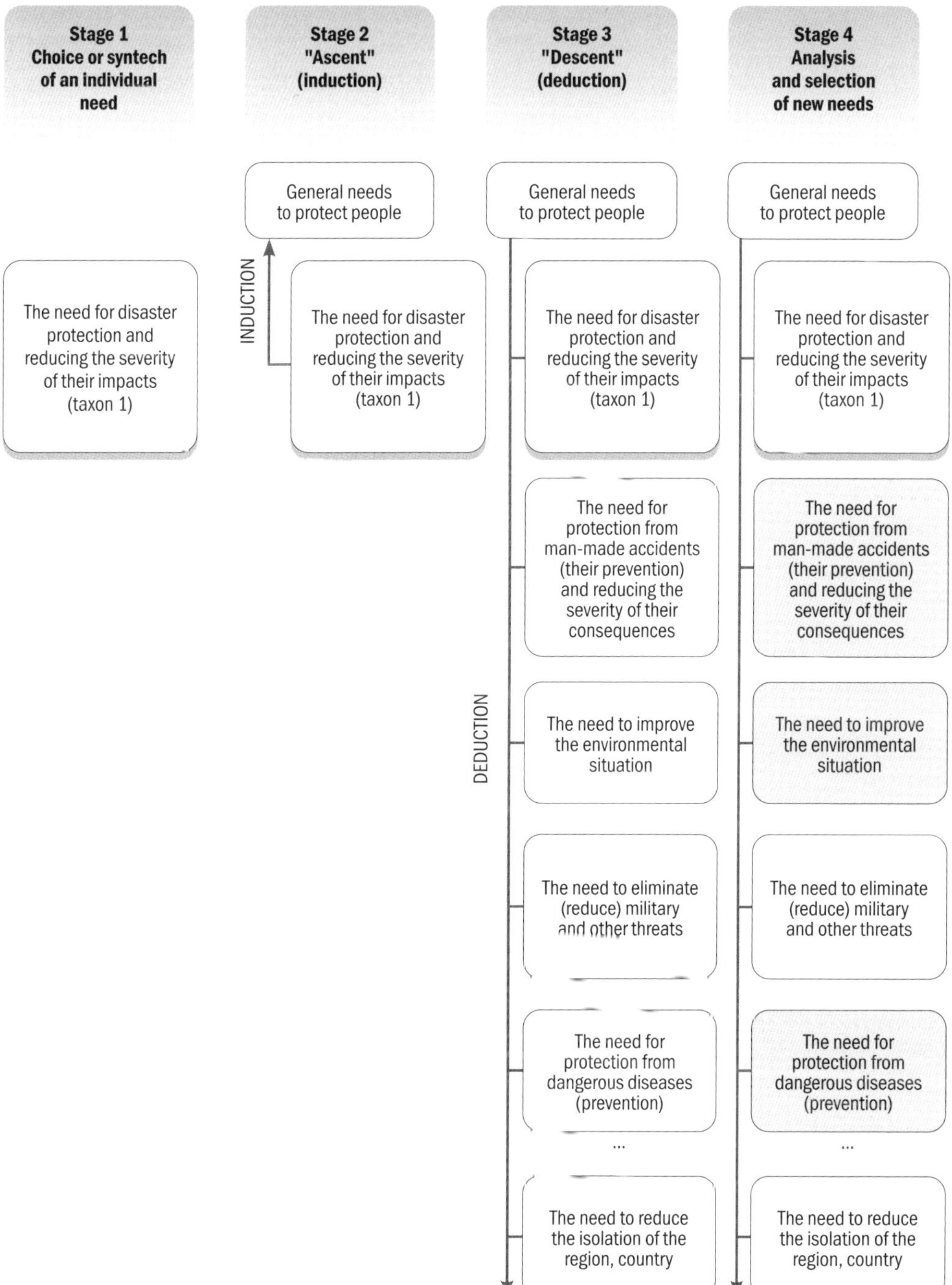

Figure 144. Example of using the "ascent-descent" method to form a variety of functions based on a given one, where ☐ – the selected new functions are marked with a shade

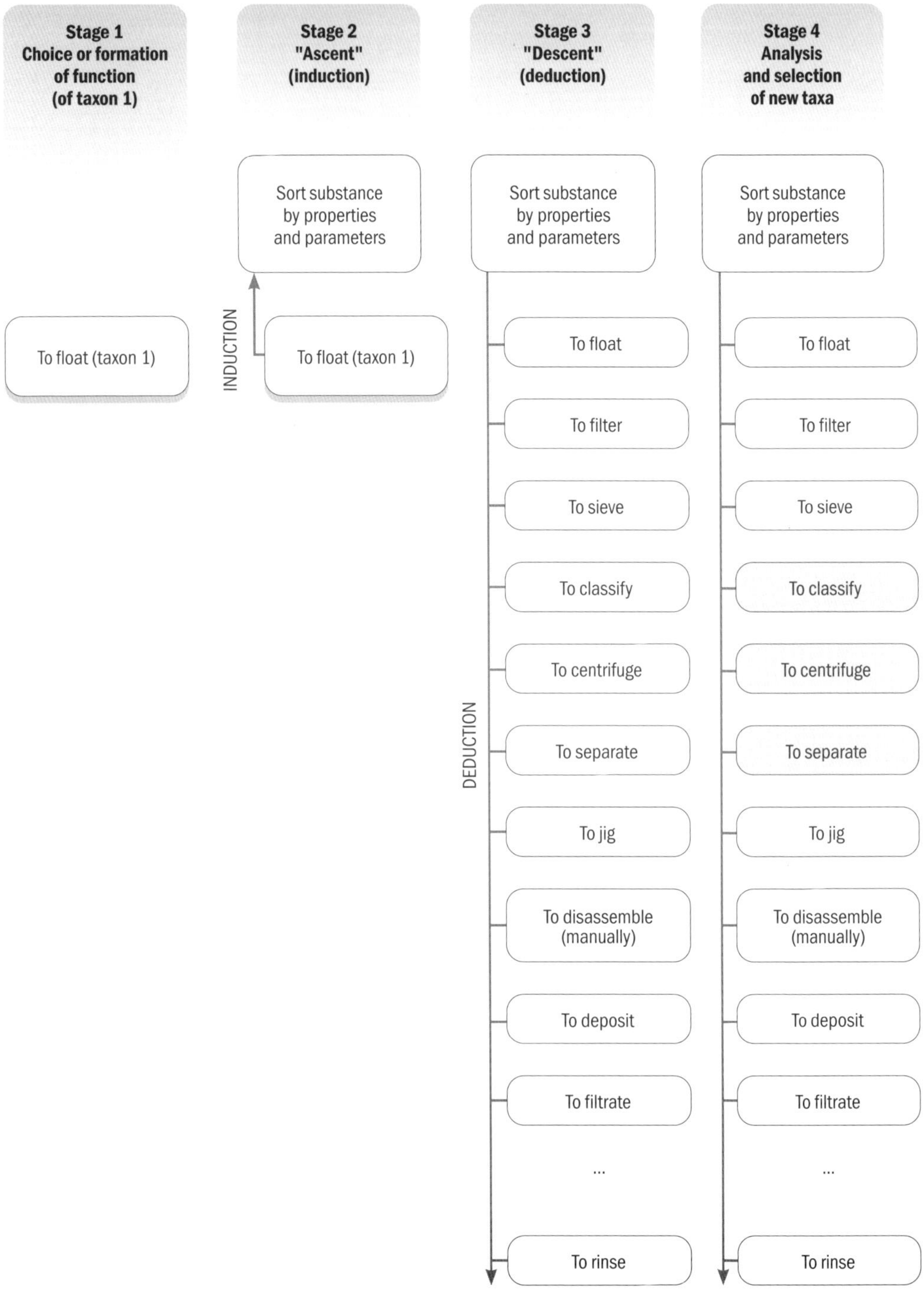

Another example of using the "ascent-descent" method at the stage of setting a problem (forming a TT for developing TS) illustrates the search for new TS functions based on a given one (Fig. 144). Taxonomies for functions can be function thesauri or variants of implementation of various generalizing functions (see Section 6.2.5.1). Fig. 144 shows that it would take much longer to obtain these results using traditional methods and means based on reproductive knowledge than using the ascent-descent method.

Examples of using the ascent-descent method at the stages of solving the problem and commercializing the results are shown, respectively, in section 6.5.

6.2.15 Generalized methodological advice when setting a task

Table 26. Methodological advice when setting a task

Stage of task setting	Methodical advice	
	General	Particular
Analysis, synthesis and selection of general human needs Analysis, synthesis and selection of technically and individual human needs	To use in the formation of general and individual technically realizable human needs, as well as functions, consumer properties and parameters of TS being designed: — criteria for progressive development, laws and patterns of TS development (especially for identifying promising directions for achieving the main design goals); — the methodology for the formation of the ideal final result and the method of focal objects (especially for overcoming barriers to thinking); — metaphors (especially for deeper and more imaginative presentation of the essence of TS being developed); — the method of "ascent-descent" (especially for the formation of a variety of human needs, functions, consumer properties and parameters of TS being designed)	1. To exclude the presentation of false needs of a man in TT: — based on phobias, religious prejudices; — contrary to morality, ethical norms, national traditions, laws of nature; — offending religious feelings; — aimed at the direct implementation of addictions to alcohol, drugs, gambling; — pseudo-aesthetic needs (for example, deafening music). 2. Use the most important human needs. 3. To form general, individual technically realizable human needs based on the antonyms of the causes of human needs. 4. To formulate individual technically realizable human needs, use the "method of sabotage" (for example, when formulating a hard-to-imagine need for comfort, it is sometimes easier to imagine TS leading to discomfort, and then, using an antonym, return to the original problem. Or, for example, when it is difficult to identify the cause of TS destruction during operation, it is advisable to try to imagine TS or external conditions leading to such destruction)
Analysis, synthesis and selection of functions of TS being designed		1. To use generalizing functions to select the types of functions included in generalizing. 2. To use "good" verbs to form new useful functions, and "bad" verbs – to identify harmful functions in the development of TS[62] and prevent their occurrence
Analysis, synthesis and selection of consumer properties and parameters of TS being designed		1. To use the typical criteria for the progressive development of TS to determine the ways of improving (increasing, decreasing, stabilizing) the parameters of TS being designed. 2. To use "good" adjectives to form new competitive consumer properties, and "bad" adjectives – to identify "bad" consumer properties, the appearance of which in TS should be prevented. 3. To use adjectives "smart" and "clean", which have high creativity, when generating consumer properties of new classes of competitive TS

When solving a scientific and technical creativity problem, the functional-cost analysis and synthesis of TS, as well as heuristic techniques, are effective.

The intersectoral fund of heuristics is constantly growing.

It should be noted that some of the mentioned methodological advice (for example, the laws and patterns of TS development, the "ascent-descent" method) can be used at the stage of solving a problem. Sometimes this leads to the adjustment to a technical task.

6.3 Problem solution

6.3.1 Functional cost analysis and synthesis of technical systems

Let me remind you that in the system design methodology, the solution of a scientific and technical creativity (design) problem includes stages 5–8 (see Section 3.2.1, Fig. 45, as well as Section 6.1).

After the formation of TT for the design of TS (made on the basis of the selected TS analogue (prototype)), it is advisable to conduct a systematic feasibility study of functions in TS analogue (prototype) in order to determine the usefulness of each implemented function and the cost of achieving it (charge for utility). The ultimate goal of the study is to develop (synthesize) solutions aimed at reducing costs while maintaining or even increasing the usefulness of TS functions. This feasibility study is called functional cost analysis and technical systems synthesis (FCA).

FCA is especially effective if an analogue (prototype) of TS being designed is produced in large quantities. When carrying out FCA of such an analogue (prototype), it is not uncommon for a specialist-expert to formulate a useful function for some elements of TS being analyzed, which this element performs (often even the aesthetic properties of this TS element are assessed as very low). Although the manufacturing cost of this element may be low, it can be substantial in mass production. By excluding such an element from TS, it is possible to exclude these costs without loss of TS utility.

If desired, the reader can easily find more detailed information about FCA on the Internet.

6.3.2 Heuristic techniques

Many creators of TS being improved or TS of new generation do not realize that along with new TS, an even more important result can be obtained — a new heuristic (principle, strategy, tactics, method, technique, etc.), with the help of which this new TS has been developed with the best (in comparison with the prototype) consumer properties.

Most often, such a result is a heuristic technique (HT) — a rational creative action that should be implemented in the process of developing TS in the formation of human needs, functions, properties, functional structure, the principle of action (or existence) of TS, a technical solution for resolving a contradiction effectively. These actions are formulated in the form of proven methodological advice, essentially *imperatives*.

Various sources indicate that heuristic techniques (HT) can be *particular*, that is, the possibility of their use is limited to one industry or even one TS class, and *general* (cross-industry), which is more valuable. Various researchers have formed (and are constantly supplemented) intersectoral funds for HT. One of the variants of such a fund, which I have formed on the basis of the adjusted and supplemented intersectoral fund of prof. A. I. Polovinkin[63], is presented in Appendix 3. However, in none of the sources I have found examples of particular (industry) HT and I have tried to form such HT myself.

For example, the insufficient emotional impact of a work of art presented in one part has led to the creation of *diptychs, triptychs, polyptychs,* which have higher level of emotional impact (Fig. 145). Based on the analysis of this phenomenon, we can formulate a methodological advice — HT: "To increase the level of emotional impact of a work of art, use the multi-part functional structure of a work of art (diptych, triptych, polyptych) from parts connected by a common idea."

Figure 145. Virgin and Child with Saints. Stained glass windows of the Church of Notre-Dame-du-Sablon in Brussels (14th century), one of the brightest examples of Brabant Gothic architecture

It would seem that this is clearly particular HT, related only to the field of art. However, on reflection, we find many implementations of this HT — multipart functional structure — in the technical realm. For example, the transition from mono (one sound speaker) to stereo, quad sound and, finally, to surround sound (many speakers) (Fig. 146). Another goal is realized, for example, in a clerical or construction knife with the blade that consists of many parts, which are removed as they become dull (Fig. 147).

Figure 146. Functional structure of the surround sound implementation

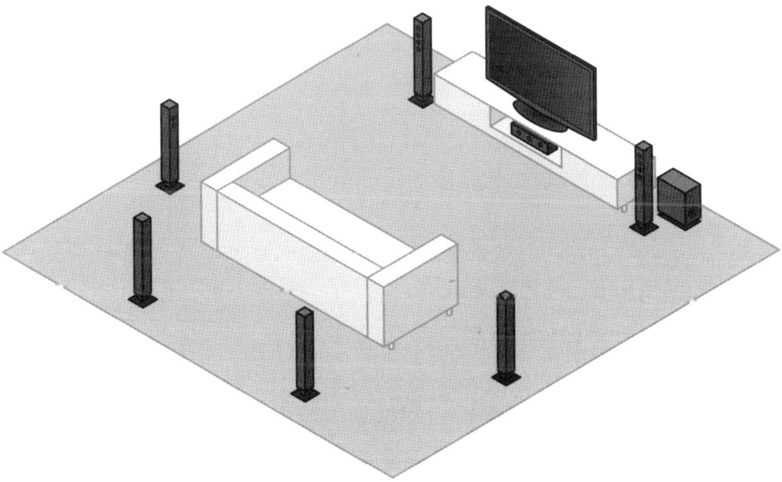

Figure 147. Knife with a multi-piece blade

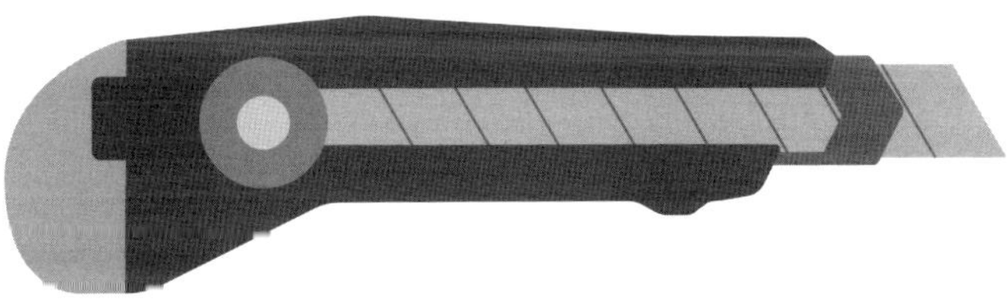

The above and other examples have led me to the conclusion that, since everything is interconnected in nature, technical and other spheres, there can be no particular (sectoral) HT in principle. And those HT that seem to be industry-specific are simply not fully understood, and they have not yet been used in other industries. By the way, this can be prevented primarily by a terminological barrier (see Section 2.3.3), when HT is described in narrow professional terms. The way out is to use generalizing functions (see Section 6.2.5) that do not indicate the specific way of implementing HT (for example, instead of the specific "separate", you should say the general "sort", and instead of "glue", "rivet", "weld" and other — "connect"). It is very important to learn how to identify and formulate HT for yourself, which rapidly develops creative abilities and ensures their effective use in *mental operations* of comparison, analysis, synthesis, abstraction, generalization and concretization, which are the basis of creative activity.

Let's look at the simple example of how you can identify, formulate HT and find new uses for it.

It is known that more than 20% of all energy produced in the world is spent on disintegration of material: rocks are crushed during mining, ore is crushed and grinded during its washing (due to separation of waste rock after grinding), ground is grinded during agricultural work, grain is grinded into flour etc. Of course, many crushing structures have been invented for these processes. One of the first of them, a jaw crusher, is shown in fig. 148.

Figure 148. Schematic diagram of a jaw crusher: 1 — material for crushing; 2 — motionless cheek; 3 — movable cheek

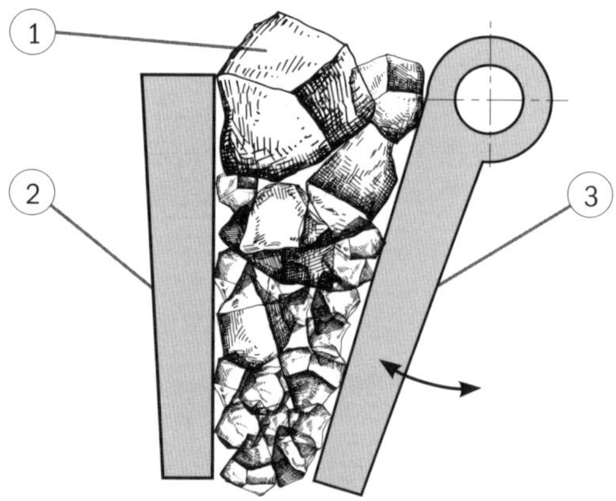

The material for crushing (1) is fed between the fixed (2) and movable (3) cheeks. The reciprocating mechanism shakes the lower part of the movable jaw (3), which, when moving towards the fixed jaw, crushes the material (1), and when moving backwards, it unloads the crusher from the crushed material. It is easy to see that the process of crushing and unloading occurs alternately (cyclically). The cycles of these processes are shown in Fig. 149.

Insufficient performance of the jaw crusher has led to the creation of a cone crusher (Fig. 150), which, with the same power as the jaw crusher, has almost twice the performance.

Figure 149. The cycles of jaw crusher operation

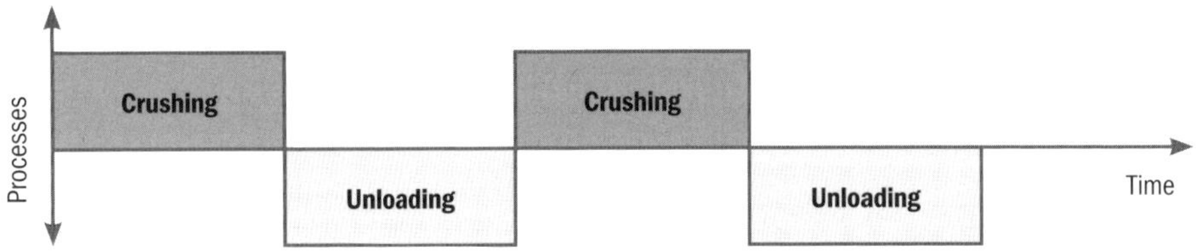

Figure 150. Schematic diagram of a cone crusher: 1 – material for crushing; 2 – cone; 3 – massive rotating shaft

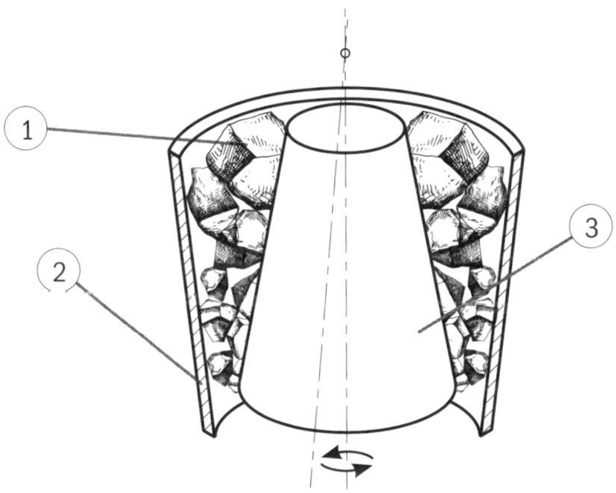

In a cone crusher, material for crushing (1) is fed into a cone (2). A massive shaft (3) is located in the cone (2), the upper end of which is suspended, and the lower end performs continuous eccentric rotation (with a deviation from the vertical axis of the cone), constantly pressing against the lower part of the cone (2). At the point of the closest approach of the shaft (3) to the cone (2), the material is crushed (1), and from the opposite side the material is constantly unloaded (1). The operating cycles of the cone crusher are shown in Fig. 151.

Figure 151. Cone Crusher Cycles

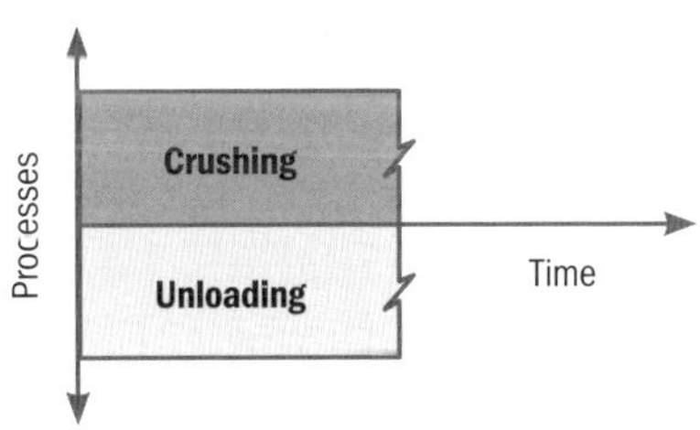

Comparing the latter achievement (cone crusher — see Fig. 150) with the prototype (jaw crusher — see Fig. 148), it can be concluded that the improvement in consumer properties (crushing and unloading performance) has occurred due to the replacement of alternating processes of crushing and unloading material due to cyclic reciprocating movements of the movable cheek to continuous processes of crushing and unloading of crushed material due to the continuous rotational movement of the shaft (3) in the cone (2) (see Fig. 150).

Thus, it is possible to formulate at least two practice-proven methodological advices — intersectoral HT.

1. To replace reciprocating motion in TS with rotary motion.
2. To replace the cyclical process in TS with a continuous one.

The transition to a continuous process can be accomplished in two ways (A and B).

A. To fill in the gaps between cycles with the same useful process (example — a more efficient cone crusher; see Fig. 150).

B. To fill in the gaps between cycles with a new or several new useful processes.

As a result, new generation TS with qualitatively new properties can be created. For example, in accordance with the USSR inventor's certificate for invention No. 343722, the cross rolling of a metal sheet is carried out in the pauses of longitudinal rolling.

In fairness, it should be noted that the continuity of processes in option B can be realized ("to fill") only partially: for example, one "old" process ("crushing") can be made continuous, the other "old" ("unloading") should be left unchanged, but at the same time several new useful processes should be started between cycles (Fig. 152).

Figure 152. Simplified diagram of the transformation of cyclic processes into continuous processes and the implementation of three new useful processes

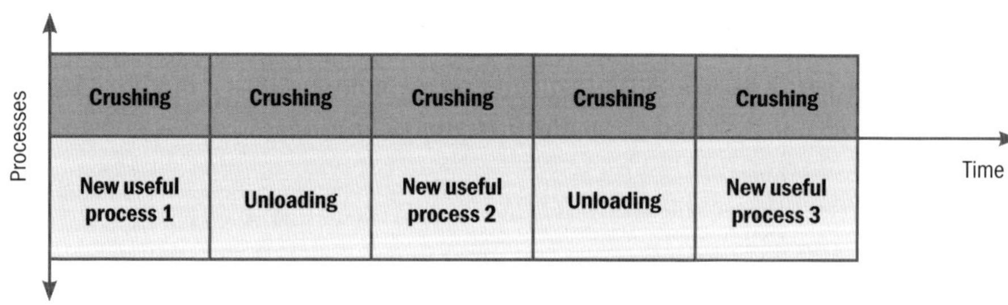

With regard to crushers, the latter HT seems exotic and, although it has not yet been implemented, it has the right to exist as proven experience.

For example, the installation of a machine gun on the wing greatly worsened the aerodynamic properties of the aircraft (Fig. 153).

Figure 153. Machine gun mounted to fire over a rotating propeller

The Dutch aircraft designer Anton Fokker, who opened an aircraft plant in Germany, in 1913 found an original solution to this contradiction — he developed a synchronizer-interrupter, which made it possible to shoot when there were no propeller blades in the path of the bullets.

History does not know how A. Fokker came to this decision — intuitively, using wrapped unconscious reasoning, or consciously.

I propose to consider the possible logic of reasoning that could lead to this effective solution.

The movement of the propeller along four sectors of the circle of rotation can be schematically represented (Fig. 154) as the alternating cyclic movement of the propeller blades over the sectors.

Figure 154. Propeller rotation circle divided into four sectors

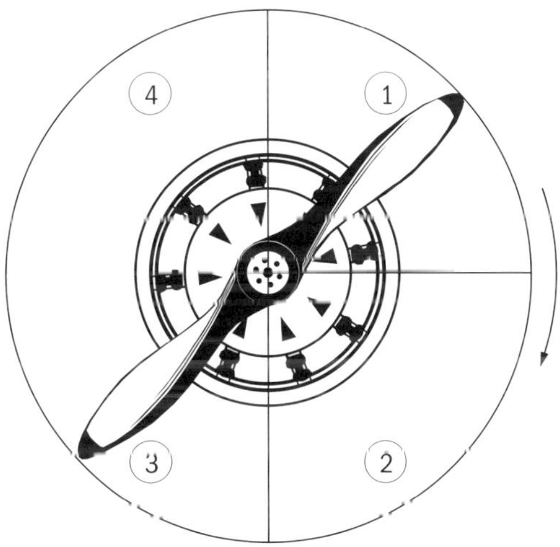

Then the scheme of the movement of the propeller blades along the sectors of the circle of rotation can be presented in Fig. 155.

Figure 155. Scheme of the propeller blades motion in the sectors of rotation circle (shaded sectors where the propeller blades are currently located)

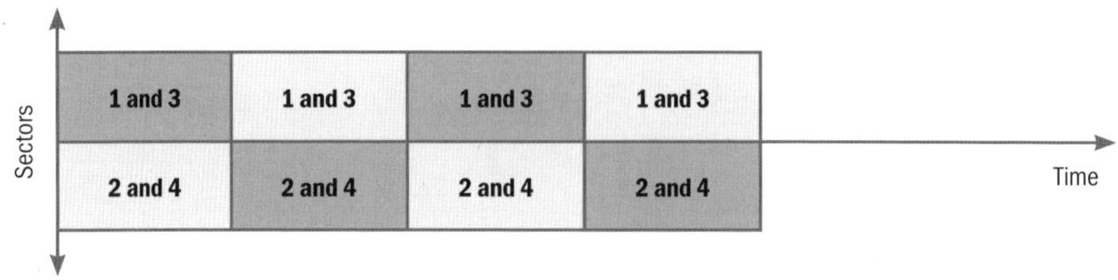

Fig. 155 shows that when the propeller rotates, sectors appear alternately in which there are no blades (first, sectors 2 and 4 in pairs, then 1 and 3). It is logical that it is these free sectors that can be effectively filled with another useful process — freely flying bullets of an aviation rapid-fire machine gun; and that has been done. Thus, the above transition to a continuous process (method B) as a methodical advice tested by experience is quite justified.

Learning how to identify and formulate HT can help the reader not only build an intersectoral HT fund, but also build their own HT fund, which may be more effective in personal creative activity than even an intersectoral HT fund.

G. S. Altshuller developed the matrix[64] that provided the formation of typical (most common) contradictions from the combination of improved and inevitably deteriorated properties. The fragment of the matrix by G. S. Altshuller is shown in Table 27.

Table 27. Fragment of G. S. Altshuller's matrix for finding HT that effectively resolve typical contradictions

What's getting worse What's improving		Capacity	...	Degree of automation	...	Loss of energy
		1	...	3	...	39
Capacity	1		...	35, 37, 1, 10	...	4, 2, 14, 1
...	...	...	...	...	...	...
Degree of automation	3	35, 37, 1, 10	...		...	36, 4
...	...	...	...	...	...	...
Loss of energy	39	4, 2, 14, 1	...	5	...	

Table 27 shows selectively improved and degraded properties (39 each). In the cells of the table (intersections of properties in rows and columns), the numbers of typical HT are indicated (there are 40 in total), recommended for effective resolution of contradictions. It is clear that typical contradictions cannot be formed at the intersections of the same properties; accordingly, HT is not indicated for such cases. I will cite the wording of some HT numbers proposed by G. S. Altshuller:

1 — Change in the state of aggregation;
2 — Preliminary action;
4 — Replacement of the mechanical environment;
10 — Copying;
14 — The use of pneumatic and hydraulic structures;
35 — Union;
36 — Feedback;
37 — *Equipotentiality.*

In my opinion, a set of properties and HT in G. S. Altshuller's matrix are in dire need of serious adjustment. Nevertheless, the original matrix is popular with many developers in different countries. The reader himself can form an effective matrix using his own experience and knowledge with the help of which he will find typical HT for resolving typical contradictions.

6.3.3 Intensified contradictions in technical systems and standards for their resolution

In Section 4.2, I talk about the effective heuristic technique of G. S. Altshuller — the formation of intensified contradictions[65] on the basis of "ordinary" contradictions in TS, leveling or even completely eliminating the negative psychological pressure on a developer, which arises when he realizes the inevitable deterioration of some TS properties during its improvement (improvement of properties). Here are some examples of the formation of intensified contradictions:

- For a crossroad (see Fig. 156). "To transfer people and goods, cars must move, and for the sake of complete safety, cars must not move";
- For a spillway dam (see Fig. 53). "Due to the flood, the inflow of water to the spillway dam increases, leading to flooding in the area in front of the dam, however, in order for the dam to cope with the discharge of water, there must not be an increased inflow of water";
- For the aircraft with wings perpendicular to the fuselage (see Fig. 54). "With the excess of the speed of sound for the aircraft with wings perpendicular to the fuselage, there should be a jump in pressure and air density at the leading edge of the wings (due to the laws of physics), and

Altshuller proposed 11 standards for the effective resolution of intensified contradictions. Three of them are universal and separate the conflicting properties of TS and the environment:

- —In time;
- —In space;
- —In time and space.

In order to generate a multitude of combinations of technically realizable human needs, functions, properties, technical decisions in TS while developing TS, use morphological analysis and TS synthesis.

there should not be such a jump in pressure and air density to protect the wings from destruction";

— For the technology of solid ore underground mining (see Fig. 55). "To ensure high productivity of blasting ore, all the fans of the blast holes should be blasted at the same time, however, since the ore can only be crushed by the blast of the fan of holes, the first from the exposed surface, all holes fans should not be blasted at the same time."

G. S. Altshuller formed 11 typical heuristic techniques (HT), which he called standards, for the effective resolution of intensified contradictions. Most of them, in my opinion, do not "manage" the standards (which should be characterized by universality, intersectoral focus) and therefore are not presented so as not to distract the reader (whoever wants, can easily find them in almost any book on TRIZ). But three of the 11 standards are brilliant, truly universal. They propose to separate conflicting properties in time, in space, in time and in space.

I note right away that all the intensified contradictions given in Section 4.2 were successfully resolved long ago (the contradiction with a glass for tea was resolved several centuries ago). However, it was not known whether developers used methodological techniques similar to those proposed by G. S. Altshuller So we will consider how the standards of G. S. Altshuller could be used to resolve these contradictions.

For safe traffic on a crossroad, the standard "to separate conflicting properties in time and space" could suggest a simple solution: some vehicles stay at red traffic light, others go green (separation in space); after some time (separation in time) the directions of movement change (Fig. 156).

To effectively resolve the intensified contradiction in the spillway dam, the same standard is suitable, which could become a hint for generating a wonderful technical decision — a labyrinth overflow spillway, implemented, for example, on the Nerl River in the Yaroslavl Region, on the Rozhaika River in the Moscow Region (Russia) and on the river in Putrajai (Malaysia) (fig. 157).

The labyrinth structure spreads (in comparison with a conventional spillway dam) the spillway in time and space. At the same time, the water overflow line with the constant dam's length significantly lengthenes, which increases the throughput of the dam and prevents excessive accumulation of water in front of the dam in case of large flood waters.

The same standard could become a hint for the generation of a very effective technical solution — swept wings of the aircraft (see Fig. 158), the design of which prevents the destruction of the wings at supersonic speeds, reducing the force effect on the leading edge of the wings from a jump in pressure and air density simultaneously.

Fig. 158 shows clearly that with the swept wing shape, increased air pressure

Figure 156. Adjustable crossroad

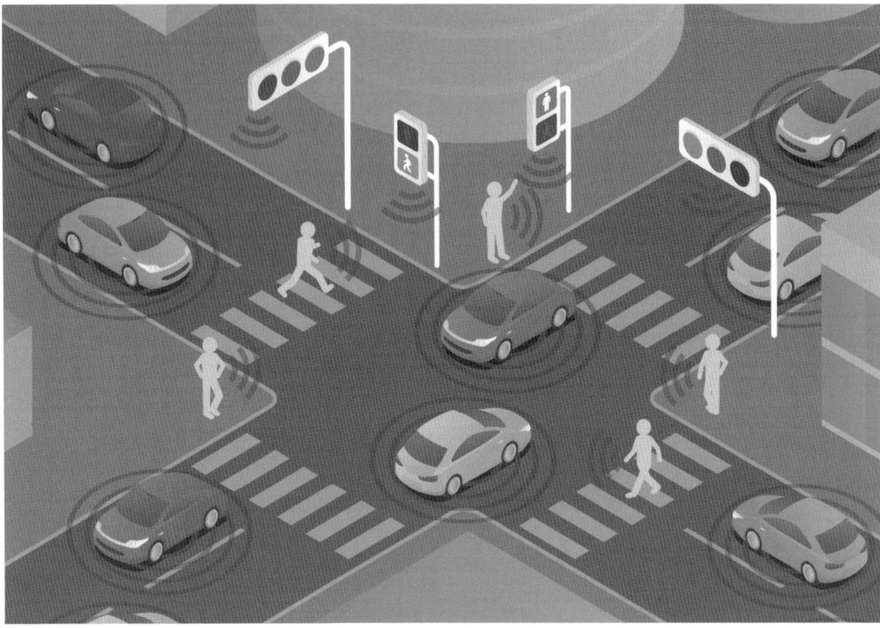

Figure 157. Labyrinth overflow spillway with a pedestrian bridge above it in Putrajaya, one of the largest cities in Malaysia

does not immediately appear on the entire leading edge of the wings (as in the wings perpendicular to the fuselage; see Fig. 54), but is distributed in time and

Figure 158. Distribution of increased force impact on the leading edge of swept wings during a jump in pressure and air density during breaking the sound barrier

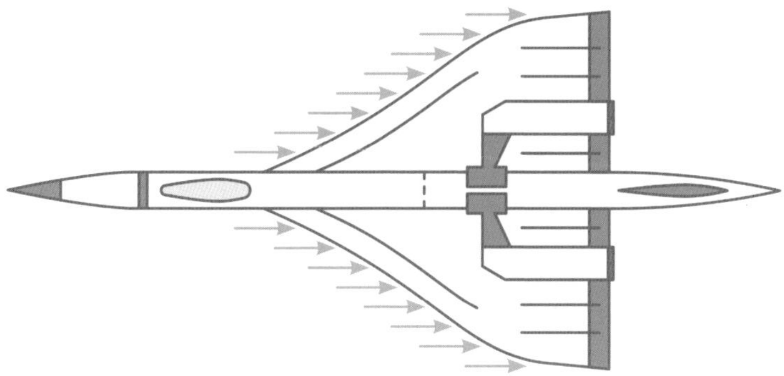

space (along the leading edge), first at the fuselage aircraft, and then sequentially goes to the rest of the leading edge of the wings.

This technical decision has not only prevented the destruction of the wings during the transition to supersonic speeds. The same standard could become (and maybe has become) a hint for the formation of optimal aerodynamic properties of an aircraft during takeoff, cruise flight and landing; in an aircraft with a variable wing geometry during takeoff and landing, the wings are placed in a position almost perpendicular to the fuselage, and in air cruise (supersonic) flight, the wings almost disappear, actually turning the plane into a rocket (Fig. 159).

To resolve the intensified controversy in the technology of underground

Figure 159. Aircraft with variable wing geometry

ore mining with the use of blast breaking, it was also possible to apply the standard of "separation of conflicting properties in time." It was implemented in the middle of the last century, when borehole ore breaking method was used. Rows or fans of wells are blasted almost simultaneously to ensure high productivity of blasting, but still separately (to ensure the process of crushing the ore), as follows, preparing and carrying out the blast. Each well in each fan is charged with an explosive with an electric detonator, which is connected to an electrical network with other detonators of the same type. When current is applied to this network, the electric detonators of the first fan of boreholes from the exposed surface (see Fig. 55) explode instantly. Each subsequent fan of wells explodes with a delay of 10–15 ms (this is provided by delay elements in electric detonators). Such successive microdecelerations of well fan explosions are sufficient for the formation and opening of cracks in the ore massif, which ensures crushing of the ore massif. In fact, each subsequent fan of wells explodes onto its exposed surface, that is, the conditions for the explosion of subsequent fans are the same as for the first.

A simple example of the separation of conflicting properties in time is chameleon glasses, which (along with optical properties that correct vision) darken during periods of bright sunlight, and at lower illumination intensity they brighten due to substances with a photochromic structure used in lenses. Electrochromic glasses have also been developed, the transparency (darkening) of which can be changed depending on the magnitude of the electric voltage applied to the glass.

To effectively resolve the contradiction in a faceted tea glass (see Fig. 56), you can use (apparently, this was done several centuries ago) the standard of "separation of conflicting properties in space." The result is the Azerbaijani vessel for tea — armudu (see Fig. 160). Its lower ball-shaped part perfectly keeps

Figure 160. Armudu

the tea hot, and the upper (expanded) part provides sufficient cooling of the drink so as not to burn your lips. Thus, the tea is hot and not too hot. There are similar vessels in other countries of the Middle East.

It can be concluded that the formation of intensified contradictions and the use of standards for their resolution is a very effective methodological technique.

6.3.4 Morphological analysis and synthesis of technical systems

The use of morphological analysis and synthesis (MA and S) of TS, developed in the 30s of the last century by the American astrophysicist Fritz Zwicky, can be simple, but effective in scientific and technical creativity at the stage of solving a problem. Morphology is the science of the form, structure of various systems.

Simplified, the use of MA and S when improving or creating a new TS generation can be represented in several steps.

Step 1 (analysis). A morphological matrix (table) is formed, where the lines indicate the names of structural elements, processes or properties of the decomposed analogue of TS being developed, and in the columns — the known implementations of these structural elements or processes that can be described in the form of their functions and properties, functional structures, principles actions, technical solutions.

At step 1, there may be synthesis elements as well. For example, along with the known structural elements and / or TS processes, new (possible) ones can be added to the lines. Accordingly, possible implementations of these structural elements and / or TS processes can also be added to the columns.

At the same time, in the last column of each line, methodologists recommend indicating as a possible option "without the name of a structural element and / or process", which can serve as the impetus for generating an original technical decision. For example, the consideration of options for creating TS that does not reflect (more precisely, weakly reflects) radar waves has served as the impetus for the creation of aircraft, ships, torpedoes — "invisible ones". Or, for example, a phone without buttons (with a touch screen)

Step 2 (synthesis). It consists in the sequential combination of all possible options for the implementation of structural elements, processes, properties of TS.

Thus, MA and S of TS, on the one hand, make it possible to quickly form all possible (within a given morphological matrix) combinations of structural elements and / or TS processes and their possible implementations, on the other hand, the requirement for the mandatory formation (without criticism) of all possible combinations allows you to include original and even fantastic options in their list, which without MA and S of TS would not have arisen in the mind of the subject of creativity due to psychological barriers to thinking.

Experience shows that manual formation of all possible combinations of structural elements, processes, TS properties and their possible implementations is extremely difficult if the morphological matrix consists of more than 5–7 rows and 5–7 columns. However, many special programs have been developed for morphological synthesis (and decision making).

Step 3. Evaluation of options (combinations) and decision making. The methods and means mentioned in section 3.3 are used to evaluate options and select an acceptable combination.

I will show the simplicity and efficiency of using MA and S of TS using the classic example with a pen, when, despite the several tens of thousands of patents available in the world, new technical decisions can be easily obtained using the formation of a morphological matrix (Table 28).

Table 28. Example of a pen morphological matrix

№	Name of a structural element, process or TS property	Implementation options					
		1	2	3	4	5	6
1	Types and kinds of fountain pens	Fountain	Ball	Capillary	Roll	Gel	Fundamentally new
2	Method of leaving a trace	Powder fusion	Burning-in	Application of ink, paste	Pressing	Scratching	Without a trace
3	Method and / or device for attaching to clothing	Hook	Clip	"Velcro"	Magnet	Chain	Without fastening
4	Writing unit pull-out and / or protection device (WU)	Removable cap	WU screw pull-out mechanism	Spring-loaded WU pull-out mechanism	Pneumatic cylinders for WU pull-out	Piezoelectric WU pull-out mechanism	Without device
5	Additional functions	Recognize counterfeit banknotes	Shoot	Accumulate information	Digitize handwritten text	Illuminate	Recognize odors (e.g. explosives)

In Table 29, the arrows connecting the cells show the formation of one of the possible options for a pen that a secret agent could use.

Table 29. Formation of one of the possible options for a fountain pen

№	Name of a structural element, process or TS property	Implementation options					
		1	2	3	4	5	6
1	Types and kinds of fountain pens	Fountain	Ball	Capillary	Roll	Gel	Fundamentally new
2	Method of leaving a trace	Powder fusion	Burning-in	Application of ink, paste	Pressing	Scratching	Without a trace
3	Method and / or device for attaching to clothing	Hook	Clip	"Velcro"	Magnet	Chain	Without fastening
4	Writing unit pull-out and / or protection device (WU)	Removable cap	WU screw pull out mechanism	Spring-loaded WU pull-out mechanism	Pneumatic cylinders for WU pull-out	Piezoelectric WU pull-out mechanism	Without device
5	Additional functions	Recognize counterfeit banknotes	Shoot	Accumulate information	Digitize handwritten text	Illuminate	Recognize odors (e.g. explosives)

In fig. 161 shows the original designs of pens, the description of which in the morphological matrix would certainly allow synthesizing many new technical solutions.

Figure 161. Original designs of pens, including new technical decisions: 1 — a pen with a laser pointer; 2 — bracelet pen; 3 — pen-vernier caliper; 4 — multi refill pen

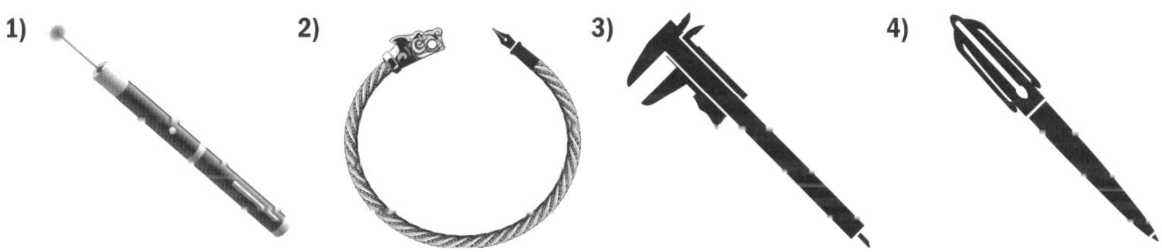

6.3.5 Description of functional structures using elementary operations

It is very heuristic to form TS functional structures using elementary operations of R. Koller, a German scientist, who proposed 12 direct and 12 reverse operations to describe any TS of different industries at the high level of abstraction. Later, other elementary operations were reasonably added to this list. The list of 16 elementary operations for describing the functional structures of TS is shown in Table 30.

Table 30. The list of elementary operations for describing the functional structures of TS

Direct elementary operation	Inverse elementary operation
1. Emission	1. Absorption
2. Conductivity	2. Isolation
3. Concentration	3. Dispersion
4. Movement	4. Restriction of movement
5. Change of direction	5. Change of direction
6. Communication	6. Interrupt
7. Increase	7. Decrease
8. Alignment	8. Oscillation
9. Consolidation	9. Disconnection
10. Accumulation	10. Output
11. Conversion of unlike streams	11. Reverse transformation of unlike streams
12. Composition of unlike streams	12. Decomposition of unlike streams
13. Conversion of streams of the same name	13. Reverse transformation of streams of the same name
14. Composition of streams of the same name	14. Decomposition of streams of the same name
15. Acceleration	15. Slowdown
16. Increase in flux density	16. Absorption

Let us consider the possibility and usefulness (creativity) of describing the functional structure using elementary operations, for example, underground mining of solid minerals (ores).

There are several hundred options for underground mining of ore, of which one can be chosen depending on many factors: the shape of the ore bodies (bed, vein, pipes, ore nests, etc.), the angle of incidence of the ore bodies (horizontal, gentle, inclined, steep), thickness ore bodies (from thin — less than 0.8 m to very thick — more than 60 m), the depth of occurrence of ore bodies and other factors.

With all the differences in development options, each of them must include elements of the functional structure that provide:

1) The organization of some conductive space from the surface of the earth to the ore body through the shafts of mines (vertical, inclined, spiral-shaped), adits, other workings, wells for supplying energy, air, water,

materials, mechanisms; this process corresponds to the elementary operation of *conductivity*;

2) Transformation of solid mineral (crushing, melting, dissolution) into the state that allows it to be transported; this process corresponds to the elementary operation of *decomposition* or *disintegration* (sometimes this operation is associated with a change in the phase state of the object);
3) Transportation of crushed (molten, dissolved) ore to the surface of the earth ("to the top", as the miners say) — *movement* operation, ore *storage* — accumulation operation, shipment to the consumer — *delivery* operation;
4) Management of the maintenance of overlying (and not only) strata (pressure of overlying strata control), need for which arises during underground mining of ore with the formation of so-called mined-out spaces (large voids) underground; at the same time, in the environment surrounding the void (rocks), there are pressure surges of rocks (operation *oscillation*), which can (if no action is taken) fall into the mined-out space. To prevent this from happening, the overlying rocks are reinforced with lining, or they completely fill the worked-out space with filling material (even concrete), or they strengthen the overlying strata with cables, or glue cracks in the rocks, which generally leads to *alignment* (according to R. Koller) of pressure.

Thus, all variants of the functional structures of solid minerals (ores) underground mining in the form of R. Koller's elementary operations can be presented in one generalized functional structure in Fig. 162.

Figure 162. Generalized description of the functional structure of solid minerals (ore) underground mining by elementary operations

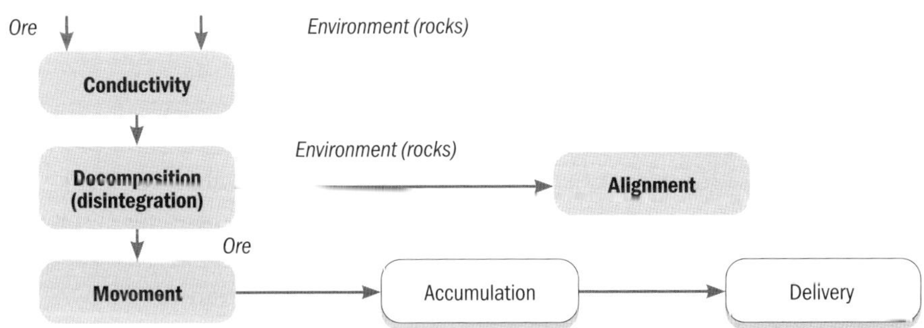

Presentation of TS functional structure creativity in the elementary operations of R. Koller is as follows.

First, the high level of abstraction of TS functional structure elements (subsystems) makes it easy to identify the main elements (subsystems) that most

To generate a set of physical principles of action (PPA) of TS efficiently, the automated system has been developed that represents PPA in the form of chains of formalized descriptions of physical effects, in which the output of energy, substance and information from the object of the effect "clings" to a similar (in quality and quantity) input of energy, substance and information of a different effect.

This approach is universal and makes it possible to substantiate the possibility of creating automated systems for the synthesis of biological, chemical principles of action, and even the principles of action of economic and other technologies.

affect the efficiency of underground ore mining. In the example under consideration, these are *conductivity, decomposition (disintegration), movement, alignment* (in Fig. 162, these elements are shaded). Indeed, the analysis shows that the increase in the efficiency of *accumulation* and *delivery*, even by an order of magnitude, changes the efficiency of underground ore mining as a whole by fractions of a percent. On the other hand, the increase in the efficiency of any of the main functional structure elements by some amount increases the overall efficiency of underground ore mining by almost the same amount.

Secondly, the high level of abstraction of the elements (subsystems) of TS functional structure reduces the negative influence of psychological barriers to thinking in the subject of creativity, in particular the most harmful of them — the barrier of terms (see Section 2.3.3), which inevitably arises when describing TS functional structure in industry terms (the names of elements or subsystems analogs) of TS. These names associatively form in a person the image of technical decisions corresponding to the names. Being "in captivity" of these decisions, it is difficult for the subject of creativity to generate a fundamentally new effective solution. To a large extent, for this reason, a lot of technical decisions based on insignificant improvement of TS analogs (prototype) appear the results of scientific and technical creativity. At the same time, competitive TS are created very rarely.

In turn, the description of TS functional structures elements at the high level of abstraction liberates thinking. This allows the subject of creativity, using various physical, chemical, biological and other effects, phenomena and their combinations, on the basis of a *mental experiment* or using an automated system for searching and synthesizing the principles of TS operation, to find or synthesize new effective options for the implementation of TS functional structure elements (subsystems). For example, to implement the "*conductivity*" element, one can consider the possibility of supplying energy and materials through capillaries in rocks and ore, and affecting the ore (for its "*decomposition*") by physical fields. To carry out the "*decomposition*" of ore when its phase state changes to gaseous (this has been used in coal mining for a long time), while the "*movement*" also changes. Underground voids can be widely used for "*accumulation*".

6.3.6 The method of forming the principles of technical systems operation based on physical, biological, chemical, geometric and economic effects and phenomena

The energy-informational model of the formation of physical effects chains has been developed by Russian Professor M. F. Zaripov and his students, based on the following logic.

In a simplified way, the effect can be represented (Fig. 163) as a certain input action on an object, which, due to a certain transformation, forms an output response (action on another object or the external environment).

Figure 163. Simplified representation of the effect

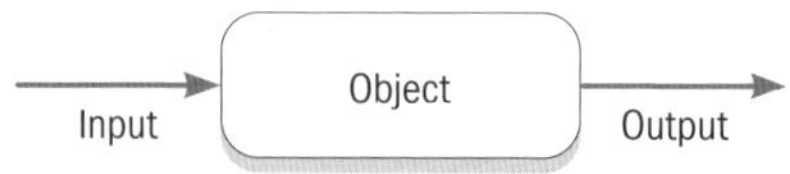

The example of the description of the effect "volumetric expansion of bodies upon heating" known from a school physics course is shown in Fig. 164.

Figure 164. Expansion effect of metals with the heating

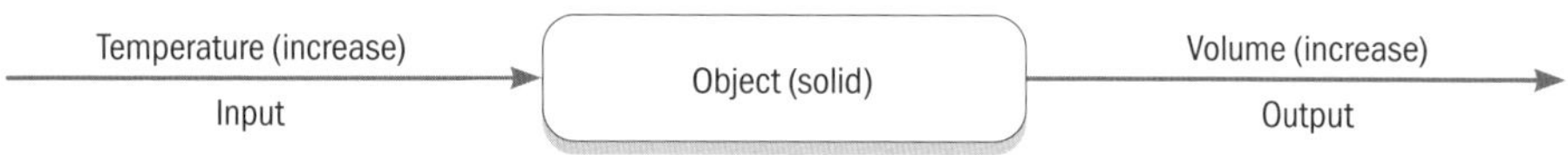

Operating principles reflecting concepts (intentions, basic ideas) of any, even the most complex, TS are chains of effects, in which the output of one effect "clings" to the input of the same name of another effect.

For example, the principle of operation of a thermal power plant can be simplified as the combination of generalized scientific and technical effects, where each of them can have many realizations: the effect of converting thermal energy into mechanical energy (Fig. 165) and the effect of converting mechanical energy into electrical energy (Fig. 166).

Figure 165. The effect of converting thermal energy into mechanical one

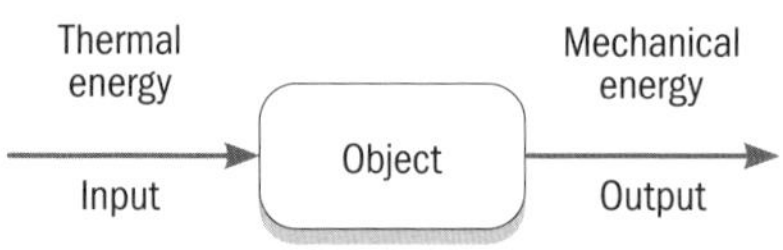

Figure 166. The effect of converting mechanical energy into electrical energy

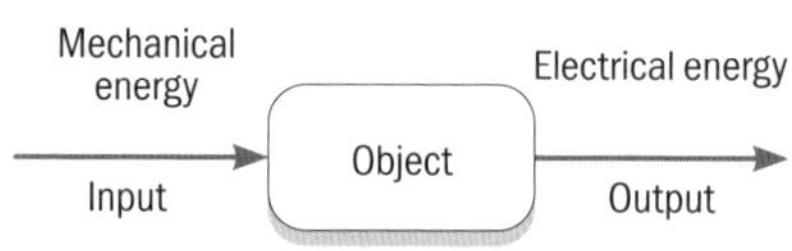

Mechanical energy is the unifying of these two effects, in the principle of a thermal power plant operation (Fig. 167).

Please note that the output and input characteristics of the effects properties must match together (the qualitative characteristics of the properties — completely, and the quantitative characteristics — completely or in certain intervals).

The simplified algorithm for the automated synthesis of operating principles,

Figure 167. Principle of a thermal power plant operation

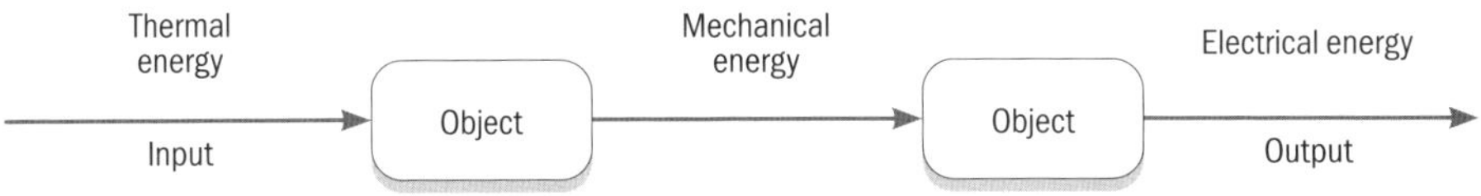

based on the above rule of combining the output characteristics (qualitative and quantitative) of the previous effect with the compatible input characteristics of the subsequent effect, can be represented as follows.

Let's form the database of effects in the form of a matrix, in which, for simplicity and clarity of descriptions of inputs and outputs of effects, we will replace them with numbers (Fig. 168).

Figure 168. Effects matrix

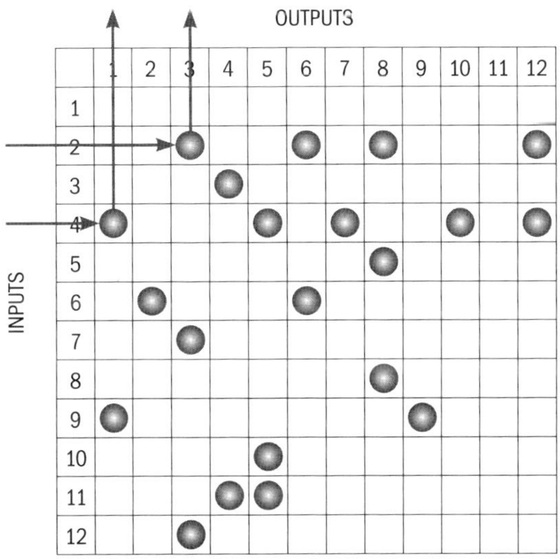

In the matrix shown in Fig.168, the effects are shown by circles (for example, the top left effect has input 2 and output 3; another effect with input 4, output 1, etc.).

If it is necessary to form the principle of TS operation from the chain of compatible effects, in which, for example, the input flow of energy, substance and information is indicated by the number 7, and the output flow by the number 12, then the path (simplified algorithm) of forming the chain will be as follows (Fig. 169).

The chain of compatible effects formed in this way, representing the desired principle of TS operation, is shown in Fig. 170.

Figure 169. An example of a simplified algorithm for the formation of the TS operation principle

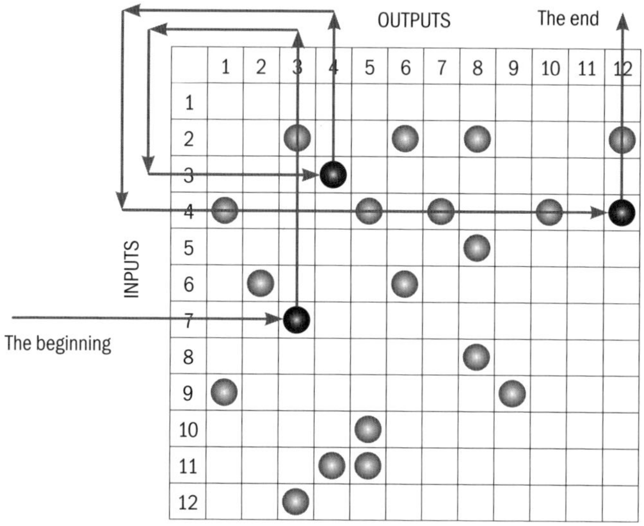

The chain of compatible effects formed in this way, representing the desired principle of TS operation, is shown in Fig. 170.

Figure 170. Functional principle of TS with a given input 7 and output 12

The approach versality has led me to the idea of the automated synthesis of not only the physical, chemical, biological principles of TS, but also economic technologies, the principles of which can also be represented in the form of chains of compatible business processes based on transformations of flows of material and non material assets, liabilities and information. To maintain uniformity in the descriptions of operation principles, I will call such business processes economic effects. In this case, objects are changed to subjects.

Let's consider the example of implementation of the simplified algorithm based on eight arbitrarily generated economic effects (business processes), Fig. 171–178. For simplicity, we omit the subjects.

To graphically represent the database of economic effects, we will indicate their inputs and outputs with verbal descriptions (Figs. 171–178).

Let's replace the verbal descriptions of inputs and outputs with numbers:

Figure 171. "Purchase organization" business process

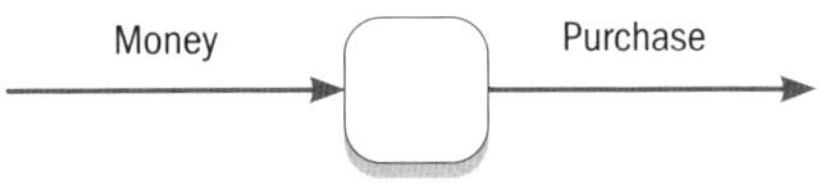

Figure 172. Business process "organization of production"

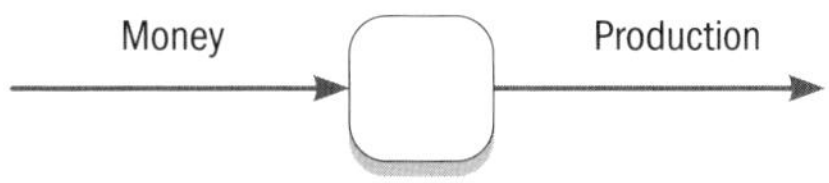

Figure 173. Business process "purchase of equipment"

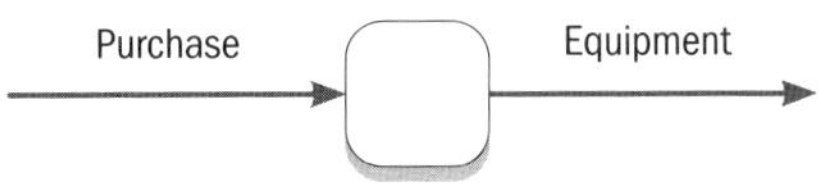

Figure 174. Business process "equipment production"

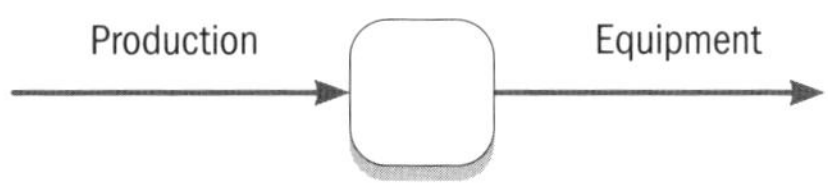

Figure 175. Business process "equipment sale"

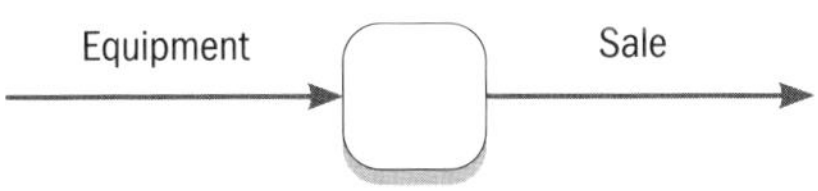

Figure 176. Business process "equipment rental"

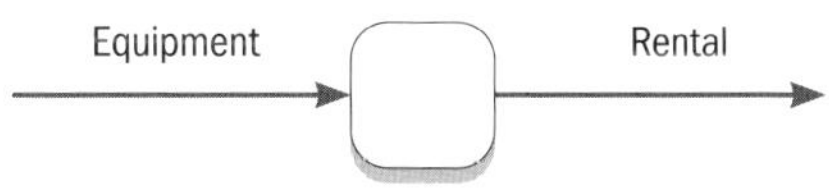

Figure 177. Business process "receiving rental payments (money)"

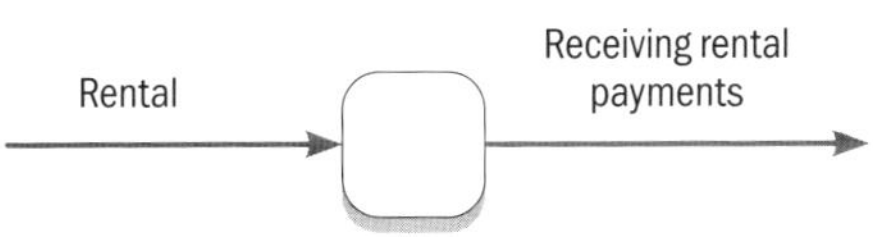

Figure 178. Business process "receiving money from sale"

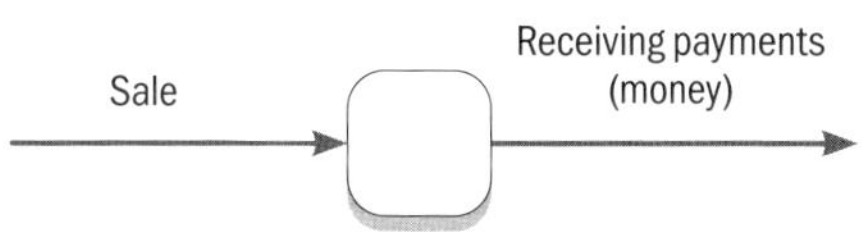

1 — cash;
2 — purchase;
3 — production;
4 — equipment;
5 — sale;
6 — lease.

Economic effects with inputs and outputs indicated by numbers are presented in Fig. 179–186.

Figure 179. "Purchase organization" business process

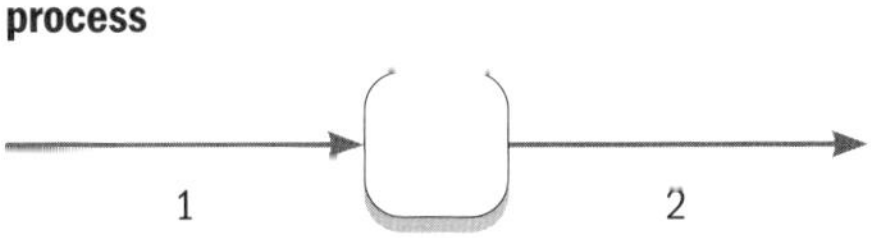

Figure 180. Business process "organization of production"

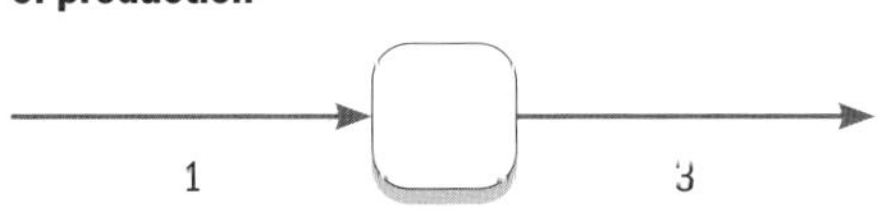

Figure 181. Business process "purchase of equipment"

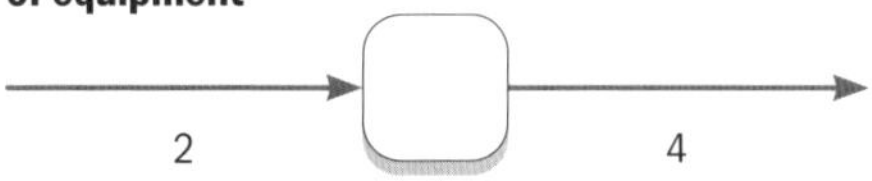

Figure 182. Business process "equipment production"

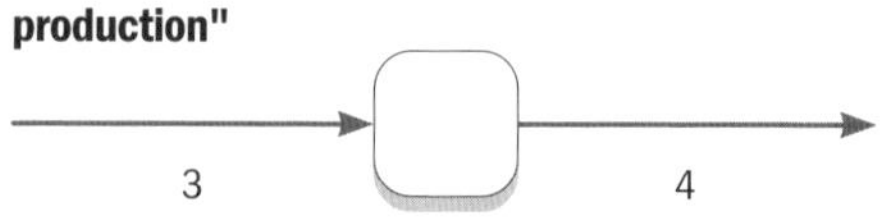

Figure 183. Business process "equipment sale"

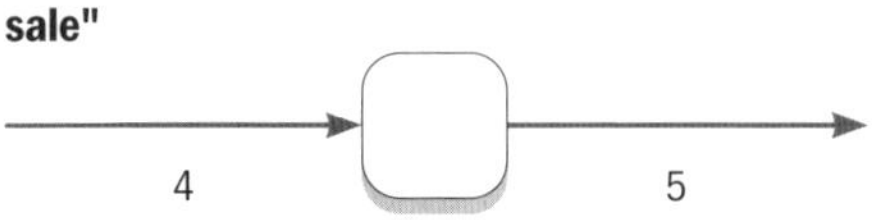

Figure 184. Business process "equipment rental"

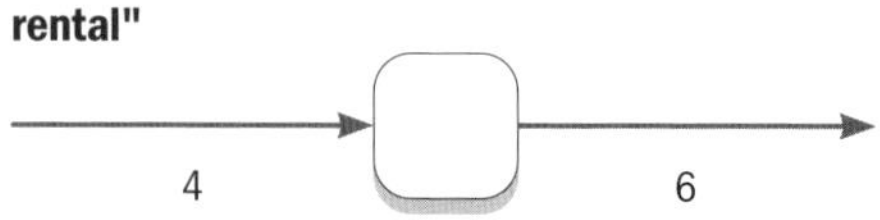

Figure 185. Business process "receiving rental payments (money)"

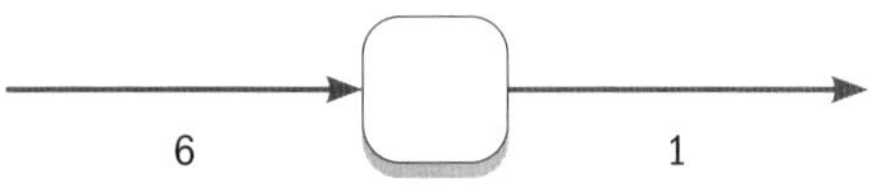

Figure 186. Business process "receiving money from sale"

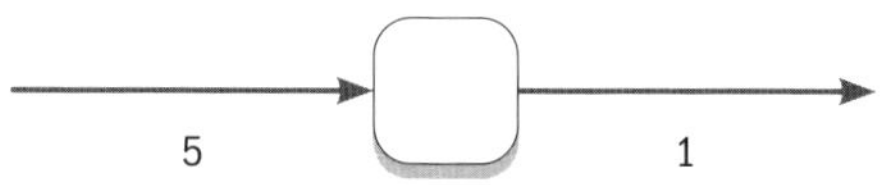

Then the database of economic effects, built according to the descriptions of inputs and outputs of effects in the form of a matrix, can be represented in the following form (Fig. 187).

Figure 187. Matrix of economic effects (business processes)

INPUTS \ OUTPUTS	1	2	3	4	5	6
1		●	●			
2				●		
3				●		
4					●	●
5	●					
6	●					

For example, it is necessary to synthesize the operation principles of economic technologies from business processes (economic effects), which have money at the input (1), and also have money at the output (1), but with increase. Synthesis options are shown in Fig. 188.

Thus, out of a total of eight economic effects, we obtain four alternative technologies, among which options "a" and "b" are rental. It is clear that with the large number of various effects, along with known technologies and processes, there can be synthesized and unique ones.

It is even more important that the descriptions of effects in a single information

Figure 188. Synthesis of operation principles of economic technologies

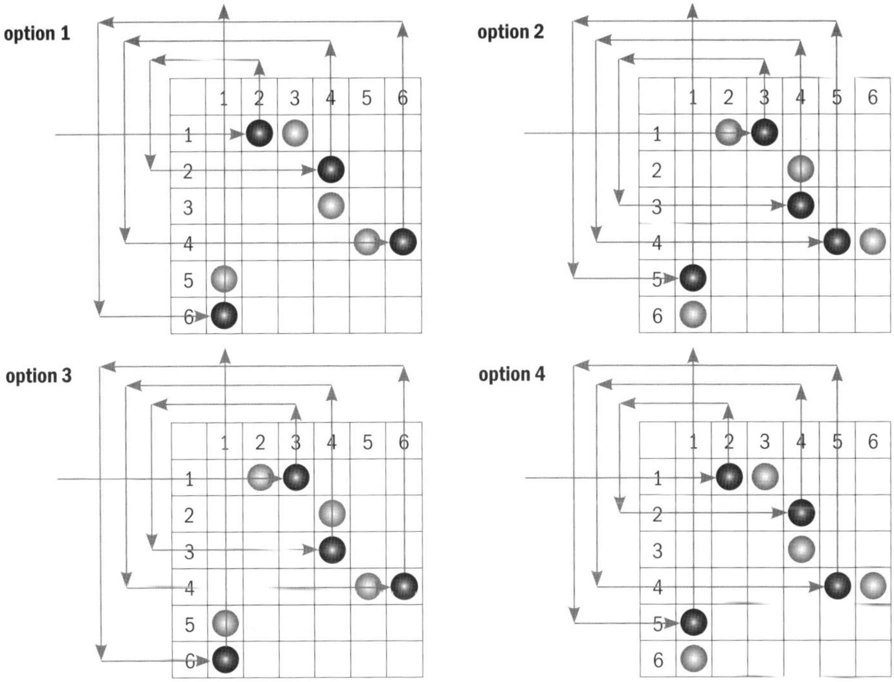

Figure 189. Synthesized principles options of economic technologies operation: a – option 1; b – option 2; c – option 3; d – option 4

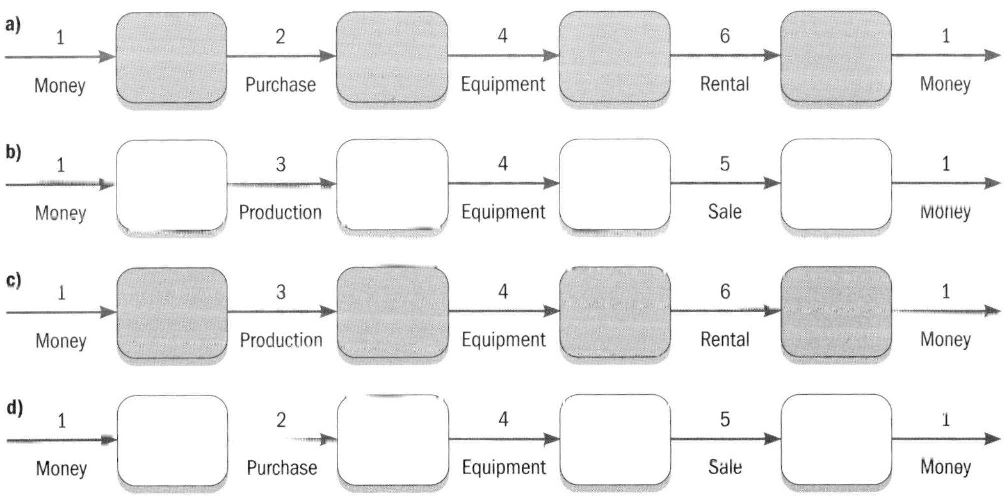

Traditional resources (natural, material, labor, financial, informational, intellectual) can be used in scientific and technical creativity, but in general they have weak heuristic capabilities. Nontraditional resources hidden in TS or in the environment of TS are more heuristic:

- — material;
- — functional;
- — informational;
- — spatial;
- — energy;
- — phenomena and processes;
- — time;
- — field;
- — systemic.

basis make it possible to create effects databases that combine the knowledge of all branches of physics, chemistry, biology, economics, and that are real contribution to the creation of meta-knowledge (mega-science) and the increase in the efficiency of knowledge convergence in new TS development.

The automated system for searching and synthesizing physical TS principles has been developed under my leadership.

6.3.7 Traditional and non-traditional (hidden) resources and their use in scientific and technical creativity

6.3.7.1 Traditional resources

In difficult life situations, including in scientific and technical creativity, many of us have often heard the so-called "very smart" advices, for example, "use resources." On the one hand, such advices are certainly correct, on the other hand, it is difficult to use them, as, for example, the statement-advice "It is better to be healthy, smart and rich than poor, stupid and sick!", especially when it comes to traditional resource groups.

The Great Soviet Encyclopedia defines such resources as money, items of value, reserves, opportunities; sources of income in the state budget[66]. In economic and financial dictionaries, resources are sources and premises for obtaining material and spiritual benefits that people need, which can be realized with existing technologies and socio-economic relations[67].

The traditional approach to resource classification is illustrated in Fig. 190.

Figure 190. Traditional resource groups

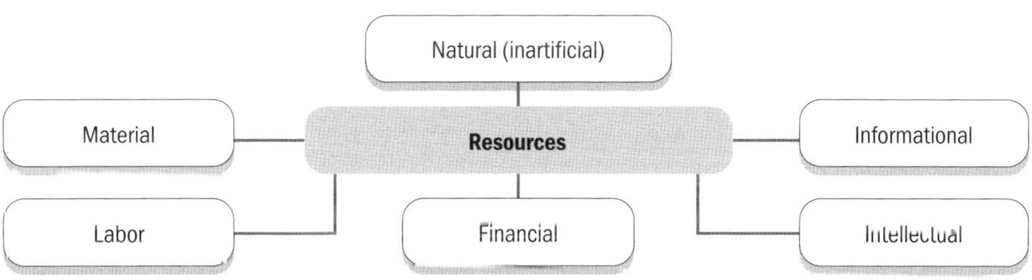

The characteristics of traditional resources groups, selected from various encyclopedias and reference books, are presented in Appendix 6.

These classifications form the idea of the essence of traditional resources. They, of course, can be used in scientific and technical creativity as various provisions, called by the names of the resource groups indicated in Fig. 190. At the same time, the idea of concentration of traditional resources for the creation of important objects and the implementation of promising (or, as they are also called, "critical", "breakthrough") directions of scientific and technical creativity may be quite heuristic. However, in general, such resources have weak heuristic capabilities for solving problems of scientific and technical creativity, including those for improving TS and creating new TS generations.

That is why the outstanding methodologist of scientific and technical creativity G. S. Altshuller made attempts to expand the traditional concepts of resources and the possibility of their use by adding time, space, the fund of physical and technical effects, Vepol analysis, physical and technical fields: mechanical, acoustic, thermal, chemical, electrical, magnetic, electromagnetic[68] to the list. And although these attempts were not systematic enough and included erroneous concepts, they were very creative in their essence and therefore became the basis for the formation and development of new heuristics, including the identification, systematization and use of non-traditional (hidden) groups of resources.

Shown below is a brief description of the results of systems research and development in this area, carried out under my leadership.

6.3.7.2 Non-traditional (hidden) resources

The heuristic classification of non-traditional (hidden) resources defines them as the type of productive knowledge (that is, effectively generating new knowledge) and includes:

- Two types of *resources* in terms of their presence in TS or the environment — *ready-made* (available) and *derived resources* (absent, but can be obtained by transforming ready-made resources);
- Nine classes of *ready-made* and *derived resources*, depending on their origin (occurrence, manifestation) or use in TS (Fig. 191).

Figure 191. Classes of non-traditional (hidden) resources

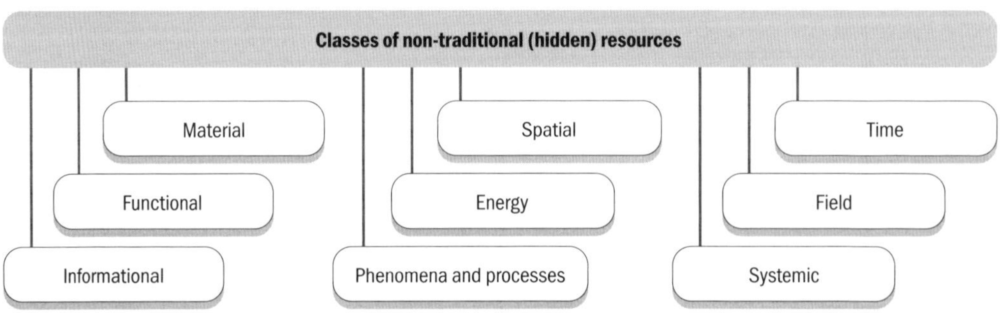

As can be seen from Fig. 191, the classes of non-traditional resources differ significantly from the traditionally distinguished groups of resources (see Fig. 190).

These resources are named hidden because most of them are not obvious, hidden in TS and TS environment, and it is necessary to carry out a functional-physical (biological, chemical) analysis of TS and TS environment to reveal them.

The concept of using non-traditional (hidden) resources is based on the following postulates:

- All TS, the environment of TS, the phenomena and processes occurring in them, any production and consumption, as well as the waste of human life and activity and of other wildlife representatives are resources and can be used to reduce costs at various stages of the life cycle created by TS, while improving others TS, creating new TS generations and pioneering TS;
- Resources are available or may arise not only during the operation (TS operation), but also at other stages of TS life cycle — during storage, utilization, disposal;
- A lot of resources arise from harmful and neutral functions in TS or impact of the environment on TS;
- *Derivatives* can be obtained from *ready-made* resources as follows:

1) *To modify* a ready-made resource (modify TS, phenomenon or process without affecting their essence) to obtain new properties of the resource;
2) *To transform* a ready-made resource (with a change in essence, phase state, etc.).

For example, air in TS or TS environment is a ready-made material resource. It can be modified (compressed), resulting in the formation of a derived material resource — compressed air — with new quantitative parameters (in particular, the density is increased, the volume is reduced). Compressed air can also be used as the resource of another class — energy.

Another example: a ready-made material resource in TS or TS environment — water — can be modified (magnetized) or transformed (transferred to another phase, or aggregate, state)[69]. The resulting derivative resources — magnetized water, steam, ice — have new properties (see Fig. 192). So, irrigation with magnetized water contributes to the yield of agricultural crops. In addition, magnetized water can flow upward as well as downward. In turn, ice and steam are used in technologies in which liquid water is not applicable.

Sometimes, not only the resources of other TS arising from their harmful and neutral functions, as well as environmental influences, can be a hint in the development of options for the implementation of useful TS functions but also some of the useful functions, space, energy, material and material basis of other TS, already used as useful. For example, a highway can be temporarily used for

Figure 192. Examples of modifying and transforming resources

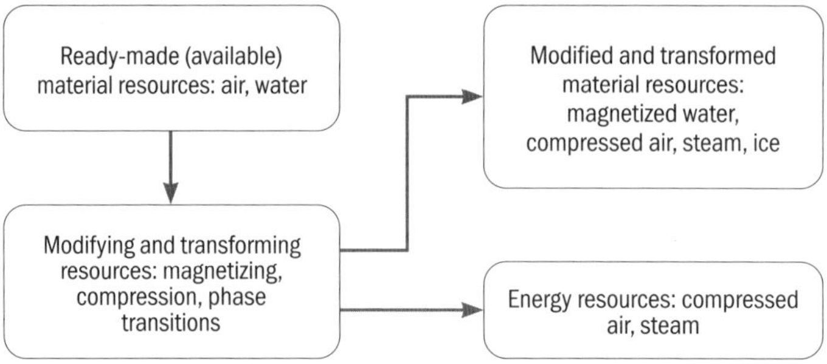

emergency landing of aircraft in peacetime, or serve as an alternate airfield in wartime. There are known cases of emergency landings of a common (non-hydro) aircraft on arable land, even on water.

TS analysis shows that certain types of potentially useful resources arise during the operation of certain TS classes (or their elements), united by design features or used processes. In particular, in the class of TS, with functioning of which a potentially useful ready-made resource — thermal energy, arises, it is possible to collect:

— Tribotechnics (technology related to friction processes);
— TS, in which mechanical processing of materials is carried out (cutting, drilling, milling, grinding, etc.);
— TS, in which processes are used, accompanied by the release of heat (combustion, stretching, compression, blow, etc.);
— All electrical systems, etc.

This TS class, for example, includes compressors, with the operation of which thermal energy arises due to a useful function — gas compression, as well as due to a harmful function — friction of the moving parts of the compressor. The resulting energy (water heated during the cooling of the compressors) is used to heat greenhouses, in which the function "generate heat energy" is useful (Fig. 193).

Figure 193. Example of using a ready-made energy resource: a — it was; b — became

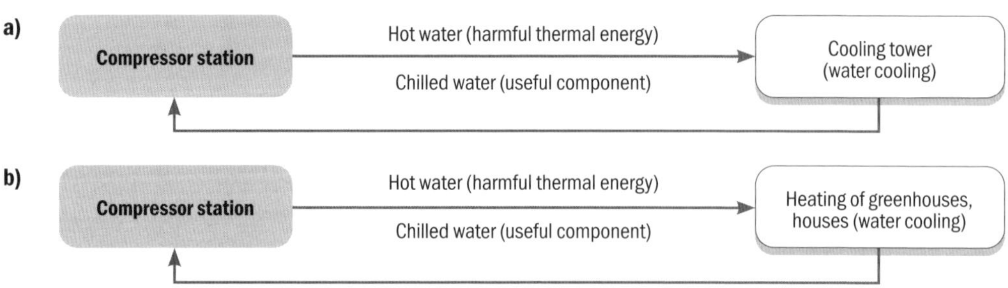

Another example characterizes the rational use of waste as material resources (Fig. 194). Metallurgical slag, ash from thermal power plants and boiler houses used to be production waste, and they were buried. However, instead of burial, they have begun to be used as a useful component in the manufacture of concrete. The result is double savings by eliminating the cost of producing the useful component and waste burial.

Figure 194. Example of the use of ready-made and modified material resources: a – it was; b – it became

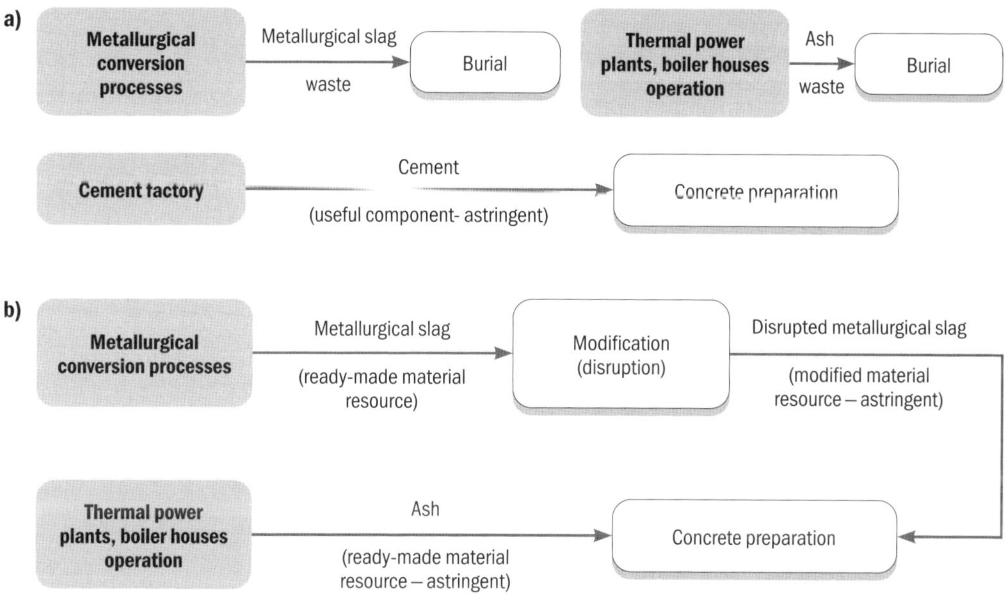

However, it should be noted that, although in many countries of the world they have begun work on accounting for production and consumption waste, their disposal long ago (for example, in Russia, in accordance with the federal law "On production and consumption waste", the federal classification catalog of waste, the state register of waste disposal, the databank of waste and technologies for the use and disposal of various types of waste are being created)[70], there is still NO full understanding of the fact that absolutely ALL waste of production and consumption, as well as human and animal life waste, can be used as resources. To a large extent, this explains the excess of the volumes of storage and disposal of waste (and often criminal emissions of waste into the environment) over their utilization.

The reasons for the occurrence of electrostatic fields in TS and the surrounding environment of TS, which can be used as resources, are shown in Fig. 195.

Figure 195. The reasons for the occurrence of electrostatic fields in TS and TS environment

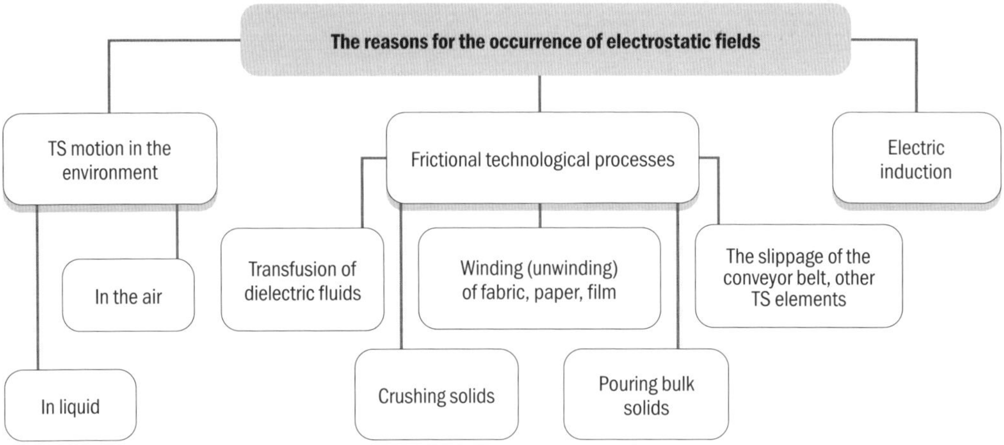

The above and many other examples have served as the basis for the formation and repeated adjustments of the detailed characteristics of the classes of non-traditional (hidden) resources.

6.3.7.3 Characteristics of non-traditional (hidden) resource classes

The characteristics of non-traditional (hidden) resources classes, developed by me, are presented in table 31.

The examples of existing and potentially possible implementations of non-traditional (hidden) resources are given in Appendix 7.

Table 31 and Appendix 7 indicate a potentially large number of non-traditional (hidden) resources types, many of which have not yet been described and systematized, which means that they are not involved in scientific and technical creativity.

Table 31. Characteristics of non-traditional (hidden) resources classes

Resource class	Class characteristics
Material	Any wastes from production and consumption, as well as from the life of humans, representatives of wildlife, substances of natural origin in the environment of TS, playing a harmful or neutral role in the performance of useful functions of TS. Material parts of TS and the environment that can be used as part of other TS
Spatial	Free spaces in TS or its surroundings (including at the microlevel), filled with liquid, gas or other filler, or being a vacuum. A part of the used TS space that can be used (usually temporarily) to implement new useful functions and properties. Spatial changes and movements of TS elements

Resource class	Class characteristics
Time	Time not used in TS before or after the end of a technological cycle, process, operation, reception, action. Execution time TS of a technological cycle, process, operation, reception, action, which in parallel can be used to perform other technological cycles, processes, operations, receptions, actions
Informational	Any information that characterizes the state of TS and the environment. Possible changes in TS properties during operation, storage, utilization, burial of TS, as well as TS properties indicating the presence of other TS properties. The presence of certain environmental objects or their characteristics, indirectly indicating the presence of other environmental objects or possible changes in the environment
Functional	Harmful and neutral functions and properties arising in TS during its operation, storage, utilization, burial or under the influence of the environment. Useful TS functions and properties that can be used (usually temporarily) as the principal, main and auxiliary functions and properties of another TS
Energy	All types of energy arising from operation, storage, utilization, disposal of TS or under the influence of the environment as harmful or neutral factors. Surplus energy or energy of natural origin not used during operation, storage, utilization, disposal of TS. Part of the energy used in TS that can be used (usually temporarily) to implement new useful functions and properties
Field	All types of physical fields arising during operation, storage, utilization, disposal of TS or under the influence of the environment as harmful or neutral factors. Physical fields in TS environment, acting as harmful or neutral factors. Physical fields due to the environment used in TS, which can be used (usually temporarily) to implement new useful functions and properties
Phenomena and processes	Phenomena and processes occurring in the atmosphere, space, earth's crust, as well as the consequences of these processes (volcanism, emissions of hot water, steam, rock falls, tsunamis, avalanches, etc.). Abiotic and biotic cycles of substances on the Earth. Phenomena and processes occurring during the operation, storage, utilization, disposal of TS, including those caused by harmful and neutral functions, as well as by the interaction of the environment with TS
Systemic	Complexes of interrelated harmful and neutral functions and properties arising during operation, storage, utilization, disposal of TS, as well as under the influence of the environment. Complexes of interrelated useful functions and properties of TS that can be used (usually temporarily) in another TS

6.3.7.4 The technique for identifying and using non-traditional (hidden) resources

The technique is presented in the form of the following two heurorhythms (heuristic algorithms).

The first *heurorhythm* can be characterized as an attempt to use the resources of "own" TS (or its environment) in "foreign" TS to implement their useful functions (see Fig. 196). In the first heurorhythm, "own" TS is TS1, and "foreign ones" are TS2 and TS3.

The second heurorhythm, in essence, is aimed at solving the inverse problem: to implement useful functions in "own" TS, you need to find "foreign" resources and, accordingly, TS that generate them (see Fig. 197).

Figure 196. Heurorhythm of options searching for using "own" ready-made, modified and converted resources in "foreign" TS

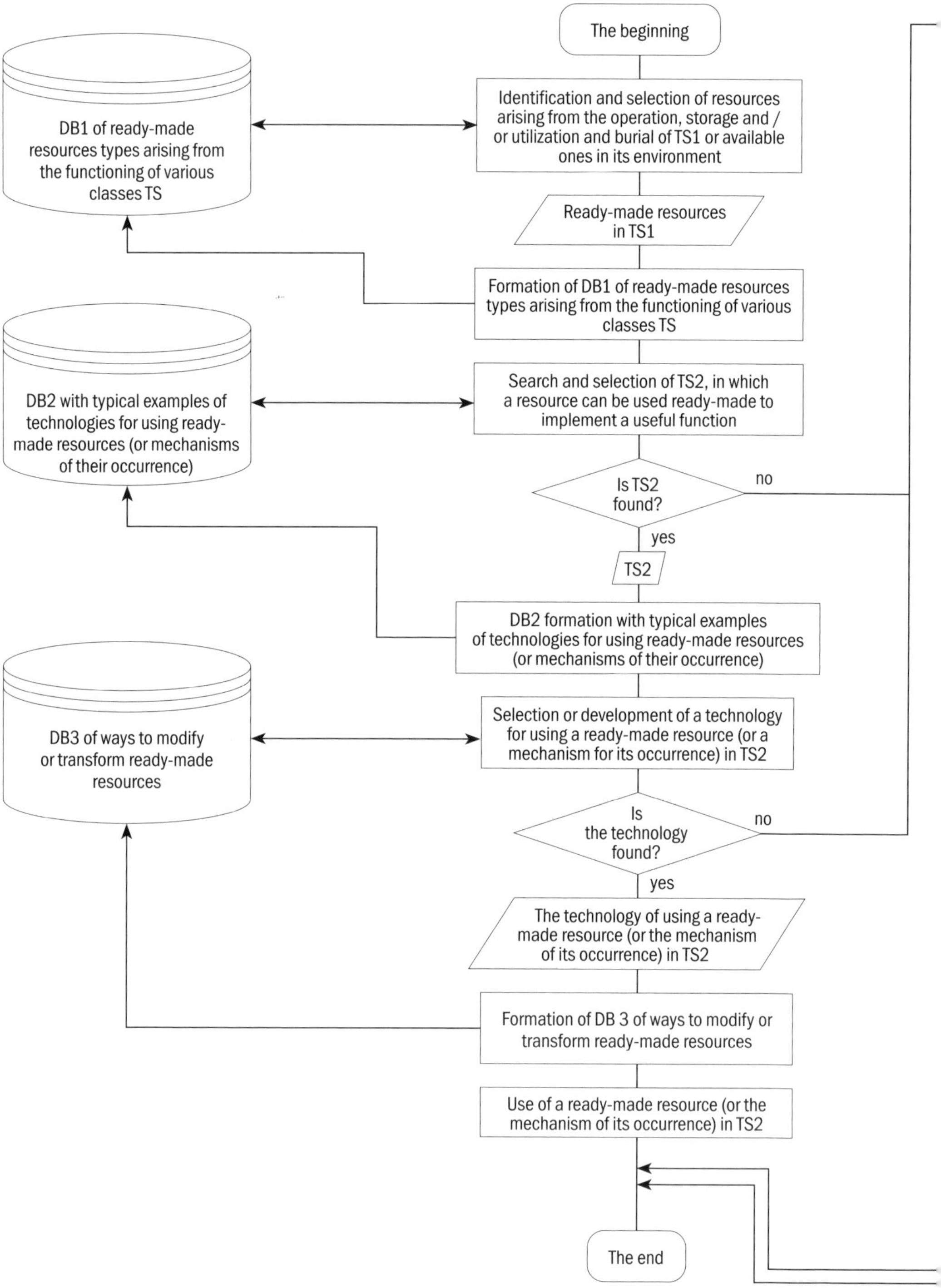

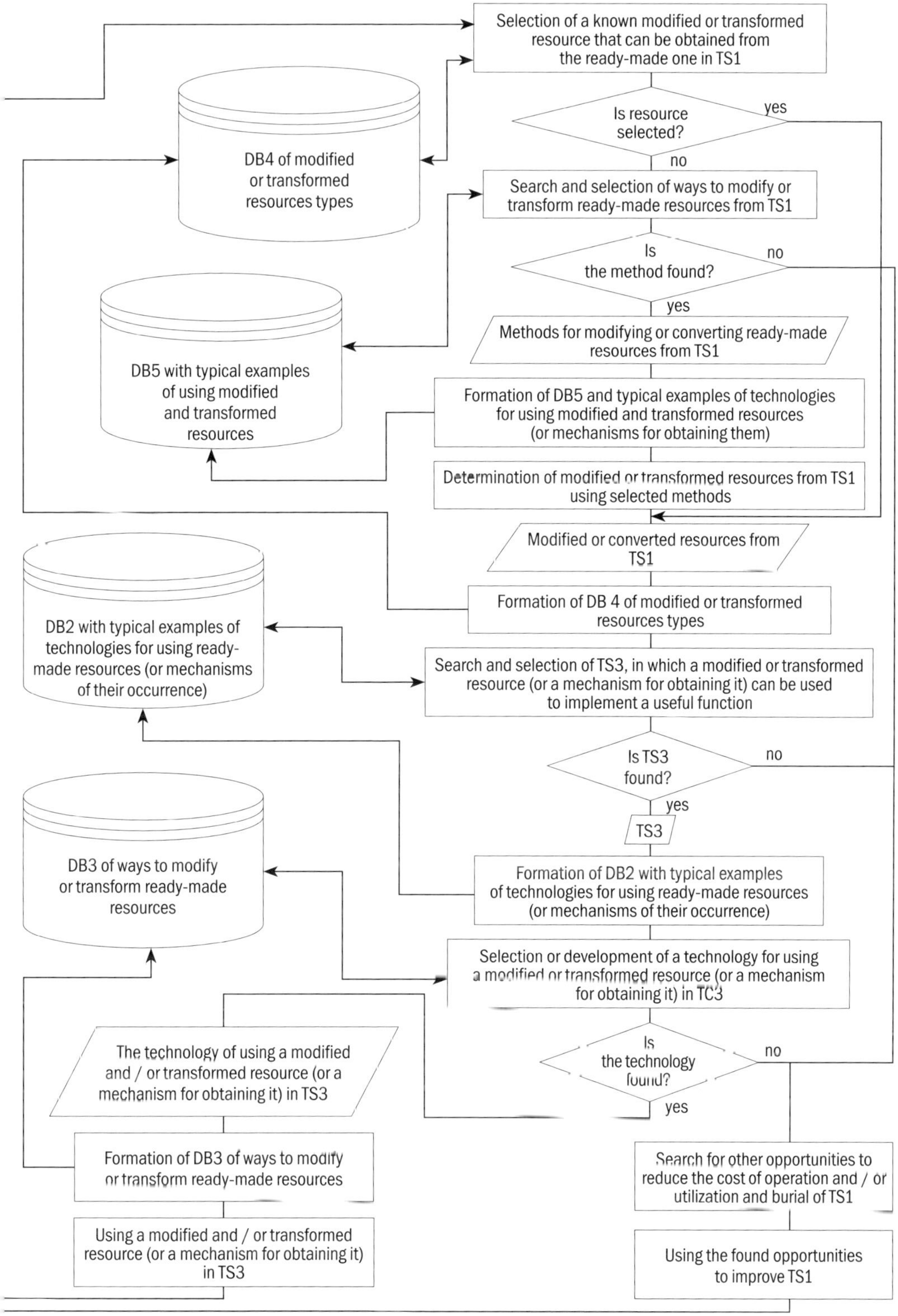
Selection of a known modified or transformed resource that can be obtained from the ready-made one in TS1
Is resource selected?
yes
no
DB4 of modified or transformed resources types
Search and selection of ways to modify or transform ready-made resources from TS1
Is the method found?
no
yes
Methods for modifying or converting ready-made resources from TS1
DB5 with typical examples of using modified and transformed resources
Formation of DB5 and typical examples of technologies for using modified and transformed resources (or mechanisms for obtaining them)
Determination of modified or transformed resources from TS1 using selected methods
Modified or converted resources from TS1
Formation of DB 4 of modified or transformed resources types
DB2 with typical examples of technologies for using ready-made resources (or mechanisms of their occurrence)
Search and selection of TS3, in which a modified or transformed resource (or a mechanism for obtaining it) can be used to implement a useful function
Is TS3 found?
no
yes
TS3
DB3 of ways to modify or transform ready-made resources
Formation of DB2 with typical examples of technologies for using ready-made resources (or mechanisms of their occurrence)
Selection or development of a technology for using a modified or transformed resource (or a mechanism for obtaining it) in TC3
The technology of using a modified and / or transformed resource (or a mechanism for obtaining it) in TS3
Is the technology found?
no
yes
Formation of DB3 of ways to modify or transform ready-made resources
Search for other opportunities to reduce the cost of operation and / or utilization and burial of TS1
Using a modified and / or transformed resource (or a mechanism for obtaining it) in TS3
Using the found opportunities to improve TS1

Figure 197. Heurorhythm of searching for "foreign" ready-made, modified and transformed resources for use in "own" TS

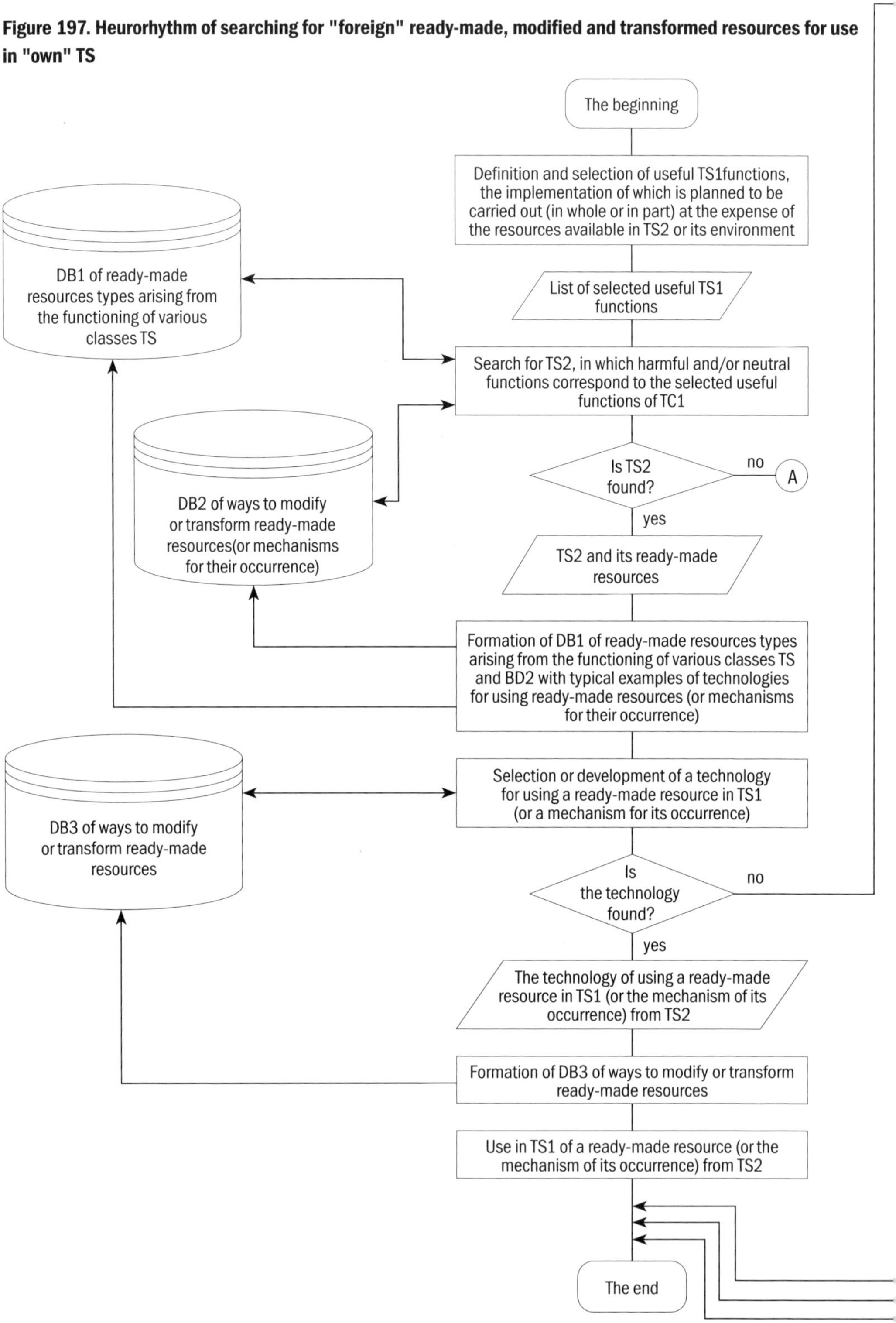

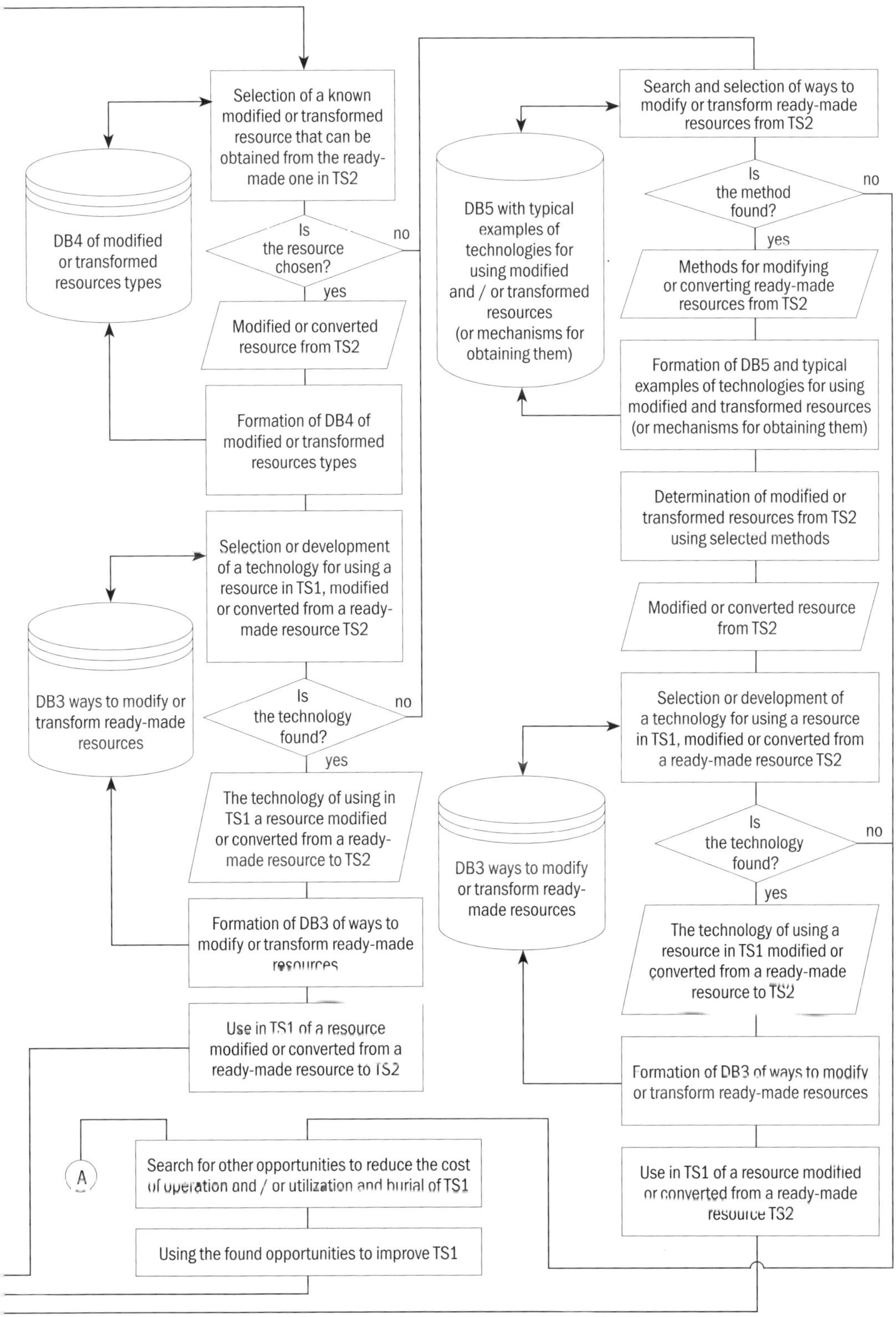

Selection of a known modified or transformed resource that can be obtained from the ready-made one in TS2
DB4 of modified or transformed resources types
Is the resource chosen?
no
yes
Modified or converted resource from TS2
Formation of DB4 of modified or transformed resources types
Selection or development of a technology for using a resource in TS1, modified or converted from a ready-made resource TS2
DB3 ways to modify or transform ready-made resources
Is the technology found?
no
yes
The technology of using in TS1 a resource modified or converted from a ready-made resource to TS2
Formation of DB3 of ways to modify or transform ready-made resources
Use in TS1 of a resource modified or converted from a ready-made resource to TS2
A
Search for other opportunities to reduce the cost of operation and / or utilization and burial of TS1
Using the found opportunities to improve TS1
Search and selection of ways to modify or transform ready-made resources from TS2
DB5 with typical examples of technologies for using modified and / or transformed resources (or mechanisms for obtaining them)
Is the method found?
no
yes
Methods for modifying or converting ready-made resources from TS2
Formation of DB5 and typical examples of technologies for using modified and transformed resources (or mechanisms for obtaining them)
Determination of modified or transformed resources from TS2 using selected methods
Modified or converted resource from TS2
Selection or development of a technology for using a resource in TS1, modified or converted from a ready-made resource TS2
DB3 ways to modify or transform ready-made resources
Is the technology found?
no
yes
The technology of using a resource in TS1 modified or converted from a ready-made resource to TS2
Formation of DB3 of ways to modify or transform ready-made resources
Use in TS1 of a resource modified or converted from a ready-made resource TS2

There are two heurorhythms (heuristic algorithms) for using non-traditional (hidden) resources.

The first one — the resources of "own" TS (or its environment) are used in "foreign" TS to implement their useful functions.

The second one implements useful properties in its "own" TS with the help of "foreign" resources and TS that generate them.

Competitive TS can be implemented on the basis of materials with unusual properties or mechanisms of their occurrence, disappearance, stabilization, etc.

At the same time, in both cases, the use of either ready-made, modified and transformed resources or the mechanisms of their emergence, which, on the basis of convergence (closing-in, transfer) of knowledge, can be used in the operation, improvement or creation of new TS.

The resources arising from the harmful effect of the environment on TS should be identified when studying the causes of TS destruction or deterioration of their consumer properties due to environmental influence.

To move from single examples of solving TS conceptual design problems associated with the rational use of resources in TS (found by the "trial and error" method) to a large-scale and effective application of the proposed technique, it is necessary to create a tool kit that includes systematized knowledge about resources and based on modern information technologies.

Under my leadership, the concept of the automated system for finding resources in TS and the environment and technologies for using resources has been developed.

6.3.8 Constructive mathematics, or the "STC +" operator

To activate scientific and technical creativity, to exclude or level out psychological inertia and barriers to thinking, to bring the moment of enlightenment (insight) closer, the methodologists of scientific and technical creativity recommend taking measures aimed at time gaps, on the one hand, albeit beautiful, but on the other hand, the obsessive chain of thoughts and emotions inherent in creativity.

It was G. S. Altshuller[71] who became the popularizer of one of those rather effective measures — the use of STC operator. In practice, this is conducting thought experiments to increase (to infinity) and decrease (to 0) the characteristics of the investigated (being developed) object, for example: size (S), time (T), cost (C).

The STC operator changes the usual idea of an object, liberates thinking, which can lead to non-trivial, even fantastic ideas.

The STC operator is, in a simplified form, the long-known *"constructive mathematics"* — the branch of mathematics that studies (as indicated in the Great Russian Encyclopedia) constructive processes, the possibility of their implementation and the results of these processes — constructive objects[72].

In constructive mathematics, two abstractions are systematically applied: *the abstraction of potential feasibility* and the *abstraction of identification*. Abstraction of potential feasibility is used, abstracting from the practical limitations of design possibilities in space, time and material. The abstraction of identification is used when one speaks of two, in one sense or another, identical objects as of one and the same object[73].

Since, in general, the idea of considering TS parameters of possible characteristics in a wide range during TS development is very fruitful, I would propose to endow TS being developed (along with STC) with **any** (including non-inherent) characteristics estimated by quantitative parameters (for example, temperature, strength, brightness), and then mentally change these parameters as much as possible (value "0" will indicate the exclusion of this characteristic from TS) and try to generate ideas and technical decisions for their implementation under such extreme conditions. It is better to call such a heuristic the "STC + operator".

6.3.9 Use of laws and patterns of technical systems development

Let me remind you that in accordance with the systematic methodology of design activity (see Section 3.2), the problem solution includes the stages (their numbering continues the numbering of the stages of the task setting): 5) analysis, synthesis and selection of the functional structure of TS being designed; 6) analysis, synthesis and selection of the operating principle of TS being designed; 7) analysis, synthesis and selection of technical decisions for TS being designed; 8) optimization of property parameters in the technical solution of TS being designed. Recommendations for using the laws and patterns of TS development at the stages of solving a problem are given in Table 32.

Table 32. Recommendations for the use of laws and patterns of TS development at the problem solving stages

Name of the law or pattern of TS development	Recommendations for the use of laws and patterns of TS development
The law of stage TS development	Use to determine the promising direction of TS development based on the implementation of the next (unrealized) fundamental labor function (FLF) or on the basis of a new implementation of technological FLF (see Fig. 97)
Progressive TS evolution law	Use to determine the most probable way of hierarchical exhaustion of the capabilities of TS being developed (see Fig. 98)
Correspondence law between functions and functional structure of TS	Use for the design of TS elements that prevent the occurrence of discrepancy between the functions (F) and the functional structure (FS) when the exclusion of an element from FS, changes in external conditions or human needs
Patterns of TS development	Use to identify trends in changes in the characteristics of the functioning and / or structure of TS, leading to the development of TS technical decisions that potentially have new useful properties and / or improved property parameters

To optimize the property parameters in a TS technical decision, TS parameter optimization methods are used.

6.3.10 Use of materials with unusual properties

The experience in the development of technologies, equipment, products (technical systems — TS) shows that often the basis of TS with high consumer properties, and sometimes TS of a new generation, are materials with unusual properties or mechanisms of occurrence, disappearance, preservation, stabilization, change, concentration, isotropy, anisotropy and other manifestations of such properties. Let me remind you that the properties of systems represent their characteristics, which are usually divided into two types:

- Qualitative;
- Quantitative (parameters).

Analysis of the properties of natural and artificial materials shows that there are many unusual ones among them. This unusualness consists in the presence or absence of a characteristic, characteristic value or range of values, in the connection between a pair of characteristics, a form of existence or another manifestation that is atypical for materials of this class.

Systematized information about materials with unusual properties can be the important type of productive knowledge, the skillful application of which can lead to original solutions to the problems of scientific and technical creativity. There are cases when the use of the material with unusual properties has initiated the development of an entire industry. For example, fluorescent materials are used in the manufacture of clothing, shoes, road signs and other items, in advertising.

There are examples of a multiple increase in TS efficiency in the industry due to the implementation of materials with properties that are fundamentally different from the properties of traditional materials.

However, many materials with unusual properties have not been identified, have not been systematized, and the technologies for their use have not been developed. One of the reasons is the absence of criteria for unusual properties; therefore, the appearance of materials with unusual properties has been most often explained by chance.

Under my leadership the studies have been carried out, as a result of which the following have been developed[74]:

The criteria for unusual properties (given in Appendix 9);

- The description form (passport) of the material with unusual properties, including its formalized presentation;
- Dictionaries and classifiers used for the formalized description of materials with unusual properties (classifier of industries in which materials

with unusual properties are created and used; classifier of material properties; ontology of scientific and technical characteristics; dictionaries of unusual properties' features);

— Databases (DB) structure of materials with unusual properties;
— Software implementation of the automated information system (AIS) for materials with unusual properties;
— Examples for filling the database of materials with unusual properties (Appendix 10).

Approbation of AIS and DB of materials with unusual properties has shown the potentially high efficiency of their use in scientific and technical creativity, as well as for improving the quality of specialists training.

6.3.11 The use of illusions in technical and artistic creation

The word "illusion" comes from lat. illusio — mockery, ridicule, deception, delusion. Now, illusion means distorted perception — the mental process of holistic reflection of objects in the human mind and phenomena of reality in a variety of their properties with direct impact on the human sense organs.

Perceptual distortion can be natural, unintentional and it can occur in mentally healthy people due to the manifestation of physical effects. For example, the appearance of mirages in front of travelers in the desert under the certain state of the atmosphere — imaginary visible displays of various objects and phenomena. The distortion of perception can inadvertently occur with mental disorders leading to *hallucinations* — delirium, visions, senses deception, arising in consciousness without external stimuli.

But you can deliberately distort perception in various ways, which leads to illusions, which are widely used in architecture, design, household items, art, military affairs, advertising and other fields in order to create the false idea of objects, phenomena, their states, as well as their properties and properties parameters. Most often this is done to increase the number of positive properties, improve their parameters or hide flaws. In military affairs, illusions are used to disguise real or to create false military objects, and in fine arts and music — to form a new emotional human state or to strengthen an existing one.

The types of illusions used in technical and artistic creation that arise due to deliberate distortions of perception are shown in Table 33.

Along with the above illusions, there are also tactile, physiological, functional etc.

Table 33. Types of illusions used in technical and artistic creativity

<table>
<tr><th>Types of illusions and their characteristics</th><th colspan="2">Examples of illusions manifestations</th></tr>
<tr><td rowspan="2">Optical (visual) – due to physical laws and / or unconscious correction of the visual image</td><td>For example, a teaspoon put in a glass of water is perceived as broken (Fig. 198).

Figure 198. Illusory representation of a teaspoon in a glass of water</td><td>There are classic examples of the trompe l'oeil technique – the art of manipulating visual perception.
For example, a distorted perception of size (the Mueller-Lyer illusion) arises from the fact that the upper line, bounded by diverging segments, seems longer than the lower one, bounded by converging segments (Fig. 199), although these lines themselves are alike.

Figure 199. Distortion of size perception</td></tr>
<tr><td colspan="2">The Müller-Lyer illusion can distort the perception of perspective. In Fig. 200a, the diverging segments from the line representing the corner of the building generally give the illusion that Fig. 200a is further away than Fig. 200b, although both are at the same distance from the reader, the lines representing the corner are alike as well.
a) b)
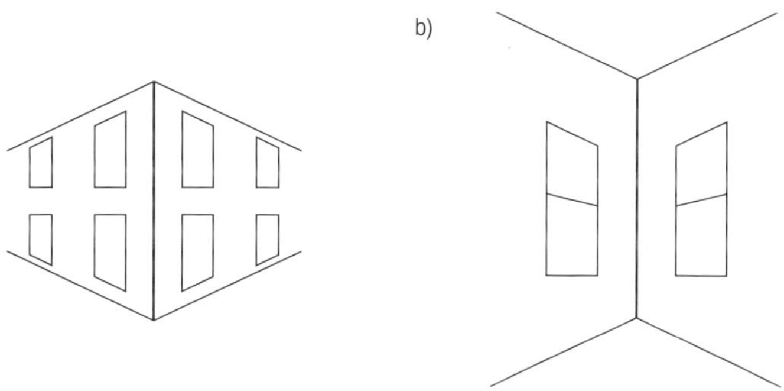
Figure 200. Perspective distortion: a – outer corner; b – inner corner</td></tr>
<tr><td>Optical (visual) – due to physical laws and / or unconscious correction of the visual image</td><td>By changing the slope of the shapes, you can create the illusion that the parallel lines appear to be intersecting (Fig. 201)
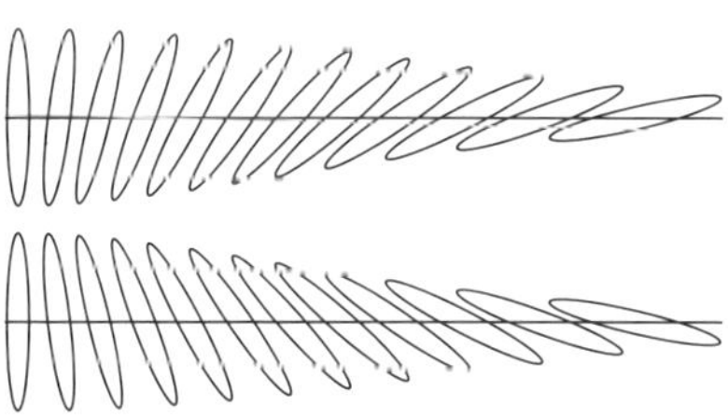
Figure 201. The illusion of intersecting straight lines</td><td>Latent images (apparent shapes) can be made with the help of creative display of other shapes and even letters in words. For example, in the well-known FedEx logo, you can unexpectedly see a forward arrow between the letters Ex, which vividly symbolizes the company's mission (speed and accuracy of delivery) (Fig. 202).
Figure 202. FedEx logo</td></tr>
</table>

<table>
<tr>
<td rowspan="2">Optical (visual) – due to physical laws and / or unconscious correction of the visual image</td>
<td>Psychologists at the University of Bristol (UK) have found that glasses with curved and tapering walls, equal in volume to glasses with straight walls (Fig. 203), create the illusion of a smaller volume, which leads to faster emptying of such glasses and, in general, to the increase in drinks consumption.

a) b)

Figure 203. Glasses with different walls: a – curved; b - straight

In addition to these types, there are optical illusions of movement, impossible shapes, dual images, aftereffect, perception of depth, gravity of objects (A. Charpentier illusion), perceptual readiness (the eye "sees" what the brain is prepared to see based on the general context of the image) etc.</td>
</tr>
<tr>
<td>Illusions of rich color and contrast are created, for example, by the fact that color areas are outlined with black frames. In Fig. 204 a, the color fields appear brighter and more contrasting than in Fig. 204 b, although they are actually alike.

a) b)
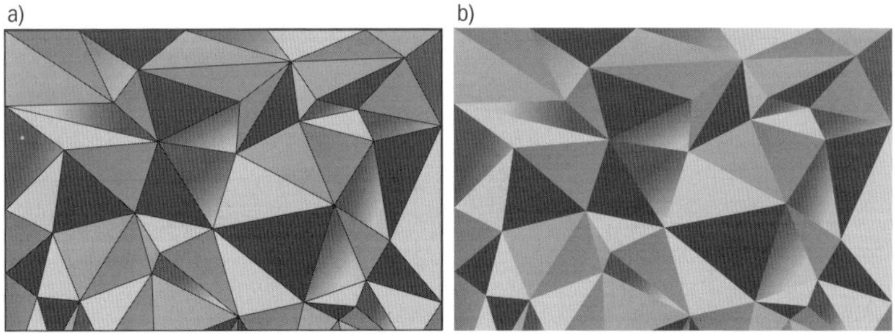

Figure 204. Illusion of color and contrast: color fields with black frames (a) and without them (b)</td>
</tr>
<tr>
<td>Auditory – due to distorted perception of the characteristics of tone: strength, rhythm, distance to the tone source</td>
<td>In music, for example, the illusion of a continuous change in the pitch level of sounds is used, called the glissando (smooth sliding from one sound to another) Shepard-Reese, which forms the perception of a continuously rising or falling tone. Reese also created a similar illusion with a continuously accelerating or decelerating rhythm. All this forms new emotional states of a person</td>
</tr>
<tr>
<td>Olfactory and gustatory – appear in the form of a subjective change (inversion) in the quality of sensations</td>
<td>These illusions are achieved in various chemical-biological and physical ways. Then, for example, sweet can seem sour, the stench is felt like perfume (and vice versa).
It has become fashionable to create fish-flavored meat dishes and vice versa. Even the so-called molecular cuisine has emerged.
To enhance the taste, dynamic contrast of tastes is widely used (for example, as in Raffaello sweets, when you can first taste a crispy shell with coconut flakes, then a delicate filling, then a crispy nut kernel). By the way, several million dollars have been spent researching the crunch of chips to make a crunch that creates the illusion of maximum palatability, salivation and the desire to buy certain foods. Knowing this information is necessary at least in order to assess the realities correctly</td>
</tr>
</table>

6.4 Formation of problem solving results

6.4.1 Formation of a variety of technical and other solutions

Full satisfaction with the results of scientific and technical creativity can be achieved only if these results are competitive. The high probability of obtaining a competitive result can be realized due to the extensive or intensive formation of a variety of results (technical and other solutions), using the methods and means, respectively, of the "old" and "new" innovation culture (see Section 5.2.4). Let me remind you that the extensive path of innovative development in many countries at the present time and in the near future, due to a lack of financial resources, is possible only for solving certain acute problems. For intensive formation of a variety of technical and other solutions, you should use the entire set of heuristics that are presented in this book. The "ascent-descent" method is the most suitable for this purpose, with the help of which the "family" of variants of the obtained solution[75] should be formed, intended for different objects and conditions of use, or based on different quantitative parameters of the properties of TS being developed.

From the formed "family" of technical and other solutions (results of intellectual activity and means of individualization of legal entities, equated to the results of intellectual activity), the most promising ones should be selected from the point of view of commercialization or other use, and the selected results should be analyzed in terms of the necessity and possibility of copyright protection.

6.4.2 Protection of intellectual activity results and means of individualization

For example, in accordance with Article 1225 of the Civil Code of the Russian Federation, "the results of intellectual activity and the means of individualization equated to them for legal entities, goods, works, services and enterprises, which are granted legal protection (intellectual property), are:

1) Works of science, literature and art;
2) Programs for electronic computers (computers);
3) Databases;
4) Performance;
5) Phonograms;
6) Communication over the air or by cable of radio or television broadcasts (broadcasters or cablecasters);
7) Inventions;

Before commercializing and / or using the results of intellectual activity (RIA), select the best ones and provide them with legal protection. And then either sell the rights to use, or organize production and / or provide services yourself.

For successful scientific and technical creativity, it is useful to know about the outstanding achievements of mankind.

8) Utility models;
9) Industrial designs;
10) Breeding achievements;
11) Topology of integrated circuits;
12) Production secrets (know-how);
13) Company names;
14) Trademarks and service marks;
15) Appellations of goods origin;
16) Commercial designations"[76].

Let me remind you that in the Russian Federation the objects of invention are *"a device; way; substance; strain of microorganism, cell culture of plants and animals; application of previously known inventions for a new purpose"*[77].

If you plan to patent an invention abroad, it is better to do it first in your own country. Then it is your own country that acts as the defender of rights. The reader can easily find more detailed materials on patent law on the Internet.

6.4.3 Commercialization or other use of problem solving results

After a preliminary assessment, selection of promising results of intellectual activity (RIA), providing them with legal protection, it is necessary to consider the possibility of commercialization of RIA (obtaining benefits) or other use (for example, for internal needs).

The commercialization of RIA is carried out in two main ways, each of which has many forms.

1. Transfer of rights to use RIA on the basis of contractual relations.
2. Organization of production (and sale) of RIA objects and / or provision of RIA-related services.

The issues of RIA commercialization can be systematically worked out in a business plan, the structure and methodology of which can be easily found by the reader on the Internet.

6.4.4 Determination of the direction for further improvement of a technical system

The directions for further improvement of the developed TS can be prompted by new needs identified in the process of additional marketing research, consequences from the law of stage development, as well as thought experiments using other heuristics presented in Table 5.

6.5 Example of generating an idea and forming its effective technical decision using heuristics

In section 6.2.2.1 "Continuous and periodic the most important natural needs", the history of the emergence of the most important natural need to have protective and rescue equipment for critical situations is described, on the basis of which, using various heuristics, I have developed and patented nine variants of individual device for descent from high objects in critical situations (fire, man-made, ecological disaster, terrorist act, etc.). These devices differ from many others by the possibility of unburdensome constant wearing and immediate use (such as a mobile phone). The device can be realized, for example, in the form of a beautiful belt, which is normally used for its designated purpose (worn with trousers and a dress). The belt is woven from a thin strong thread ($D \approx 1$ mm) or a rope 100–150 m long and can be unraveled with effort. If it is necessary to descend from a high object (for example, a building), a thin thread should be pulled out several meters, fastened to something: large pieces of furniture, a radiator, a pipe, a door, etc. — andyou can start descending. In this case, the effort of unraveling the thread (rope) can be fully or partially compensated for the action of the brake mechanism, the role of which can be played by the belt buckle.

The history of the generation of ideas in the formation of effective technical decisions using heuristics for the creation of permanently wearable individual descenders in critical situations is shown in Table 34.

Table 34. History of creation and development of TS technical decisions using heuristics

Heuristic used	Stages of creation and development of TS technical decisions
1. Based on the the most important natural need to **have protective and life-saving equipment for critical situations,** a new urgent need for the **constant availability** of a life-saving equipment has been formed	Formation of a current need for an individual emergency rescue device **that can be carried with you at all times without burden or kept nearby ready for immediate use**
2. Initially, the design of a drum (with minimum dimensions) was developed, used for winding (unwinding) a thin strong thread. However, in accordance with the rules of **functions** formulation, the traditional functions of the drum **"wind"** and **"unwind"**, indicating specific implementation options, have been **replaced by generalizing (abstract) "lay"** or **"pack"** and **"unpack"**	Development of a technical solution with qualitatively new properties – a rope-launching device without **a drum.** This made it possible to abandon traditional constructive implementations of rope-launching devices and significantly reduce the size and weight of the device
3. With the help of the most important natural need to **arrange** the spacing of objects and their structural elements (including signs, symbols, sounds); the shape of objects (transform in the form of regular geometric shapes); types and characteristics of movements; connections between objects and their structural elements or to **replace the existing order** with another, **the idea of a new laying (packing) of the launching** rope is formed	Development of a technical decision with qualitatively new properties – the launching rope, unraveled from a belt and / or other part of clothing, shoes, accessories

Heuristic used	Stages of creation and development of TS technical decisions
4.With the help of the "ascent-descent" method, a variety of fabric products is formed	Development of a technical decision with qualitatively new properties – the launching rope, unraveled from a belt and / or other part of clothing, shoes, accessories
5.With the help of the technique of forming the ideal final result (it seems that there is no technical system, but the function is being performed), a technical decision close to ideal has been formed	Development of a technical solution with qualitatively new properties – a fabric in which, due to weaving, gluing and other bonds, **the rope is unraveled** with a given **force (there is no braking device** in the traditional sense)
6. With the help of the "ascent-descent" method, a variety of options for using the device has been formed	Variants of use in case of man-made disasters, terrorist acts, in tourism, mountaineering, speleology, emergency towing, in military and other special clothing

6.6 One hundred outstanding achievements of humanity

For successful practice of scientific and technical creativity, it is useful to know about the outstanding achievements of mankind, which have had the greatest impact on its formation and development. One list of achievements alone makes a creative person think about the reasons for the emergence of outstanding technical systems indicated in the list, their main functions and properties, the history of TS development.

There are many such lists (e.g. including 100 achievements), in which, as a rule, only 20–30% of TS overlap. This indicates great subjectivity in the assessment of achievements. Moreover, the smaller the list of outstanding TS in terms of the number of TS, the less likely it is to meet the same list compiled by another author or other authors. That is why it is almost impossible to objectively name one TS that has had the greatest impact on the formation and development of mankind.

Appendix 11 contains the list of 100 outstanding achievements of mankind, formed by A. I. Polovinkin and significantly corrected by me.

6.7 Promising directions for the development of technical systems

I have formulated 10 promising directions for technical systems development, including research and development activity related to them, the results of which are likely to be in demand in the world in the next 5–15 years.

In the list of directions, which is given below with comments, the criterion of the significance of the directions is not taken into account.

Promising directions for TS development and scientific developments in the next 5–15 years

1. Robotization of all sectors of the economy.
2. Revealing and substantiation of new patterns of TS development in order to improve TS or create new TS generations.
3. Reducing the cost of operating and improving TS or creating TS of new generations at the expense of non-traditional (hidden) resources.
4. Creation of TS based on analogs and analogies of TS in animate and inanimate nature.
5. Creation of a meta-knowledge base (mega-science) for effective convergence of different fields of knowledge.
6. Development of materials with unusual properties and technologies for their use.
7. Realization of the most important human needs (especially permanent ones) while improving TS or creating new generations of TS.
8. Creation of efficient energy converters based on new operating principles.
9. Implementation of the concept of "smart" in any objects.
10. Creation of the Internet of Things — a system of computer networks and physical objects with remote access to them.

1. *Robotization of production processes in all sectors of the economy and industry.*

The inevitability of this is due to the law of the staged TS development during the transfer from a person to TS of all fundamental functions of labor (FFL): technological, energy, management and planning. If this condition is met, transition to the promising direction of TS development based on a new implementation in TS of the technological FFL will take place.

The implementation of the state program "National Technology Initiative", which, in particular, provides for the creation of unmanned aerial vehicles, cars, sea vessels confirmes the need for large-scale robotization of production processes in Russia.

2. *Conducting scientific research to identify and substantiate new patterns of TS development and use these patterns to improve or create new generations of TS.*

Taking into account the possible large number (several tens of thousands) of new TS development patterns and the potentially high efficiency of their use in scientific and technical creativity, the conduct of such research is of great national importance.

3. *Revealing non-traditional (hidden) resources in TS and the TS environment, mechanisms of resource generation, as well as technologies for their use to reduce the cost of operating existing TS and improve or create TS of new generations*[79].

Since there can be about a million undiscovered and undescribed non-traditional (hidden) resources in various sectors of the economy, the industry, the conduct of such research and development is of great national importance.

4. *Creation of TS based on analogs and analogies of TS in animate and inanimate nature.*

Almost half a century ago, the Nobel laureate Soviet academician N. N. Semenov said: "I think that the further development of technology largely depends on how people are able to transfer the principles of living organisms operation into it."

Currently, this direction is known as the creation of nature-like systems (based on the use of functions, properties, functional structures, principles of action and other solutions implemented in nature in TS).

5. *Creation and development of a mega-knowledge base (or mega-science), combining different fields of knowledge for the effective implementation of the knowledge convergence mechanism.*

With the creation of such a base, in addition to the sharply increased efficiency of knowledge convergence, the basal principle of heuristics will be implemented, which almost excludes the influence of the "narrow" specialization of scientists.

6. *Development of materials with unusual properties and technologies for their use.*

The creation of materials, the use of which ensures the implementation of a set of new functions, has great prospects. For example, they are discussing the creation of paint that can be applied to any object (equipment, structures), as well as clothing and accessories; the paint converts any object into a solar-to-electrical energy converter.

7. *Realization of the most important human needs (especially permanent ones) while improving or creating TS of new generations.*
8. *Creation of efficient energy converters based on new operating principles.*

Within the framework of this direction, it is promising to create portable transducers that use human motion and / or energy, as well as human waste.

9. *Implementation of the concept of "smart" in any objects.*

The concept of "smart" ("smart" properties) can be implemented in room air, water, environment, technology, machinery — from paper clips to complex systems, including a "smart" city.

10. *Creation of the Internet of Things — the system of interconnected computer networks and connected physical objects (things) with built-in sensors and software for collecting and exchanging data, with the possibility of remote control and management in the automated mode, without human intervention*[80].

Fig. 205 shows that in 2018, the number of devices (personal computers, smartphones, tablets, home appliances, industrial equipment, smart TVs, devices attached to the body) connected to the Internet of Things worldwide reached

Figure 205. The number of devices connected to the Internet in the world with a growth trend for the current year

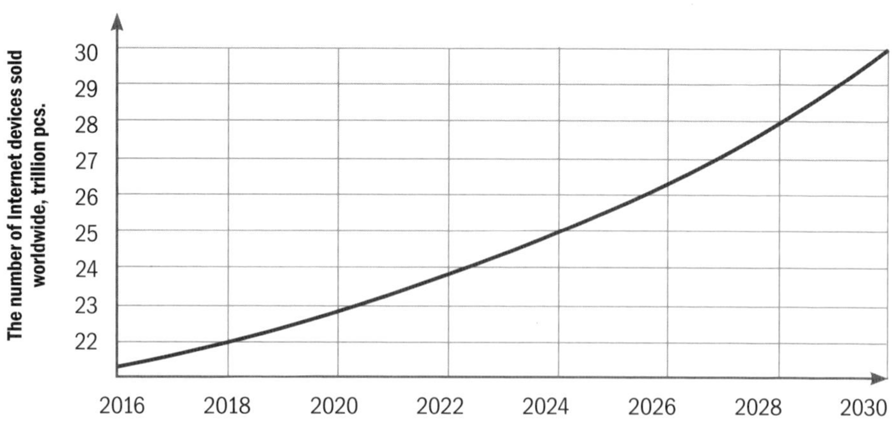

22 billion. According to expert forecasts, by 2025, 38.6 billion devices will be connected in the world, and by 2030–50 billion.

It should be noted that many promising directions for TS development and related areas of scientific research intersect, generating new qualities. Artificial intelligence, and neural networks in particular, can (and already do) make a significant positive contribution to the implementation of every promising direction in the development of TS.

Chapter 7
Training of creative professionals

7.1 The growing demand for creative professionals and the limitation of vocational education system

With the growth of the Earth's population, accompanied by the rise in the cost of technologies, raw materials, equipment, depletion of natural resources, deterioration of the environmental situation, aggravation of competition, the needs of people for protection from various threats, as well as for improving the level and quality of life of each man and society as a whole, are constantly growing. This, in turn, causes the need for training and increasing the number of creative professionals who are able to go beyond **reproductive** activity, generate many effective ideas and implement them with great productivity, creating goods and developing services with high consumer properties.

However, the systems of professional education existing in the world, with rare exceptions, are based "on traditional (reproductive) didactics, which considers teaching as the process of objectively *determined* (strictly defined) development, provided only by the transfer of already known knowledge, which practically does not develop the creative abilities of students and allows them to prepare specialists only for reproductive activity"[1]. And although most teachers are focused on training creative professionals and often repeat after the ancient Greeks (according to some sources, Socrates, according to others — Plutarch): "A student is not a vessel that needs to be filled, but a flame that needs to be fired!" — they themselves "flame up" (that is, motivate to active cognitive activity aimed at developing the creative abilities of students) rarely, because they do not know how to do it.

One of the main reasons is the insufficient elaboration and implementation of the methodology and especially the means of computer support for creative pedagogy in the vocational education system, which effectively ensures the formation of students' creative abilities.

The methodological systems developed in the 70s and 80s of the XX century by the outstanding Soviet educators and innovators V. F. Shatalov, S. N. Lysenkova, Sh. A. Amonashvili, G. L. Ilyin, L.A. and B. P. Nikitin, training development of V. V. Davydov, the didactic system of L. V. Zankov, the stage-by-stage formation of mental actions by P. Ya. Galperin, the system of enlarged didactic units by P. M. Erdniev, school of "dialogue of cultures" V. S. Bibler, contextual training by A. A. Verbitsky and others made a great contribution to improving the effectiveness of training[2]. There are many outstanding educators and innovators in other countries. However, the methodological systems developed by educators-innovators have not yet become prevailing in the vocational education system, since (in my opinion) they have not been supported by the systemic methodology of design activity, where design is understood as the third culture[3] that unites traditionally non-intersecting humanitarian, artistic and scientific technical cultures, and has been also not equipped with heuristics that not only develop motivation to learn, but also provide students with significant scientific and practical results.

This pedagogical technology, based on the systematic methodology of design activities and heuristics, called creative (projective) pedagogy, has been developed and successfully tested under my leadership. The results of many years of work were published in the monograph[4], which became a pedagogical bestseller.

7.2 Formula for creative pedagogy

Creative (projective) pedagogy is developed on the basis of the creative pedagogy formula formulated by A. G. Aleinikov (last years lived in the USA) according to the rules for drawing up *formulas of inventions*, which, with any complexity of the described objects, most often consist of one sentence (single-link) and include:

— The name of the proposed object, for example, technology (method), equipment (device), material (substance);
— A pre-distinctive part (that is, the common thing that unites the proposed object with analogues and prototype);
— The goal (as a rule, formulated in a very standard and general way);
— A distinctive part (that is, features that make up the essential differences of the proposed object from analogues and prototype, the implementation of which assumes a positive effect)[5].

The formula of creative pedagogy (according to A. G. Aleinikov)[6]: "Pedagogy of creative orientation, containing pedagogical influence on the subject for mastering a certain learning material (learning subject) and characterized by the fact that, in order to increase the effectiveness of teaching, pedagogical influence is carried out against the background of centrifugal supercritical interaction, in this case, the student is transferred from the rank of the object of influence to the rank of the subject of creativity, and the traditional (basic)

Vocational education systems, with rare exceptions, are based on traditional (reproductive) didactics, practically do not develop creative abilities and train professionals only for reproductive activity.

A creative (projective) pedagogical technology has been developed on the basis of the systematic methodology of design activities, interpreted using the formula of A. G. Aleinikov's creative pedagogy.

educational material is transferred from the rank of the subject of development to the rank of a means of achieving some creative goal, while the additional material contains the description and demonstration of heuristic techniques and methods action."

In other words, to achieve the effect of creative orientation in the developed pedagogy and its approbation, the following is necessary:

A. To create educational process provided with a background of centrifugal, open in meta-knowledge, mega-science (including other, in addition to a narrow specialty, knowledge) supercritical (allowing only benevolent, constructive, developing criticism) interaction, which contributes to the disclosure and development of students' creative abilities[7].

In general, this is the basal principle of heuristics implementation aimed at increasing the efficiency of the knowledge convergence mechanism (see Section 2.3.1) in the case of the subject-subjective relationship between the teacher and the student.

B. To reorganize the educational process, during which the student becomes a creator (subject of creativity), and main learning material (which is studied with increased interest) becomes a means of achieving a creative goal.

C. To introduce additional learning material, including heuristic principles, laws and patterns of TS development, strategies, tactics, methods and techniques (see Table 5), which allow the student to dramatically increase the effectiveness of creative activity.

In general, with such an organization of training, students increase their motivation for cognitive activity, their creative potential develops and, ultimately, competencies are formed that are necessary not only for reproductive, but also for productive (creative) activity[8].

7.3 The functional structure of developed creative (projective) pedagogy

Taking into account the fact that the achievement of a creative goal in any productive activity (including learning), in essence, is the solution to any problem or task, the implementation of a design or concept and is carried out in the process and according to the laws of design[9], creative (projective) pedagogy has been actually developed on the basis of overlapping (or combining) the stages of design preparation, which are presented in the systematic methodology of design activities (see Fig. 15), and the formula of creative pedagogy proposed by A. G. Aleinikov.

As a result, the functional structure of creative (projective) pedagogy has been formed (see Fig. 206), which describes primarily technical professions training.

In accordance with it, the goal of studying a certain learning material (learning subject) can be defined as a creative goal, the training itself is carried out in stages (numbering of stages from top to bottom) from abstract to specific as the

Figure 206. Functional structure of creative (projective) pedagogy

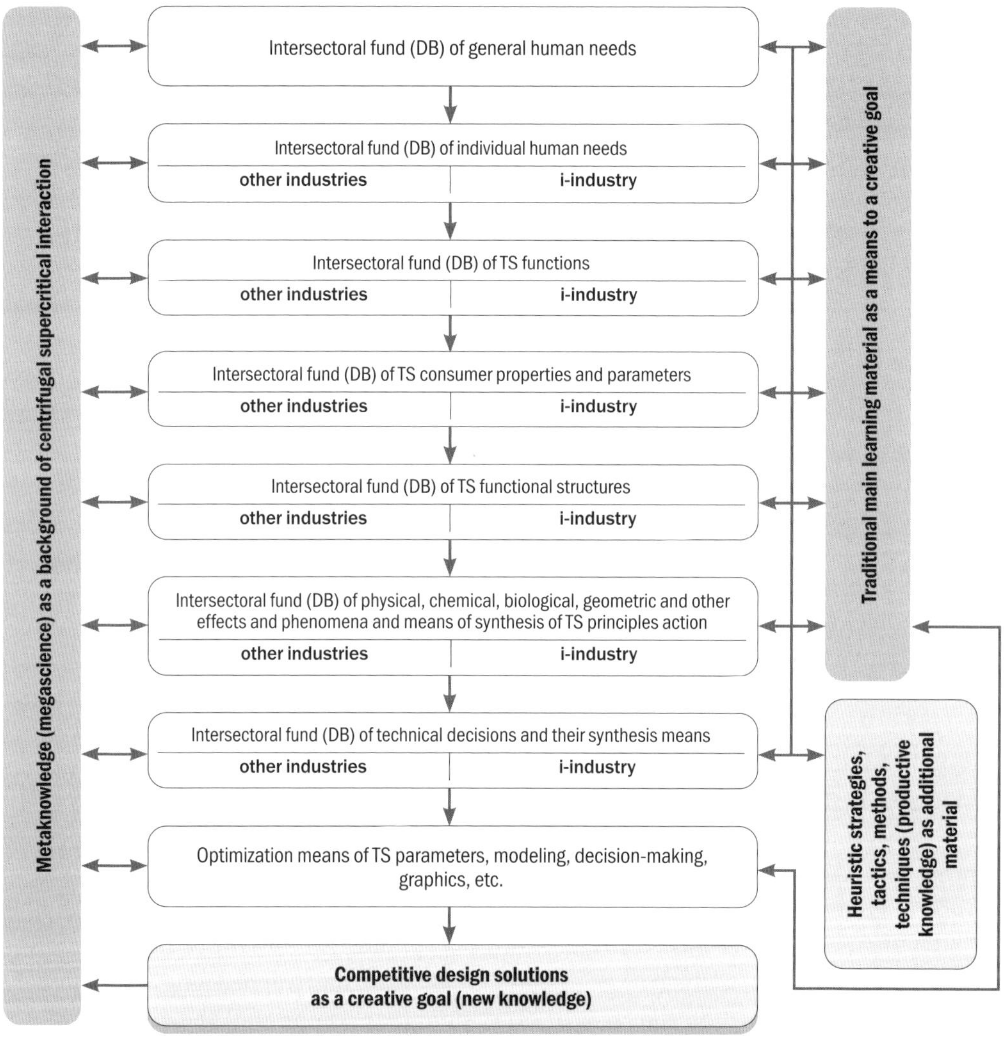

development of a competitive design solution. Metaknowledge (megascience) is used from stage 1 to stage 8 to maximize the implementation of the basal principle of heuristics. Resources "I" are industry knowledge that is mastered in stages 2–7. In the learning process, metaknowledge (megascience) is supplemented with new knowledge consolidated by students (and teachers) at the last (ninth) stage.

The sequence of stages of achieving a creative goal in creative (projective) pedagogy is as follows:

At **the first stage**, a need is selected from the intersectoral fund (database — DB) of general human needs, which must be fulfilled.

At **the second stage**, an individual human need is selected.

At the same time, a powerful mechanism for the formation of new needs is the use of the list of the most important human needs that I have developed in any analogs and prototype of TS. The analysis shows that almost all TS are created or qualitatively improved based on the implementation the most important human needs.

At **the third stage**, the search or synthesis of TS functions is carried out.

At **the fourth stage**, the search or synthesis of TS properties is carried out, while the ontology of TS physical properties developed under my leadership (includes about 1400 properties) can be used.

At **the fifth stage**, to form TS functional structures, elementary operations of R. Koller can be effectively used, allowing with the necessary level of abstraction to describe the functional structure of any TS, so as not to "drown" in the structural and technological details and easily highlight the main and secondary functions of TS under study.

At **the sixth stage**, the search or synthesis of TS operation principle is carried out, which determines the main goal of TS functioning and is the task of great combinatorial complexity. To search and synthesize TS operation principles, the automated system developed under my leadership can be used (see Sections 1.4.4 and 6.3.6).

At **the seventh stage**, the search or synthesis of TS technical decisions (TD) is carried out. In this case, you can immediately get competitive TD with the required parameters. The parameters are optimized at **the eighth stage**.

Moreover, the main learning material — traditional subjects (physics, chemistry, mathematics, biology and specialty subjects) — is used as a means of achieving a creative goal. And additional material, including productive knowledge (see table 5), is mastered at the first seven stages (stages of conceptual design) to increase the efficiency of creative processes and achieve a creative goal, a deeper acquisition of the studied material (subject), development of the creative potential of students, effective formation of competencies necessary for professional creativity and, finally, for obtaining new knowledge.

One can and should be taught scientific and technical creativity step by step from childhood in the family, and then continue learning throughout his life. 40 criteria for curriculum creativity can help this, thanks to which creative pedagogy will enter every schoolroom and every classroom.

7.4 Didactics of creative (projective) pedagogy

The development of creative (projective) pedagogy has required the formation of **a new creative (projective) didactics**[10], the difference of which from traditional (reproductive) didactics is shown in Table 35.

Table 35. Comparison of reproductive (traditional) and creative (projective) didactics[11]

Elements of vocational training	Type of didactics	
	Traditional (reproductive)	Creative (projective)
Type of training	Passive-contemplative and explanatory-illustrative	Active, based on the achievement of a creative goal and the formation of new knowledge in the learning process
Problem statement by the teacher	Object	Functional
Method or way of solution	As a rule, unambiguous and indicated by the teacher	The student chooses methods and solutions himself, and uses heuristic strategies, tactics, methods, techniques as well
Learning example	The teacher indicates the prototype	The student identifies the prototype himself, using analogies of properties and relationships in animate and inanimate nature, society
The result of the decision for the teacher	As a rule, unambiguous and known	As a rule, it is ambiguous and unknown (new knowledge is formed)

The most important distinguishing properties of creative (projective) didactics are the functional (and not the traditional object one) approach to the statement of the learning problem, as well as a change in the imperative in the relationship between the teacher and the student — from "follow me!" to "lead yourself!" (of course, the assistance of a teacher is not excluded). The creative goal is the development and / or implementation of any design that is essential in the life of the student, and not only in the educational process. At the same time, in contrast to mass education, it is becoming more and more personified.

In general, the development of creative (projective) pedagogy has shaped the formation and implementation of the following **new educational objectives:**

— The formation of a new mentality based on the belief that education not only consumes and replicates new knowledge, but is also the producer of new knowledge and information;

— The creation of new types of cognitive methodology that allows one to overcome barriers to thinking, develop the ability to critical, abstract, imaginative, creative and other types of thinking and thereby form productive knowledge that generates new knowledge in the learning process based on the achievement of specific creative goals[12].

Moreover, the **high creativity** of many **heuristics**, for example, generalizing functions (see Section 6.2.5), which determine the formation of a multitude of their implementations that provide a more effective convergence of knowledge, **leads to an important** logical **conclusion about the advisability of training broad specialists.** For example, instead of TS developers for railway, air, water, road transport, to train TS development specialists who implement the generalizing function of "moving" in all (not only transport) industries and even in the microcosm (however specialization should not be excluded). Of course, it is necessary to develop methodology and develop means of creative (projective) pedagogy taking into account their integration into existing education systems.

7.5 On the criteria for the creativity of curricula, textbooks and manuals

The high importance of the upbringing of creative individuals has led to the growth of pedagogical technologies of creative orientation, the formation and implementation of creative curricula on their basis, the creation of creative educational and methodological support of such programs. However, judging by the final results, the effectiveness of most creative educational programs and their educational and methodological support is insufficient.

This is largely due to a lack of clear criteria for the creativity of curricula, textbooks and manuals, which could be guidelines for educational researchers and practitioners in the selection, improvement and / or creation of new programs, textbooks and manuals with a creative focus. To eliminate this shortcoming, I have developed 40 criteria for the creativity of curricula, textbooks and manuals that are approved by the teaching community[13]. These criteria are given in Appendix 12.

7.6 Teaching children scientific and technical creativity

Scientific and technical creativity consists not only in the formation of new knowledge, the creation and transformation of objects, but also in (and this is the main thing) the improvement of creativity subject, that is, a man.

One can and should be taught scientific and technical creativity step by step from childhood in the family, and then continue learning at different levels of the educational system and further throughout his life. Teachers advise to develop creative thinking at the initial stages in a playful way, through various types of gymnastics (for the brain, fingerplays, articular, etc.).

Teachers should try to explain complex things to a child in simple language, from human needs to the systematic methodology of design activity — the basis of any human activity. To some extent, this confirms well-known statement of the famous physicist Ernest Rutherford: "If a scientist cannot explain what he is doing to the cleaning lady who washes the floor in his laboratory, then he himself does not understand what he is doing."

The main competencies that should be formed in children at the initial stages of teaching scientific and technical creativity are as follows:

- To identify and formulate the functions and properties of objects and phenomena;
- To find the differences and similarities of objects and phenomena;
- To classify objects and phenomena on various grounds;
- To identify and formulate contradictions in objects and phenomena;
- To make assumptions of a forward-looking nature;
- To form fantastic transformations of objects and phenomena;
- To form various implementations of the most creative adjective "smart", which characterizes the properties in objects and phenomena, etc.

When developing educational and methodological materials for teaching children scientific and technical creativity, it is useful to recall the brilliant universal answer of the great Russian writer Maxim Gorky to the aspiring children's writer, who asked: "Please advise me how to write for children?" — "In the same way as for adults, only even better!" — answered the master.

Of course, first of all, teachers, school teachers, teachers of colleges and universities should be trained in the methodology of scientific and technical creativity based on a new innovative culture.

7.7 Prejudices and delusions of humanity, protection from frustration and "collapse of the last hope"

A large number of informative and instructive articles and even books have been written about the stories of human prejudices and delusions in science, creative activity, including those among inventors. The church denied Galileo's assertion about the rotation of the Earth. Even such a great mind as Napoleon did not understand and rejected the first proposal of a steamer and a torpedo. Electric light was considered dangerous to the eyes, the telephone was destructive

There are many prejudices and misconceptions in science and creative activity. It is not without reason that Martin Heidegger said: "Humanity is heading back into the future!"

Optimism, patience, hard work, resistance to frustration are the most important qualities that support effective creative activity.

There is no "collapse of the last hope" — there is a way out of any situation.

to the hearing, aircrafts heavier than air were considered impossible, and the railroad was simply jeered. Not so long ago, the indivisibility of the atom was asserted, but it turned out that that boundary was also conditional. It is worth remembering the famous expression: "Everything that could be invented has already been invented" (see section 5.2.4).

The unjustifiably negative attitude of famous people and society as a whole towards certain scientific discoveries and inventions often led to *frustrations* of the authors of many outstanding (as life has shown) scientific discoveries and inventions, and sometimes even to the death of scientists and inventors. However, life has confirmed their greatness.

Inadequate perception of certain scientific achievements and inventions still happens. That is why the outstanding philosopher of the 20th century Martin Heidegger, with his natural paradoxical feature, exclaimes: "Humanity is heading back into the future!" — bearing in mind that psychological barriers to thinking, social stereotypes, personal failures (perceived too negatively) lead to the fact that, although humanity is moving into the future, but as if backward, constantly looking back at the past, which complicates the transition to the NEW.

Psychologists know that especially young people immediately start perceive many, even insignificant, as they say, "trifling" life troubles as "collapse of the last hope", which leads to strong and dangerous to health and even life experiences. After a while, almost all of these negative events and related experiences are reevaluated and revised. But due to the inability to prevent the state of "collapse of the last hope", one may not live to see a change in the assessment of a negative event.

With the development of the Internet, the situation has become even more dangerous, as many so-called "death sites" have appeared, which are pushing young people to commit suicide. For example in Russia cyber police shut down tens of thousands of these sites every year. But they reappear in many countries. Criminals take advantage of the insecurity of children who leave information about themselves on social networks.

That is why it is important to develop the constant sense of optimism, patience, hard work, resistance to frustration, which is the most important human quality that supports effective creative activity. For example, to prevent frustration arising from the inability to immediately fulfill any need, you should find a reasonable explanation that reduces or eliminates this need (instead of worrying about a lack of money and the inability to buy (for instance) grapes, convince yourself that "the grape is still unripe").

There are many examples from great people's experience who have found themselves in a situation of "collapse of the last hope." After overcoming despair and digging out themselves of the "hole", they have later remembered with a smile about their strong feelings on trifling occasions. In fact, there is never "collapse of the last hope" — there is always a way out of any situation. Therefore, according to the recommendations of psychologists, signs in the subway and other places with the inscription "No exit" (which have sometimes become a reason for suicide for people experiencing acute grief and being in depression) are changed to "Exit from the opposite side."

Chapter 8
G.S. Altshuller is certainly great, but ...

It is believed that Genrikh Saulovich Altshuller, whom I repeatedly mention, made a huge contribution to the development of the methodology of scientific and technical creativity. There is even a publication "Heinrich Saulovich Altshuller — da Vinci of the 20th century?" Despite the question, this is more of a statement. Maybe indeed, G. S. Altshuller is underestimated? Let's try to figure it out.

Genrikh Saulovich Altshuller (1926–1998), pseudonym Genrikh Altov, is the author of TRIZ-TRTS (the theory of inventive problems solving — the theory of technical systems evolution), as well as TRTL-ZhSTL (the theory of the development of a creative individual — the life strategy of a creative individual), inventor, writer.

In 1948 he wrote (together with R. B. Shapiro) a 30-page letter to Joseph Stalin sharply criticizing the state of affairs with invention in the USSR. In 1950 he was arrested by the USSR Ministry of State Security on charges of spying for British intelligence, without trial he was sentenced by the Special Meeting of the USSR Ministry of State Security to 25 years in prison. In 1954 he was rehabilitated by the KGB of the USSR.

In 1970, he established a school for young inventors in Baku, which in 1971 grew into AzOIIT (Azerbaijan Public Institute of Inventive Creativity) — the world's first TRIZ training center. From 1989 to 1998 — President of the TRIZ Association. On his initiative, in 1997, on the basis of the TRIZ Association, the International TRIZ Association (MATRIZ) was established[1].

Many books by G. S. Altshuller have been published many times in different countries and are still in demand.

In 1991, I interned at the University of Illinois (USA) and, since I was very keen on TRIZ, along with work on the topic of the internship, I was looking for any literature in the field of methodology of scientific and technical creativity. Already at that time there were electronic catalogs in the library of the University of Illinois, and American colleagues, at my request, found many publications on the

G. S. Altshuller, the creator and popularizer of the theory of the inventive problem solving (TRIZ), is certainly great, but ...

request of "methodology of scientific and technical creativity", "methods and techniques in invention", "how to create an invention", etc. Due to my poor English, with great effort it took many days for me to look through those materials using a dictionary. Much to my disappointment, it turned out that absolutely all (except one) of the books were completely devoted to the stories of the creation of inventions, often fascinating and to some extent instructive. But these were just stories. And only one book contained elements of the methodology of scientific and technical creativity: Altshuller G. S. Creativity as an Exact Science[2] ("Creativity as an exact science: The theory of the solution of inventive problems"). It was probably the eighth edition — a phenomenon extremely rare for monographs.

Much credit goes to G. S. Altshuller who **focused** on a little-known and little-understood (even by scientists and specialists) **paradigm** in all his books and speeches: **all TS do not develop chaotically, but on the basis of objective laws that are cognizable and can be used to resolve contradictions** in technical and in other areas **much more effective than** the widespread enumeration of options (most often unsystematic and known as **the method of "trial and error"**). In the final form, this paradigm was implemented by G. S. Altshuller in the form of the theory for solving inventive problems (TRIZ), including *heuristic algorithm* — the algorithm for inventive problems solving (ARIZ).

In general, a heuristic algorithm is a practical method that includes techniques and their sequence for solving a problem. The correctness of the method for all possible cases has not been proven, but in most cases it accelerates the solution of the problem and improves the quality of the result.

Initially, ARIZ was the sequence of rational steps in creative activity based on the use of heuristics, consisting of methods, standards, and special systematized information resources. Since 1959, for almost 40 years, G. S. Altshuller developed 16 ARIZ modifications. One of the latest modifications of ARIZ-85V contains 42 steps.

Here are the most important heuristics from the TRIZ point of view:

— The concept of the ideal final result (IFR) and justification of the feasibility of the formation of IFR as the first step in solving the problem;

— The concept of an aggravated contradiction (for GS Altshuller, the contradiction has another, more common name — "physical"), recommendations for the formation of "physical" contradictions and 11 standards for their resolution, three of which — to share contradictory properties in space, in time, as well as immediately in space and in time — in my opinion, are brilliant;

— Su-field analysis (the system of substance and field), including fepoli, tepoli, full su-field, incomplete su-field, complex su-field, harmful su-field, chain su-field, double su-field and 76 standards for their construction, destruction, development, folding su-field models;

- Laws of systems development, including the law of displacement of a man from the system, the law of the completeness of parts of systems, the law of energy conductivity, the law of coordination of systems parts rhythm, the law of increasing of systems ideality degree, the law of uneven development of systems parts, the law of transition to a supersystem, the law of transition from the macrolevel to the microlevel, the law of increasing the degree of su-field;
- TRIZ widely uses the concepts of "coordination" (direct, reverse, homogeneous, heterogeneous, internal, external, direct, conditional, buffer, collapsed, resource, dynamic) and "mismatch", as well as "collapse" and "deployment" of systems;
- The matrix (table) has become widespread, which makes it possible to form typical contradictions in technical systems and find heuristic methods (there are 40 of them) aimed at effectively resolving these contradictions;
- The method of little people is an original find that reduces the possible negative consequences of empathy (man's ability to identify him, for example, with a technical system);
- STC operator (size, time, cost) helps to overcome the psychological barrier to thinking;
- The proposal to use non-traditional resources (in the interpretation of G. S. Altshuller): time, space, fund of physical and technical effects, su-field analysis, physical and technical fields (mechanical, acoustic, thermal, chemical, electrical, magnetic, electromagnetic) was the creative basis for the formation and development of new heuristics;
- The theory of the development of a creative personality (TRTL), or the life strategy of a creative personality (ZhSTL) developed by G. S. Altshuller (together with I. M. Vertkin) is the extension of TRIZ[3].

It should be especially noted that the most important achievement of G. S. Altshuller (still not surpassed) — the introduction of millions of people (of different ages and social status) in many countries of the world to scientific and technical creativity. G. S. Altshuller, his incredible ability to work, great pedagogical gift and talent of a bright popularizer contributed to this, resulting in the emergence of schools and centers for the study of TRIZ, as well as domestic and international associations of TRIZ.

But (or however) there are quite a few concepts in TRIZ and TRTL that contradict the concepts generally accepted in physics, design, psychology, and even logic, which greatly reduces the usefulness of these theories.

A logical question: is it worth talking about this against the background of the impressive final result of G.S. Altshuller's activities? Moreover, he cannot respond to criticism. However, one day, I will not be able to answer the criticism in his and my address as well.

I am deeply convinced that constructive criticism can be an impetus for the further development of the methodology of scientific and technical creativity. That is why, being guided by the modern trends in the development of education, which consists in **changing the imperative** that characterizes the relationship between the teacher and the student, from **"follow me!"** to **"lead yourself!"** I will try to explain the existing contradictions clearly, so that the reader (whom I consider as a student in this case) can independently and impartially choose a point of view (mine or GS Altshuller's one) or form his own assessment of TRIZ.

1. The title "Theory of Inventive Problem Solving"[4] has no right to exist, not only because in the modern science, the recognition of the theory, as a rule, is the defense of theses that reveal and develop the theory (for almost 50 years there has not been a single thesis on TRIZ), but also because **inventive problems do not exist!**

I will try to prove the latter. Imagine that there is a technical problem (any of them) that students, housewives, workers, engineers, and academics have undertaken to solve. Someone will not be able to get any solution, others will have a variety of solutions. And only **solving a problem**, but after the following steps:

- Conducting a search for analogs of solutions to the problem in patent funds of different countries (by the way, such a search is also recommended when setting a problem to clarify and correct it);
- Analysis of the solution analogs to a problem;
- Revealing significant differences of the obtained problem solutions from the found patented analogues;
- Revealing the compliance of significant differences with the definition of inventions or utility models under domestic and / or foreign patent law;
- Patents application filing — **it is the solution of a problem that can be protected by patents for inventions, utility models, etc. However, everyone has had one and the same problem while solving it!**

Therefore, there is no so called "inventive thinking" as well. Otherwise, then you should admit that there is "useful model", "rationalization", "school", "household", "academic" and other types of thinking.

It's amazing that no one has said this before. This is a common misconception. At one time, I had to argue with a senior official of the regional education management body, who demanded to replace in the report on research work carried out

TRIZ is not a theory, and there are no inventive tasks!

under my leadership by order of the education management body, the "inventive tasks" available there (according to the official) with "rationalization" ones. But most officials are functionaries rather than scientists, so it is especially distressing to listen to arguments (with reference to TRIZ) about "inventive problems" and "inventive thinking" from scientists (including professional psychologists) with high academic titles and, it would seem, who know all about the types of thinking.

Of course, all this does not add glory to TRIZ, but introduces serious confusion, since many, preferring to engage in "inventive tasks", will look for something that does not exist. The same applies to "inventive thinking", which is demanded by educational authorities, but many people want to develop in their own way.

2. The concepts of "ideal" and "contradictions" were formed by the ancient Greeks. However, formulating the concept of IFR, G. S. Altshuller wrote: "An ideal system is no system, but its function is preserved and executed."

But what is the way to create something that does not exist?! If we put mysticism off, then there is always a system, but made in its own way. I consider it important to say it ones again that this contradiction can be easily resolved by adding words "**as if**". That is, one should read: "... it seems **as if** there is no system, but ... (you can further keep to the text)".

3. The term "physical contradiction"[6] is improper, since according to GS. Altshuller, it has a special creative meaning of an aggravated contradiction (in this book I call it "intensified"), which differs from the classical "physical contradiction". All this brings confusion, especially since G. S. Altshuller's classical biological, chemical contradictions are also called "physical".
4. Su-field analysis[7] has become my pain, I have been looking for the appropriate approach five times! But I have not been able to understand the logic and methodology of this analysis in order to accept and use or criticize it. At the same time, I have constantly recalled the humorous statement of one of my teachers — the outstanding scientist in the field of mining, Doctor of Technical Sciences, prof. Vladimir Rafailovich Imenitov (I was counted as his 41st candidate of technical sciences): "If I don't not understand this, then anyone will hardly understand it, and I am not the last fool in this world!" It should be added that for half a century I have happened to meet many "TRIZ members" who enthusiastically have talked about Su-field analysis, but I have not met a single one who would have actively used it in practice.
5. Regarding the laws of systems evolution[8], I would like to note the following. The fundamental law of the stage TS development, formulated in 1972 by S. S. Tovmasyan and logically explaining the emergence and TS development, G. S. Altshuller presented exactly the opposite — as the law

of displacing a man from the system. At the same time, it is not clear how to explain the displacement of primitive man from TS when there is no TS.

In the books of G. S. Altshuller, there is no TS taxonomy, or at least the list of TS types representing their diversity (as, for example, in Fig. 1), so it is difficult to imagine the operation of such laws of system evolution as the law of completeness of systems parts, the law of energy conductivity, the law of harmonization of systems parts rhythm, the law of uneven development of systems parts, the law of transition from the macrolevel to the microlevel, in such TS, as, for example, hand tools, technologies, processes, operations, artificial biological objects, structural materials (see Fig. 1). And in the absence of a clear concept of "su-field", it is impossible to comprehend the law of increasing the degree of su-field.

So far one has not managed to meet a scientist (in the fields of general physics, in particular mechanics) who would confirm that the mono-bi-poly-mono transition law for TS objectively exists. Moreover, even its name makes many specialists dumbfounded.

6. The methodological usefulness of all ARIZ modifications is greatly reduced by the fact that G. S. Altshuller did not operate with such important concepts in scientific and technical creativity as human needs, TS design (including TS design stages). Instead of these concepts, he used the terms "physical solution", "technical solution", "computational solution", "mini-problem", "maxi-problem"[9] "stage", "operational stage", "synthetic stage". If the reader sees objectively existing stages of TS design in these ARIZ stages, then he is lucky.
7. The theory of creative personality development (TRTL), or the life strategy of a creative personality (ZSTL), consists, in essence, of six correct, but rather trivial statements about what distinguishes a creative personality[10]:
   - A worthy goal;
   - Plan for achieving the goal;
   - High efficiency, persistence;
   - Excellent problem solving technique;
   - The ability to take the heat;
   - Effectiveness.

That's the whole theory. In fairness, it should be noted that the book "How to Become a Genius" contains many fascinating and instructive stories that illustrate these claims. However, many conclusions are perceived with difficulty, as, for example: "... using the TRIZ terminology, the secret of great efficiency and versatility of achievements is in complete closing down of creative work"[11].

8. Taken as a whole, it should be noted once again that in all his books, G. S. Altshuller, probably wishing to popularly and clearly explain his own

methodology (desire for good), introduced many concepts and terms that contradicted the modern scientific terminology and the scientific picture of the world. I give a far from complete list of such concepts and terms, judge for yourself: "information conductivity"[12], "the first level — the smallest inventions"[13], "scattering fields", "mechanical fields", "chemical fields", "sound fields", "deployment" and "closing down", "coordination" and "mismatch" of systems[14], "suprateen", "x-element", "superficial contradiction (administrative)"[15], "x-factor"[16], "anti-function"[17], "monosystem", "bisystem", "polysystem"[18], "supereffect" , "complete" and "incomplete" TS[19].

However, I am convinced that even a child (the books "TRIZ in Kindergarten" have already appeared) should not, for the sake of clarity, say, for example, "bibika" instead of "car".

Unfortunately, the followers of G. S. Altshuller during his lifetime and after his departure, have not added something useful to TRIZ. Basically, those were bad "rehashes" of TRIZ, including MATCEM (an abbreviation for memorizing, in essence, a false pattern of TS development due to changes in fields).

9. And finally, the last — personal admission. When I read the book by G. S. Altshuller's "Creativity as an Exact Science"[20], I was just delighted with such a striking comparison of creativity with the exact science and I retained this admiration for many years until I met the collection of stories by O. Henry, named after one of his famous stories **"The Exact Science Of Matrimony"**[21]. It was not the first edition in Russian; in 1959, a two-volume edition of O. Henry's Selected Works was published. G. S. Altshuller probably read it. And although no one can say for sure, my sense of admiration has diminished a little.

And yet, returning to the huge positive contribution of G. S. Altshuller in the development of scientific and technical creativity in the world, it should be stated: **G. S. Altshuller, of course, is great!** Without any "buts".

Noting the enormous contribution of G. S. Altshuller in the development of scientific and technical creativity, involvement in active creative activity of millions of people around the world, it should be stated:

G. S. Altshuller, of course, is great! Without any "buts"

Conclusion

The book presents the results of many years of work on the development of the methodology of scientific, technical and other creativity in any field, which essentially consists of choosing (sometimes instantaneously and unconsciously) a design, but more often — in the formation of a design, and then in its implementation.

To rationalize the processes of creative (design) activity, a system methodology of design activity has been developed, including 8 stages, 3 strategies and 3 basic tactics. The first 7 stages relate to conceptual design, in the course of which qualitative characteristics of the design object are formed, determining the success or failure of the design as a whole.

The book convincingly demonstrates that a sharp increase in the efficiency and effectiveness of conceptual design is ensured by the use of heuristics (principles, strategies, tactics, laws and patterns of technical equipment development, technologies, products — technical systems, methods) — productive knowledge, supplied with methods for their use, instead of the "trial and error" method that is widespread in the world.

Particular attention is paid to the creative (designed) pedagogical technology developed by the author, aimed at training creative specialists who are able to go beyond the scope of reproductive activity, generate many effective ideas and implement them with high productivity, creating goods and developing services with high consumer properties. The following 10 criteria of creativity of educational programs, textbooks and teaching aids will help in this; they can serve as guidelines for educational researchers and practitioners when choosing, improving or creating new programs, textbooks and teaching aids with a creative orientation.

The author's patented device for personal growth training "Find an idea!", attached to the book, made in the form of a bookmark, is of great interest and practical value for the book's readers (adults and children).

The book provides biographical information and an objective description of the activities of the outstanding methodologist of scientific and technical creativity

Genrikh Saulovich Altshuller, the author of the Theory of Inventive Problem Solving (TRIZ), whose books, published in many countries, have had a great influence on the development of scientific and technical creativity in the world.

However, after 50 years, these books, in the opinion of many methodologists, have become morally outdated and require serious adjustments.

Therefore, one of the main goals of the book "Mental Karate: Methodology of Scientific and Technical Creativity" is to replace the books of G. S. Altshuller, while preserving and developing all the best of his methodological achievements, which have stood the test of time.

I am sure that this book with heuristics and the "Find an idea!" device will be a unique instrumentarium for the effective growth of creative potential, the formation of the foundations of a new innovative culture in each reader, which will also ensure an increase in the quality of their lives.

Appendix 1
Causes of occurrence and main groups of protection needs

Causes of occurrence of protection needs		Protection needs		
general	individual	general	individual	technically implementable
Insufficient protection from natural disasters and increasing severity of their consequences	The unexpectedness of a natural disaster. Destruction of buildings due to tsunamis, earthquakes, etc. Large human losses associated with the inability to free people from the rubble quickly, deliver them to a safe place, etc.	The need for protection from natural disasters and reducing severity of their consequences	The need for early warning of a natural disaster. The need to prevent the destruction of buildings and structures. The need to release people from the rubble quickly and send them to a safe place, extinguish fires, etc.	Creating of: — effective methods and means of forecasting and early warning of natural disasters; — high-strength materials, resistant structures; — perfect means of communication; perfect technologies, — equipment and products for the rapid elimination of the consequences of natural disasters
Growth in the number of dangerous man-made accidents and the severity of their consequences	The unexpectedness of a man-made incident. Large human losses associated with the inability to free people from the rubble quickly, deliver them to a safe place, extinguish fires, decontaminate the territory, etc.	The need for protection from man-made accidents (their prevention) and reducing the severity of their consequences	The need for prevention or early warning of a man-made accident. The need to quickly release people from the rubble, send them to a safe place, extinguish fires, decontaminate the territory, etc.	Creating of: — perfect means of control and protection of existing TS; advanced technologies, equipment and products for the elimination of the consequences of hazardous man-made accidents; — new generations of TS with increased reliability

Causes of occurrence of protection needs		Protection needs		
general	**individual**	**general**	**individual**	**technically implementable**
Sharp deterioration in the environmental situation	Natural disasters, man-made accidents, unnatural migrations of insects, animals, ecological imbalance, etc.	The need to improve the environmental situation	The need for prevention or early warning of natural disasters, man-made disasters, unnatural migrations of insects, animals and possible consequences, restoration of the ecological balance	Creation of environmentally safe TS, as well as TS for eliminating negative environmental consequences
Military threats	Local conflicts between countries. Internal political and / or economic instability in the aggressor country	The need to eliminate (reduce) military and other threats	The need to create or strengthen border troops. The need for early detection of the enemy and his actions	Development and use of technologies and means for the peaceful resolution of conflicts. Quantitative increase in military potential
	Desire to test military technology and equipment in real conditions; strengthening of religious fanaticism, etc.		The need to strengthen the troops with new generation military TS, which have various purposes, etc.	Qualitative improvement of military TS and / or creation of new generation military TS, having various purposes
Outbreaks of dangerous diseases	Natural disasters, man-made accidents, military actions, as a result of which reservoirs, treatment facilities have been destroyed, epidemics, etc.	The need for protection from dangerous diseases (their prevention)	The need for clean water and non-contaminated food. The need for effective methods and means of treatment	Creation of effective methods and means of prevention and treatment of diseases (including means of decontamination and purification of water and food)
Lack of food and drinking water	Living in areas unsuitable for agriculture, cattle breeding, poultry farming (including due to a lack of water sources). Prolonged drought. Ineffective economic activity in the field of agriculture, cattle breeding, land reclamation, irrigation, etc.	The need for sufficient amounts of food and drinking water	The need for the creation and long-term storage of food and water supplies. The need for early drought forecasting. The need for efficient agriculture, cattle breeding, land reclamation, irrigation, etc.	Creation of technologies and means for long-term storage of food and drinking water supplies, early forecasting of drought, efficient farming, cattle breeding, poultry farming, land reclamation, irrigation, etc.

Causes of occurrence of protection needs		Protection needs		
general	**individual**	**general**	**individual**	**technically implementable**
Rise in crime	Poor socio-economic conditions and upbringing. Imperfection of laws. Late detection of crime signs. Imperfect locking devices or their absence. Large number of unsuccessful attempts to neutralize, arrest criminals. Mass escapes of prisoners during their transportation, as well as from prisons	The need to reduce crime	The need for social crime prevention. The need for early detection of signs of crime and criminals. The need for reliable protection of objects and people's lives. The need for effective neutralization, detention, transportation and maintenance of criminals	Creation of effective methods and means of crime prevention, neutralization, detention, transportation and keeping criminals
Increase in terrorist attacks and severity of their consequences	Poor socio-economic conditions, insufficient education and upbringing of people. Imperfection of laws. Aggravation of racial and religious conflicts. Absence or imperfection of methods for detecting terrorists and impending terrorist acts. Large human losses due to the inability to release the hostages quickly, rescue people from the rubble, deliver them to a safe place, extinguish fires, etc.	The need to exclude terrorist acts, as well as reduce the severity of the consequences	The need for social prevention of terrorist acts. The need for early detection of terrorist acts and terrorists. The need to release hostages, people from the rubble quickly, transport them to a safe place, extinguish fires, etc.	Creation of perfect TS for the prevention, detection and elimination of the consequences of terrorist acts
Unwanted climate changes	Natural disasters, man-made disasters. The negative impact of human activities on the climate: unjustified deforestation, irrigation or reclamation, etc.	The need to preserve the climate, as well as high standard of living	The need to prevent man-made disasters and natural disasters (for example, the fall of a large meteorite on the Earth). The need for safe technosphere	Creation of TS, preventing unwanted climate changes, as well as adequate to new conditions

Causes of occurrence of protection needs		Protection needs		
general	**individual**	**general**	**individual**	**technically implementable**
Negative consequences associated with the geographic isolation of the country, region	Large human and material losses due to a lack or imperfection of communications and transport infrastructure, providing communication between the region, the country with neighboring regions, countries	The need to reduce the isolation of the region, country	The need for modern communications and transport infrastructure, providing communication between the region, the country with neighboring regions, countries	Creation of telecommunication technologies and means, transport infrastructure, reducing dependence on the isolation of the country, region

Appendix 2
Methodology for carrying out functional-physical (chemical, biological) analysis and synthesis[1]

To provide information support for the conceptual design of TS, it is necessary to form a fund of descriptions of the most effective TS of various industries, which can serve as prototypes for solving various problem situations. At the same time, each TS, before being included in the fund, must be subjected to functional-physical (chemical, biological) analysis and synthesis to identify various types of productive knowledge, orienting a designer in the direction of the most probable ways to improve the prototype or create new TS generations.

The developed methodology for such analysis and synthesis assumes the following sequence of actions.

1. To give a short verbal description of TS, a diagram (indicating sources of information).
2. To identify the principal, main, auxiliary, harmful and neutral functions.
3. To identify and describe realizable TS consumer properties, which characterize the measure of its usefulness, perfection, technical level and competitiveness. To indicate quantitative properties characteristics.
4. To form TS functional structure, highlighting new (in comparison with the prototype) elements in it.
5. To reveal CPD of TS.
6. To reveal the heuristic techniques implemented in TS.
7. To reveal physical, chemical, biological, geometric and other effects used in TS. To give examples of TS in which the identified effects are realized. To indicate sources of information.
8. To identify technical and formulate intensified contradictions in TS.

9-11. To formulate the conclusions (consequences) of laws of stage manifestation of TS development, the correspondence between the functions and structure of TS, progressive evolution of TS.

12. To identify the patterns of TS development, which can be used for improving TS.
13. To identify non-traditional resources hidden in TS and TS environment, and formulate recommendations for their use.
14. To reveal TS analogies in another TS, animate and inanimate nature.

Appendix 3
Intersectoral fund of heuristic techniques[1]

1. Form transformations

Name	Examples of using
1.1. To go from symmetrical form to asymmetrical one. Technique inversion	In the hopper, the axis of the outlet of the lower conical part is offset relative to the axis of the cylindrical part, which prevents free-flowing material from hanging
1.2. To shape the object under load to resist the load (for example, convex form)	The arched shape of the elements of bridges, dams, fulcral walls and other structures provides effective resistance to loads
1.3. To use a circular shape, a ball and its parts (sector, segment, layer, Euler triangle)	Ball-shaped wheels provide moving the vehicle in any direction
1.4. To make holes, cavities, recesses in the object. Technique inversion	Building blocks, bricks with cavities have a lower mass, increased heat and sound insulation properties
1.5. To turn the shape of an object (element) inside out (change the traditional shape)	Ceiling of buildings, made in the form of saddle structures (instead of domed)
1.6. To use film coatings instead of spacious structures	Electric heater made of conductive film (instead of spiral, rod heater, etc.)
1.7. To move from flat surfaces to curved forms	The convex shape of the bow of the ship below the waterline (Yurkevich's bulb) reduces the wave drag when moving at high speed
1.8. To shape load bearing element concave or trough	Scoop water-lifting machines. Belt conveyor for moving bulk materials
1.9. To use the property of elastic shells to change shape under the influence of external loads	Airbag in the car
1.10. To give a line object a helical shape	Screw timber flume through the dam (instead of an elongated hump)
1.11. To perform elements connected to each other to transfer movement, in the form of a Reulcaux triangle, "equiaxial contour", hypo- and pericycloids	Shaft profile connection. Cam mechanisms for various purposes
1.12. To change the shape of an element at different stages of work	Dozer blade on an "elastic" base that changes shape as the load changes
1.13. To find the globally optimal shape of the main object (element)	The main element of the orbiting space station, made in the form of a torus
1.14. To harmonize the shape of various elements with the environment	Lamps on trees are made in the shape of their fruits

2. Structure transformations

Name	Examples of using
To exclude element with retaining basic functions by the object	A car with a semitrailer almost fully performs the functions of a car with a trailer
2.2. To exclude an element from the process temporarily when there is no load on the element	In an empty car, some of the axles with wheels are raised above the road.
2.3. To remove "extra elements" so that one element would perform the functions of several, even if part of the effect will be lost	Thermal air curtain of a doorway instead of a double-sided side one
2.4. To attach an additional specialized tool, instrument, etc. to the base object.	A bow saw, which is additionally equipped with a lower row of teeth, provides a cleaner cut in fibrous materials
2.5. To perform a rigid connection with the possibility of changing the rigidity	Antenna of solid bushings connected with a cable gains a certain flexibility
2.6. To combine objects that have an independent purpose into one whole	Combine harvester
2.7. To combine or unite traditionally incompatible objects, eliminating emerging contradictions	Air conditioner – a device that acts as a heater and air cooler
2.8. To use the principle of aggregation. Create a basic structure on which to "hang" the working bodies, units	Offshore fixed or floating platforms for well drilling and oil and gas production
2.9. To make the object removable, easily detachable	Ejection pilot seat
2.10. To split the whole object into smaller homogeneous elements that perform similar functions	The hold of the ship is divided into compartments, as a result of which the unsinkability of the ship is ensured
2.11. To divide the object into parts; process each part separately	Lighter carrier
2.12. To divide the object into parts: "hot" and "cold"; isolate one from the other	The electric motor of the refrigerator, which heats up during operation, is placed outside its body (mounted on the back wall)
2.13. To make an object as a compound structure	Accumulator battery of several accumulators for obtaining electric current with specified parameters
2.14. To bring energy sources closer to executive bodies and working areas	Heavy-duty mining dump truck with independent drives for each wheel (motor-wheel)
2.15. To divide a frequently damaged object into parts	Knife blade, consisting of separate cutting sections, which are removed one by one when they are abraded or damaged
2.16. To introduce safety devices or blocking	"Safe" electrical outlet with offset plug sockets
2.17. To equip the object with pre-prepared emergency equipment	Mine lift equipped with a parachute emergency braking system in the event of a cable break
2.18. To discard the element that has fulfilled its purpose or has become unnecessary (unhook, burn, dissolve, evaporate, etc.)	External aircraft fuel tanks discharged after running out of fuel

Name	Examples of using
2.19. To go from discrete to continuous volume shaping. Technique inversion	Use in construction of in-situ concrete instead of bricks, blocks, etc. Transition to pipeline transport
2.20. To divide the object into parts, move only the lighter and more compact part	Aerial cableway: the drive is stationary, only light trolleys can be moved
2.21. To replace several specialized objects with one universal one	Universal electric motor for both direct and alternating current
2.22. To use a "mediator" object (carrier of energy, force, mass, etc.)	Driver in impact machines (construction copra, forging and stamping hammers)
2.23. To create an easily used stock of working bodies and elements	Drill rods cassette feeding

3. Transformations in space

Name	Examples of using
3.1. To change the traditional orientation of an object in space	Storage of boats on a sports base placing them in an upright position
3.2. To use free space between elements of an object (one element goes through cavities in another element)	The cable for supplying current to electric drill motor is located in the hollow space of drill pipes
3.3. To place objects (elements) one inside the other according to the principle of "nesting dolls"	Loader buckets stacked one into another, etc.
3.4. To go from point contact to line contact, from line contact to surface contact, from surface contact to bulk contact. Technique inversion	Propeller transformation: wheel – caterpillar – air cushion. Heating devices: soldering iron – frying pan – microwave oven
3.5. To bring the working bodies of the object closer to the place where they perform their functions without moving the object itself	Dewatering plant with submersible pump
3.6. To arrange objects in advance so that they can come into action from the most convenient place and without wasting time on their transportation	Fire safety system, the elements of which are distributed throughout the premises and begin to function when a fire signal is received
3.7. To move from sequential placement of elements in space to parallel or mixed. Technique inversion	Arrangement options for ore washing machines: sequential-selective flotation; collective flotation; collective flotation with subsequent selection
3.8. To select "volumetric" and "non-volumetric" parts in the object; move the "non-volumetric" part outside the boundaries, limiting the volume	Aircraft with external fuel tanks
3.9. To move elements exposed to harmful factors outside their areas of action	Position the air intakes of the aircraft engine in areas of stabilized air flows, for example, under the wings
3.10. To change the relative position of elements that perform different functions	Change the location of the tug and barge (the tug becomes the pusher)
3.11. Use constructions that include expandable or variable elements in space	Extended column of drill pipe

4. Transformations in time[2]

Name	Examples of using
4.1. To go from a stationary in time regime to a changing one.	The area and configuration of the elements for capturing the air flow by the air intake of the aircraft engine are changed at different stages of flight.
4.2. To go from continuous supply of substance (energy) or continuous action (process) to discrete periodic or pulsed one. Technique inversion	Impulse rocket engine for orientation and stabilization of spacecraft flight, in which fuel injection and heating of the working fluid are performed discretely
4.3. To eliminate useless (harmful) time intervals. Use pauses between periodic actions to perform another action	See section 6.3.2 for examples with jaw and cone crushers in fig. 148–152
4.4. To move from sequential execution of operations to parallel (simultaneous). Technique inversion	The bucket wheel excavator simultaneously carries out the operations of breaking (separating), loading and transporting the rock mass
4.5. To combine technological processes or operations. Technique inversion	Concrete mixer truck: combined operations of mixing and transporting concrete
4.6. To use the principle of coordinating the operation of object elements in time. Move from an asynchronous process to a synchronous one. Technique inversion	Shooting through the propeller with synchronization of the frequency of shots and the speed of blades rotation
4.7. To use the principle of time (frequency) separation of the operating conditions of object elements (flows, liquid and granular bodies, signals, etc.)	Frequency, time, code and combined methods of signal separation in telemechanics
4.8. To use the principle of multiplexing (separation and sharing resources)	Different liquids are pumped through the same pipeline using separators (piston, ball, etc.)
4.9. Go from a stationary physical field to a time-varying one. Technique inversion	Pulse soldering. System of recording and reproduction of information by means of a changing magnetic field

5. Transformations of motion and force[3]

Name	Examples of using
5.1. To replace a forward or backward motion with a rotary motion. Technique inversion	Circular saw instead of strip saw. A cone crusher in comparison with a jaw crusher
5.2. To eliminate or reduce reverse and intermediate motions	Milling processing in comparison with planning one. Rotary line instead of conveyor
5.3. To replace the traditional motion trajectory with another, more efficient one	Rotary line instead of conveyor
5.4. To change the direction of motion, including to the opposite one	Reverse thrust propeller or jet engine reduces landing distance the of aircraft
5.5. To change the working conditions so that you do not have to raise or lower the object being processed	Processing an object with a "movable" tool: with flexible drive, chemical, thermal, beam (laser) methods, etc.

Name	Examples of using
5.6. To replace sliding friction with rolling friction. Technique inversion	Roller conveyor – a device that makes it easier to move the load compared to dragging
5.7. To divide an object into parts that can move relative to each other	Flaperon (aileron – flap) of the aircraft is performed in the form of two rotary elements to increase the controllability and bearing capacity of the wing
5.8. To make moving elements stationary ones and stationary elements – moving ones	The gas burner of a melting furnace with a variable torch orientation, compared to a stationary burner, provides a more uniform and faster heating of the metal charge
5.9. To change working conditions so that hazardous or "harmful" operations are carried out at high speed. Technique inversion	The turbine on-speed is carried out skippingly when passing through resonant frequencies
5.10. To compensate for the action of the mass of an object by connecting it with an object with a counter-force (in particular, a lifting force)	Counterbalance lift
5.11. To change the direction of the operating force and environment	When breaking rocks with a wedge cut, the property of rocks is used to collapse with lower efforts for separation and shear compared to compression
5.12. To replace mechanical operation in a way without metal removing	A method for removing a metal layer by local heating a workpiece (part) with a plasma jet
5.13. To pre-expose the object to a strain opposite to unacceptable or undesirable strains arising during operation	The roof of the mine, subject to stretch deformations, is pre-strengthened with anchors with their preliminary stretching, which increases the stability of the roof
5.14. To use methods to intensify motion. Technique inversion	The production of metal powders is carried out by eddy spraying of a stream of metal melt, the intensity of which is enhanced by the rotation of the eddy flow around its axis
5.15. To use methods for intensifying mixing, unloading bulk materials, tilting products, etc.	A mixer in which the blades are not made in the form of a propeller, but in the form of a Mobius strip, as a result of which, within one revolution, the point of contact of the blade with the product changes by 360 degrees
5.16. To apply methods of moving material by pushing, pressing out (intrusion, extrusion, plugging, etc.)	Extruder – a machine for the production of products from polymeric materials by depression through a device with a hole of a certain section (head)
5.17. To use the principle of backpressure, neutralization (full or partial) of pressures and loads in adjacent elements of the object	Gas storage tanks of equal size and increasing capacity, placed coaxially into one another
5.18. To change the load distribution scheme: distribute the concentrated (point) load along the line, linear – along the plane, planar – along the volume	Tracked vehicles, air cushion
5.19. To use combinations of loading and multidirectional motions of the tool and the target object	Methods of flat and volumetric vibration abrasive processing of items

6. Transformations of material and substance

Name	Examples
6.1. To remove needless material that does not carry a functional load	Hollow cross section box beam instead of solid beam
6.2. To change the material properties of the object (element) surface	Surface hardening of metal parts by thermal, thermochemical and / or mechanical methods
6.3. To make the element or its surface from a porous material. Fill the pores with a substance with the desired properties	In a gas tank filled with honeycomb (cellular) material, there is less danger of gasoline ignition
6.4. To use parts made of material with post-hardening properties	Products made of concrete, reinforced concrete, steel concrete, epoxy resins (blocks, piles, anchors, insulating parts, etc.)
6.5. To replace some elements of the environment with elements having different physical or chemical properties	Artificial freezing of the water phase of the rock mass changes its filtration and strength properties, which is used when driving shafts, metro tunnels, etc.
6.6. To use flexible shells and films instead of rigid volumetric structures. Technique inversion	Mine working support, made of pneumatic shells (cushions), instead of support made of rigid elements
6.7. To replace an expensive durable element with a short-lived, less expensive one	Barrier made of hollow polymer elements filled with water instead of concrete blocks
6.8. To use materials that reduce waste in the manufacture of parts	For example, go from cutting parts method to molding parts method
6.9. To use reinforced, composite, porous and other new effective materials	Concrete products and structures reinforced with steel reinforcement (rods, wire) have higher strength, especially tension strength
6.10. To replace TS environment	Cleaning nuts and seeds from shells, making powdered sugar from granulated sugar in chambers where high gas (air) pressure is created, and then abruptly released
6.11. To switch to waste-free technologies (for example, to receive waste materials in a form suitable for the manufacture of other parts, products, etc.)	When molding parts from thermoplastics, the gating channels in the mold are made in the form of useful products (door handles, fastening brackets, letters, etc.)
6.12. To observe poorly visible objects or processes, use coloring additives, labeled atoms, etc.	Ferro-powder is added to the fluid used for hydraulic fracturing. Results are evaluated by magnetic logging
6.13. To use intermediary materials that change the conditions for the interaction of object elements	Glass plates, "glued" into a block (with glycerin, water, etc.), are subject to less risk of destruction during mechanical processing and transportation
6.14. To use the properties of a substance or material to change its state and characteristics under external influence	Chameleon glasses. Shape memory materials, alloys, polymers
6.15. To use substances that regulate the interaction of object (environment)elements	Surface active substances with wetting, emulsifying, washing and other valuable properties are used in ore washing, finishing and dyeing fabrics, etc.
6.16. Transfer (place) an object or its element into another environment that excludes the action of harmful factors	Hydroabrasive processing of parts (abrasive material is supplied with an aquatic medium, which eliminates dust formation)

7. Quantitative changes

Name	Examples of using
7.1. To change abruptly the parameters of object (elements), environment properties	"Blowing the cold" – instant cooling of a red-hot part by immersing it in liquid nitrogen – increases its strength tenfold
7.2. To change the size of the object during storage, transportation	Inflatable pontoons for shallow water pilotage. Folding bike
7.3. To allow a slight decrease in the required effect with a significant decrease in cost	Replacing cement with other binding materials, especially those to be buried, somewhat reduces the strength of the resulting concrete, but significantly reduces its cost
7.4. Increase the intensity of technological processes in the working area (in a limited area or in a confined space)	The effects of vibrations of sound or ultrasonic frequency on the working area intensify the processes of cleaning gases and waste water, drying bulk materials, medicinal substances, fruits, degassing liquids, obtaining suspensions, emulsions and aerosols, homogenizing food products, etc.
7.5. To increase by times the length of the active face or the contact area with unchanged dimensions due to the use of the properties of the Möbius strip	Double-sided sanding belt

8. Combined techniques

Name	Examples of using
8.1. To use sliding, folding, prefabricated, inflatable structures that change the shape, dimensions and capacity of the object	Telescopic antenna, retractable fire escape ladder, escalator, folding-bed, aerostat, radiosonde, etc. Pneumatic building structures
8.2. To replace elements that are dissimilar in material and shape with one standard or unified element	Velcro in comparison with traditional zipper
8.3. To use liquid or gaseous elements instead of solid ones	Aircraft jet ailerons instead of solid material ailerons
8.4. To attach properties to the object that are opposite to invalid or exclude unwanted changes that occur during operation	Wire brush capable of processing metals, plastics, rubber without getting greasy. Antifreeze is a coolant that does not freeze at low temperatures. Fuel dope
8.5. To attach to the object new properties (for example, to ensure its floatability, tightness, self-restorability, versatility)	The flight strip for emergency landing of the aircraft on the fuselage should be made in the form of a giant brush
8.6. To create closed waste-free technologies with utilization or returning pollutants to production	Regeneration of drilling fluids with the return of water and useful fillers into circulation (weighting agents, surface active substance, etc.)
8.7. To eliminate the influence of a harmful factor by compensating it with another harmful factor	Harmful acidic components of waste gases from thermal power plants are neutralized by absorbing them with alkaline and waste waters of hydroslag – ash disposal of the station

Name	Examples of using
8.8. To use or accumulate braking and other side energy	Recovering electrical energy when braking electric trains, moving downhill of open-pit trolleys, etc.
8.9. To use waste energy (substance), unnecessary element having fulfilled its purpose for other purposes	Thermal waters used to generate electricity are sent to heat greenhouses, etc.
8.10. Object specialization	For example, vehicles: cars, trucks, buses, ambulances, fuel trucks, firefighters, sweepers, etc.
8.11. To change harmful factors so that they are no longer harmful	Using recoil energy when fired to reload weapons
8.12. To introduce additional elements to ensure their correct connection when assembling an object	Electrical connector with "keys" to prevent misconnection
8.13. To use production waste for useful purposes	Waste rock and waste of metallurgical production (slags) are used for the manufacture of building blocks, dumping of road beds, etc.
8.14. To use the ability of an element to change its properties as a result of the transition from a stationary state to a mobile one	The appearance of an electric current in a conductor moving in a magnetic field
8.15. To use the ability of an object (element) to change properties with the change of the objects (elements) state or properties interacting with it	The device that brings the fuse of an underwater mine into a firing position contains a piece of refined sugar, which gradually dissolves
8.16. To replace the mechanical system with an acoustic, thermal, chemical, electrical, magnetic, odorant, or a combination of both	Filter indicator that emits a strong odor when clogged
8.17. To make friction surfaces of separate easily replaceable parts	Replaceable plain bearing shells made of thin-walled bimetallic strips

9. Conversions by analogy

Name	Analogue	Solution
9.1. To use the principles, properties and signs observed in living and inanimate nature related to the following properties or states:		
object shape	Tortoise shell; cave vault	Vaulted ceiling (dome) of spectacular, religious buildings, etc.
	Streamlined bodies of aquatic environment inhabitants (dolphins, sharks, etc.)	Streamlined hulls of submarines and vehicles (boats, torpedoes, etc.)
object structure	Bamboo trunk	Hollow structures with reinforcement elements

Name	Analogue	Solution
the arrangement of object elements in space	Repeatability of the same type of elements in the structure of cells, honeycombs, leaves, crystals, etc.	Volumetric blocks of buildings; elevator bins; storage compartments, holds of ships, etc.
	Spiderweb	Trampoline. Fishing nets. Bonded non tissue material
change of an object in time	Anabiosis – hibernation of some animals	Artificial hypothermia
moving an object and changing its position	The ability of some aquatic environment inhabitants (dolphins, whales, sharks) to control their hydrodynamic characteristics	Aerodynamic and kinematic balancing of the rudders, balancing the loads on the elements of the control surfaces of the aircraft
	Variable, energetically beneficial flight of a bird flock "wedge"	Aircraft elements providing flow "along a liquid contour" instead of flow around "along a rigid contour"
	Vortex atmospheric phenomena (vortices, whirlwinds, tornadoes)	Vortex windscreen for high-speed water and land vehicles
	Peristaltic contractions of the walls of hollow tubular organs of humans and animals (intestines, stomach, ureters, etc.)	Peristaltic pump – a device for dosing or supplying various liquids and gases
	Lizard's tail, detached in a dangerous situation (ability to automy)	Drop rocket boosters or fuel tanks
transformation of material and substance	Polishing the surface of stones with ground swell, air and water flows under sand effect	Surface treatment of metal and stone products by sandblasting and hydroabrasive methods
	Differentiation of the material of the earth's crust under the influence of temperature, pressure, gravitational forces	Zone melting – a method of refining materials (metals, semiconductors), growing single crystals
	Decomposition of plant litter by saprophytic microorganisms to easily assimilated sugars	Hydrolysis of wood to obtain glucose, xylose, ethyl alcohol, etc.
quantitative indicators of the object	Change in the volume of the swim (air) bladder of fish	Apparatus and devices, the height of rise (descent) of which is associated with the weight of the medium displaced by them (air or water)
	Optical phenomena in the atmosphere of the mirage type	Projection devices and projectors
the combined characteristics of the object	Transformation of modes of movement in the animal world from contact along the line and surface with sliding friction (reptiles) to the contact of elements at a point with less energy consumption for movement (ungulates)	Transformation of vehicles from skids to rollers, wheels and walking devices (walking excavator, some models of all-terrain vehicles, etc.)
	The process of natural reproduction of living organisms and plants	The mode of artificial reproduction in the incubation conditions of some animal world species
	Natural and artificial selection of animals, plants, microorganisms	Hybridization and selection of organisms with morphological characteristics populations

Name	Analogue	Solution
9.2. To use objects and principles of action used in other branches of technology	Aircraft using wing lift	Hydrofoil
	Aircraft turboprop engine	Compressor unit of a gas pumping station with an aircraft turboprop engine
	Thermal insulation of terrestrial objects, carried out by multilayer, porous coatings	Thermal protection of the descent part of the spacecraft due to the sublimation of high-strength coatings
9.3. To use solutions similar to ancient objects	Arched viaduct in ancient Rome	Arched railway and road bridges
	Mill water wheel	Hydraulic turbine, multi-bucket devices for dredgers, lifting machines, etc.
	Icarus Wings	Glider, hang glider, etc.
	M.V. Lomonosov's "Night vision" tube	Modern night vision devices
9.4. To apply solutions known from science fiction literature	Columbia cannon, manned projectile nucleus, booster rocket, method of splashing down the projectile nucleus in the ocean, oxygen apparatus from the novel by Jules Verne	Spacecraft launch rocket, thruster, life support systems, landing of spacecraft (with astronauts) in the ocean
9.5. To use the principle of imitation	Sugar (sucrose)	Sugar substitutes: saccharin, xylitol, etc.
	Military airfield	Mock airfield
9.6. To use children's toys as prototypes	Spinning top	Gyroscope, gyrocompass and automatic control systems based on their use
9.7. To use cheap copies instead of a complex, expensive object	Aircraft, ships, transport vehicles, ground buildings and structures interacting with the environment	Models and layouts of these objects, tested in wind tunnels, hydraulic flumes, on vibration stands in order to optimize the parameters of the original objects
	An area of the earth's crust subject to tectonic and man-made impacts	Models made of equivalent and optically active materials that simulate the action mechanisms on the earth's crust

Appendix 4
Examples of implementation of the most important human needs

Full name of the most important human need	Option of the most important human need	An example of implementing option of the most important human need
1. To protect (prevent) from diseases and pain, sense of discomfort (itching, loud sound, bright light, unpleasant odor, sweating, etc.), from the harmful effects of the environment and human actions, as well as to treat	**1.1.** To protect (prevent) disease	The **mobile seat** I have developed (Fig. A4.1) (installed on ordinary seating furniture) forces the sitting person to spontaneously (reflexively) balance, making light, almost invisible swinging of the pelvis to keep the body in balance, and, as a result, sit in a straight line, natural, healthy and beautiful pose.  **Fig. A4.1. Mobile seat and its use: a - appearance; b — use in the office, at home; c — awards and patents** At the same time, the deep back muscles along the spine, as well as the gluteal muscles, are constantly in good shape, which **prevents curvature of the spine** and the appearance of cellulite[1]. In addition, the ability to move while sitting (rocking, turning, bending) **prevents blood stagnation** in the pelvic organs.

Full name of the most important human need	Option of the most important human need	An example of implementing option of the most important hu need
1. To protect (prevent) from diseases and pain, sense of discomfort (itching, loud sound, bright light, unpleasant odor, sweating, etc.), from the harmful effects of the environment and human actions, as well as to treat		A computer keyboard is home to thousands of times more bacteria tha a toilet. **A germicidal irradiator mounted on a computer** aimed a keyboard (and hands) **inactivates (kills) bacteria** with ultraviolet ray It would seem, how can you improve a wooden toothpick – actually a s However, **toothpicks coated with a bactericidal composition** are great demand (Fig. A4.2), which to some extent **prevents diseases o teeth** and gums. Medicines can also be applied to toothpicks.  **Fig. A4.2. Bactericidal-coated toothpicks**
	1.2. To protect food from spoilage and, as a result, a person from diseases	Modern **packaging** not only protects food from spoilage, but have add useful properties. Among the technological innovations are packages th **heat** or **cool** food themselves, **regulate the intensity of microwave heating** and **even absorb substances harmful to health**[2] (Fig. A4. 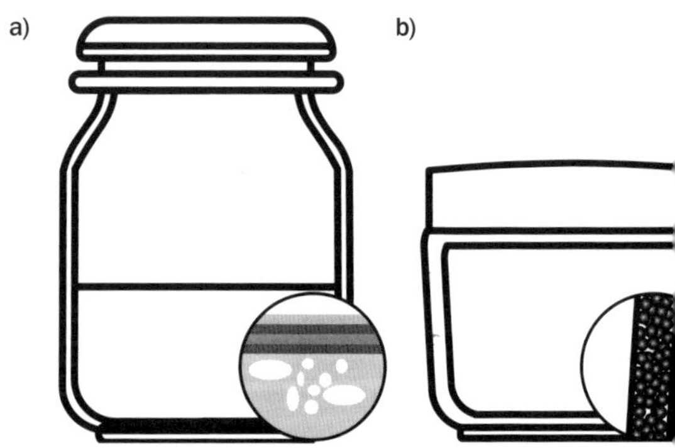 **Fig. A4.3. Modern packaging: a - absorbing harmful substance regulating the intensity of microwave heating**

Full name of the most important human need	Option of the most important human need	An example of implementing option of the most important human need
1. To protect (prevent) from diseases and pain, sense of discomfort (itching, loud sound, bright light, unpleasant odor, sweating, etc.), from the harmful effects of the environment and human actions, as well as to treat	**1.3.** Protect from harmful actions of people	On the top panel of the **biometric mouse**, developed by the American company APC (Fig. A4.4), there is a fingerprint scanner. When accessing protected data, the program prompts you to put your finger on the scanner and, if the fingerprint matches, opens the user access to sites or working folders. 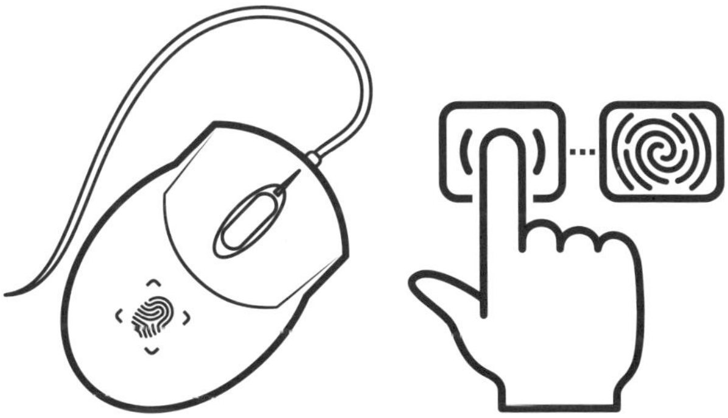 **Fig. A4.4. APC Biometric Mouse** The technology patented by Pay By Touch (Fig. A4.5) allows you **to pay** for purchases **using the customer's fingerprints**. The scanner compares them with prints from the store's database and, if the buyer is successfully identified, debits the money from his bank card. The probability that the fingerprints of two people can match is almost zero.  **Fig.A4.5. Pay By Touch technology**

Full name of the most important human need	Option of the most important human need	An example of implementing option of the most important human need
1. To protect (prevent) from diseases and pain, sense of discomfort (itching, loud sound, bright light, unpleasant odor, sweating, etc.), from the harmful effects of the environment and human actions, as well as to treat	**1.4.** To protect from sense of discomfort	The mobile phone generates high frequency vibrations that **repel mosquitoes** (Fig. A4.6) (the same for a key fob).  **Fig. A4.6. Mosquito repellent phone** The American company ExOfficio produces **clothing that repels mosquitoes** (Fig. A4.7).  **Fig. A4.7. Mosquito repellent clothing**

<table>
<tr><th>Full name of the most important human need</th><th>Option of the most important human need</th><th>An example of implementing option of the most important human need</th></tr>
<tr><td rowspan="2">2. To have protective and life-saving equipment in case of emergency</td><td>2.1. To have protective and life-saving equipment in case of an accident</td><td>The American company United Motors has built several airbags in a motorcycle jacket (Fig. A4.8). They inflate in an accident and protect the rider's upper torso.

Fig. A4.8. Motorcycle jacket with airbags</td></tr>
<tr><td>2.2. To have a life-saving device when falling into water</td><td>In the USA, a bracelet has been developed that has the functions of a lifesaving ring (Fig. A4.9). At the push of a button, the gadget turns into an inflatable pillow that allows you to stay on the surface of the water.
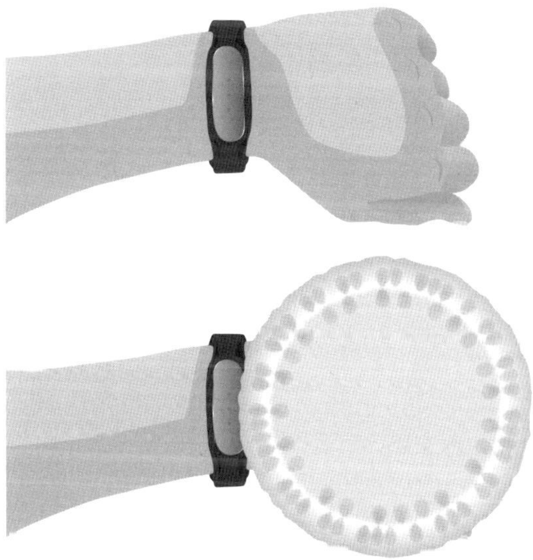
Fig. A4.9. Rescue bracelet that allows you to stay on the surface of the water</td></tr>
</table>

<table>
<tr><th>Full name of the most important human need</th><th>Option of the most important human need</th><th>An example of implementing option of the most important human need</th></tr>
<tr><td rowspan="2">3. To receive, store and transmit information</td><td rowspan="2">3.1. To receive information about the time, the state of human organs</td><td>The German company Rodenstock has developed special sunglasses with a display for athletes (Fig. A4.10). The side-mounted mini-display projects the stopwatch and heart rate directly into the athlete's eye. All the necessary electronics weigh 7 g and are hidden in the frame of the glasses[3].
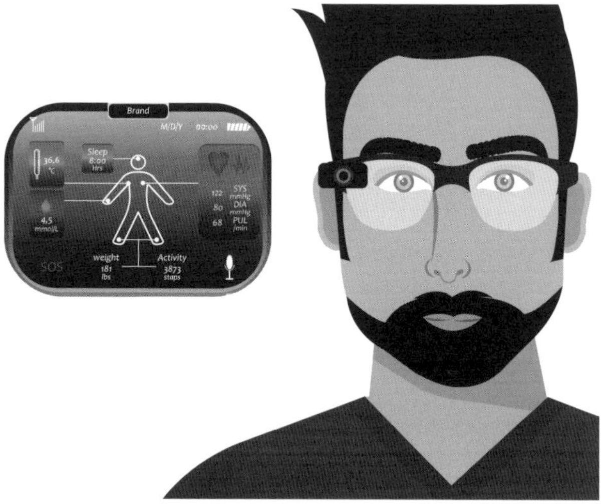

Fig. A4.10. Display glasses</td></tr>
<tr><td>The new line of Babyglow clothing, developed in the UK, signals child's temperature rise with a change in color (Fig. A4.11).
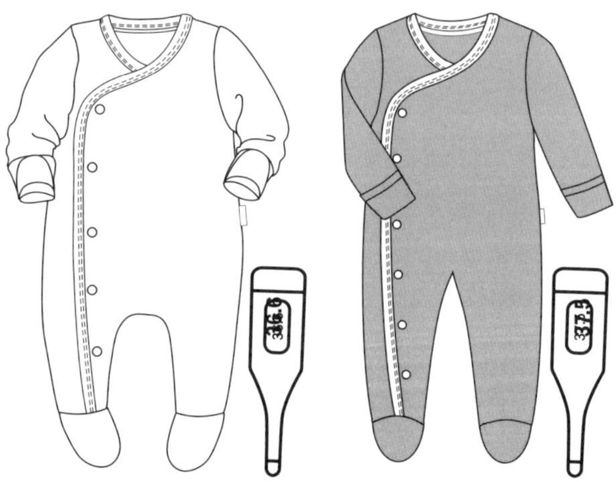
Fig. A4.11. Babyglow Clothing</td></tr>
</table>

Full name of the most important human need	Option of the most important human need	An example of implementing option of the most important human need
3. To receive, store and transmit information	**3.2.** To receive and transmit information about the state of the environment and technical system	American scientists have created a unique **T-shirt that can record the degree of cleanliness of the environment** and warn of danger (Fig. A4.12). The t-shirt is decorated with pink lungs, made of thermochromic fabric, on which blue blood vessels show through when the permissible level of air pollution is exceeded. In the future, it is planned to release clothes with the image of a liver, which will change color when alcohol vapors are in the air[4]. In the state of Texas, USA, a chemical that smells like rotten meat is added to a gas pipeline, and vultures circle over the site of the gas leak. 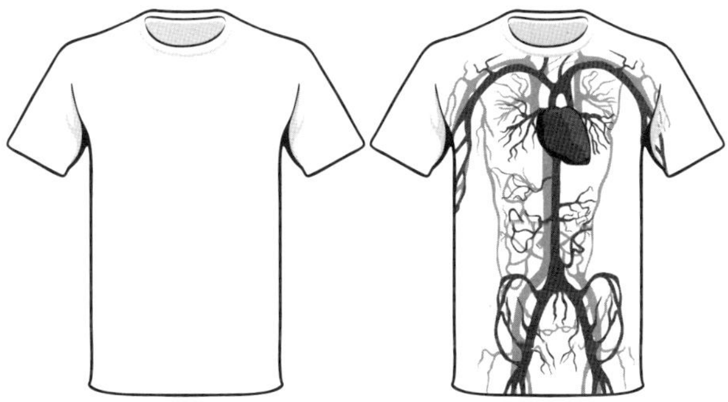 **Fig. A4.12. T-shirt that fixes the degree of air purity**
	3.3. To receive and transmit information to overcome the language barrier	Another example with a T-shirt. Three Swiss people have come up with a new way of **communicating while traveling for those** who do not speak languages: images of all things necessary in everyday life are placed **in the form of icons (pictograms) on T-shirts** (Fig. A4.13). When the owner of such a T-shirt on a trip needs something that he cannot convey in words, he simply point his finger at the corresponding icon.  **Fig. A4.13. Icon T shirt**

Full name of the most important human need	Option of the most important human need	An example of implementing option of the most important human need
3. To receive, store and transmit information	**3.4.** To receive and transmit information in case of emergency	French engineer Christophe Carolainen has proposed **a mine detector** that not only signals the presence of a mine in the ground, but **stops a man** walking through a minefield[5]. Highly sensitive sensors are built into the boots, the signal from them goes to the microprocessor on the belt. Electrodes are attached to the leg muscles. The microprocessor delivers a small discharge to them, sufficient to temporarily paralyze the muscles, which prevents them from **taking a step on the mine**. This happens in 20 microseconds – much faster than a man could react to the sound of a mine detector[6]. For more prosaic purposes, another invention is intended - **flip-flops**, which **signal the metal** in the sand (as a rule, these are lost jewelry) (Fig.A4.14)  **Fig. A4.14. Flip-flops with a metal detector**
	3.5. To receive and transmit information for the adaptation of people with disabilities	Ryan Patterson, a high school student, received the Intel International Science and Technology Fair (2001) Grand Prix and an Intel scholarship (2002) of $ 100,000. His invention is a glove that converts the signs of the sign language (Fig. A4.15), used by the deaf and dumb, into letters and displays the words on the computer screen[7]. Patterson received a provisional patent and refined his invention to include sound reproduction of words after they were written. 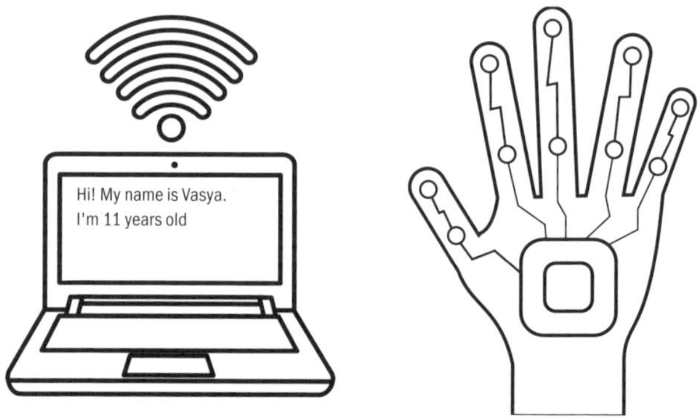 **Fig. A4.15. Glove that translates sign language into words**

Full name of the most important human need	Option of the most important human need	An example of implementing option of the most important human need
3. To receive, store and transmit information	**3.6.** To receive and transmit information about the strength properties of the material	Christoph Coplin from the Fraunhofer Institute for Integrated Circuits (Germany), having introduced microcapsules with an odorous reagent into polypropylene, has obtained **a plastic that begins to stink when destroyed**, which is very important for many technical systems. Microcapsules (with a diameter of 1–50 micrometers) are resistant to elastic loading and are destroyed only upon fatal deformation. They burst, emitting a specific odor (Fig. A4.16). 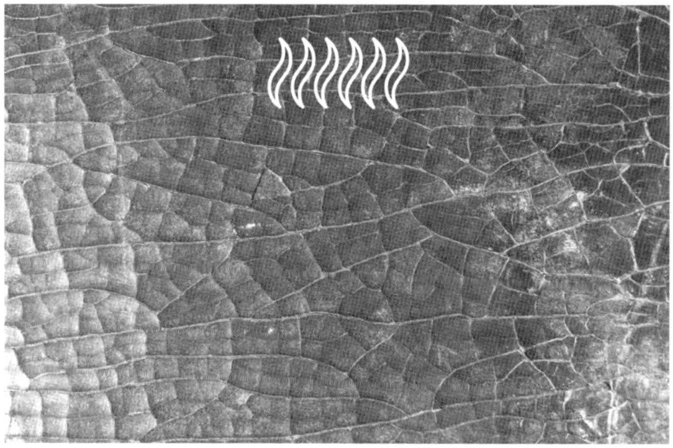 **Fig. A4.16. Cracked plastic structure**
	3.7. To receive and transmit information, including information about the interlocutor	Unusual **glasses** have been developed in Finland. In appearance, these are simple sunglasses, but they can **establish the identity of the interlocutor** and display all available information about him on the inside of the lens (Fig. A4.17). Once you focus on the person, the analysis program starts. On the outside of the frame is a small camera that is connected to the Internet. Photography quickly "makes its way" through the available databases, such as social networks.  **Fig. A4.17. Personality identification glasses**

<table>
<tr><th>Full name of the most important human need</th><th>Option of the most important human need</th><th>An example of implementing option of the most important human need</th></tr>
<tr><td rowspan="2">3. To receive, store and transmit information</td><td rowspan="2">3.8. To receive and transmit information about the presence and / or absence of goods</td><td>In some Japanese cafes, chips are embedded in the glasses, which remotely inform the waiters that the customer is running out of drink[8] (Fig. A4.18).

Fig. A4.18. Glasses with chips</td></tr>
<tr><td>In many countries, the technology of radio frequency identification of goods has been implemented: all goods are equipped with chips (Fig. A4.19), and store shelves are equipped with sensors. When a customer places an item of goods from a shelf into a cart, the information goes straight to the computer system. This allows for real-time tracking of inventory (both on the shelf and in the store)[9] as well as the rate of sales.

Fig. A4.19. Goods chips</td></tr>
</table>

Full name of the most important human need	Option of the most important human need	An example of implementing option of the most important human need
		American company Dr. Fresh has implanted a timer and **a tiny LCD screen** into **the cap of the toothpaste tube** (Figure A4.20). After the toothpaste is applied to the brush and the cap is closed, **the screen flashes** with colored lights[10] for the time necessary to brush your teeth. The cap signals that you can finish brushing your teeth by stopping blinking. 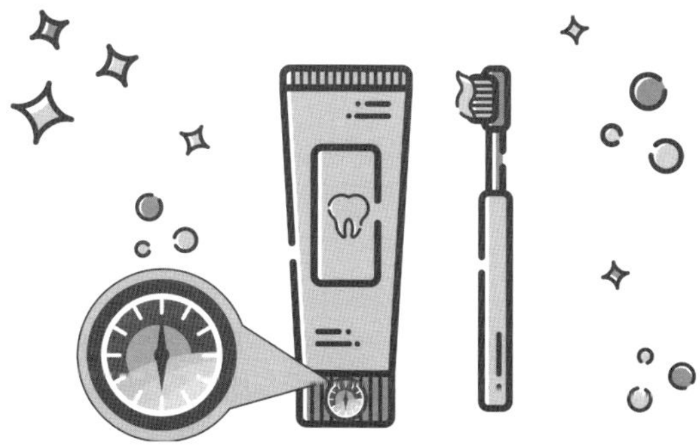 **Fig. A4.20. Tube with a timer in a cap**
3. To receive, store and transmit information	**3.9.** To receive and transmit information about the start and / or end time of actions, as well as about the actions themselves	Some drug **packages inform** about the need (time) of taking[11].
		The Japanese watch manufacturer Citizen Watch has demonstrated **a watch with** an iVERT Bluetooth **adapter**[12] (Fig. A4.21). It is synchronized with a smartphone and notify the owner of incoming messages and calls (**the watch vibrates, blinks and displays** the name of the caller and his number on the display). Using the buttons on the watch, you can answer a call[13] or make a forward.  **Fig. A4.21. Bluetooth watch**

Full name of the most important human need	Option of the most important human need	An example of implementing option of the most important human need
3. To receive, store and transmit information	**3.10.** To receive information about product properties	Specialists from the New Zealand Institute for Plant and Food Research have developed a high-tech **label** that allows you to **determine the ripeness of fruits** (Figure A4.22). A tiny sensor is built into the sticker that analyzes the composition of gases emitted by the fruit as it ripens[14].  **Fig. A4.22. Smart label** When gases properties parameters reach the desired level, **the label changes color**, which means that the fruit is ripe.
	3.11. To keep information	One of the American firms has begun producing **a ballpoint pen with a roll of paper** in the body (Fig. A4.23). The length of the paper is 60 cm. The notes can be torn off or rolled back[15]. 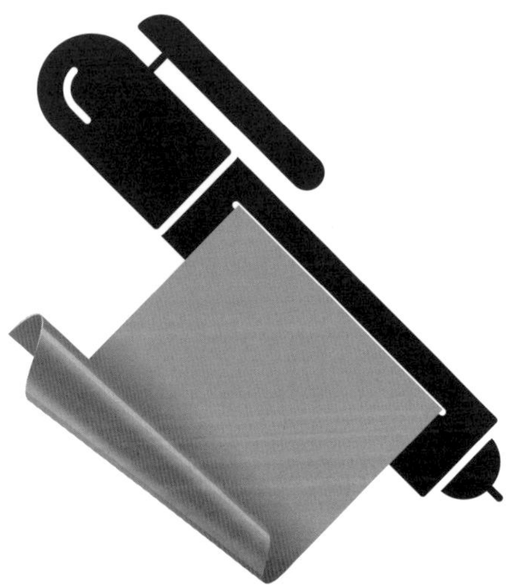 **Fig. A4.23. Pen with paper roll**

Full name of the most important human need	Option of the most important human need	An example of implementing option of the most important human need
3. To receive, store and transmit information	**3.12.** To transfer information	If a man without **means of communication** turns out to be cut off from civilization, he will be able to **transmit a distress signal in Morse code** using a flashlight built into a dress watch (Fig. A4.24). The alphabet itself is printed on the dial.  **Fig. A4.24. Dress watch with flashlight and Morse code**
4. Provide TS versatility, including through the implementation of household functions (cut, puncture, open, unscrew or screw, fasten, heat, cool, enlarge, clear, illuminate, etc.)	**4.1.** To heat, to cool	In modern cars, the steering wheel can be **heated** and **cooled** (Fig. A4.25). In Germany, the production of clothing with a heat-insulating air "cushion", which regulates the degree of insulation, has begun.  **Fig. A4.25. Temperature-controlled steering wheel**
	4.2. To implement household needs	The Swiss record-breaking **folding knife** includes 87 **tools** with 141 functions (see fig. 39). Recently, a USB flash drive has been added to a folding knife.

Full name of the most important human need	Option of the most important human need	An example of implementing option of the most important human need
4. Provide TS versatility, including through the implementation of household functions (cut, puncture, open, unscrew or screw, fasten, heat, cool, enlarge, clear, illuminate, etc.)	**4.3.** To open	Bottle cap **openers** are **built** into a handle, a lighter (Fig. A4.26), glasses and even ... the sole flip-flops (a type of beach shoes).  **Fig. A4.26. Multifunctional lighter**
	4.4. Illuminate	The **umbrella** with a built-in mini-lamp to **illuminate the road** (Fig. A4.27).  **Fig. A4.27. Umbrella with flashlight**

Full name of the most important human need	Option of the most important human need	An example of implementing option of the most important human need
4. Provide TS versatility, including through the implementation of household functions (cut, puncture, open, unscrew or screw, fasten, heat, cool, enlarge, clear, illuminate, etc.)	**4.4.** Illuminate	A similar need to **illuminate** is realized in the **key fob** (there are also keys that "respond" to the voice, whistling, **informing** about their location), as well as in **wrenches** (Fig. A4.28). 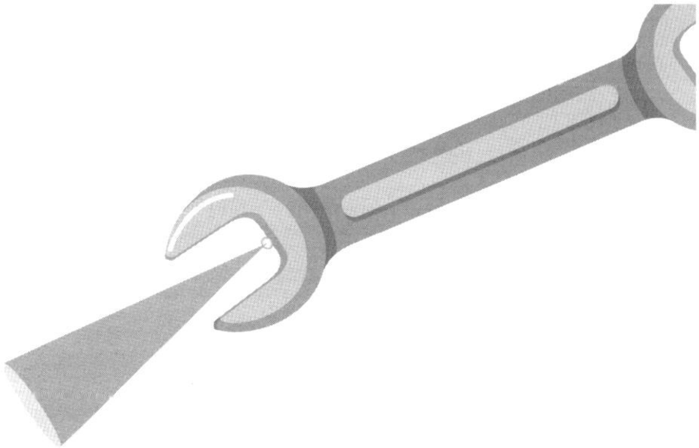 **Fig. A4.28. Wrench with with flashlight**
	4.5. To clean	**Clothes** that do not require washing and **can be self-cleaned** (Fig. A4.29) have been created by a group of scientists from China and Austria.  **Fig. A4.29. Self-cleaning clothing** A layer of anatase-modified titanium dioxide, five atoms thick, is applied to the surface of the fabric using nanotechnology. This composition is actively used in the space industry due to its ability to decompose pollutants when exposed to sunlight. Tests have shown that in just 20 hours the suit, drenched in red wine, completely restores its salable condition[16].

Full name of the most important human need	Option of the most important human need	An example of implementing option of the most important human need
5. To ensure compatibility of functions, properties, relationships (and their parameters) in different TS, society	**5.1.** To ensure compatibility of functions, properties, relationships (and their parameters) in the economy, education system, science, culture, etc.	To a large extent, this most important human need is aimed at the implementation of the **fundamental function of labor** – management[17] – **in the law of staged** TS **development**, and its **importance increases with the development of globalization** of labor markets, economics, political systems, religious cultures, etc. development of many sectors of the economy. For example, with a great delay, the USA and the USSR (and then other countries) agreed on **the compatibility of docking mechanisms of spaceships and stations**[18] (Fig. A4.30).  **Fig. A4.30. Docking of spaceships in open space** A lack of uniform data transfer protocols delayed the development of **the Internet** for years, and a lack of uniform MARC standards delayed the exchange of electronic resources between **libraries**[19] for many years, since library databases in different countries turned out to be incompatible. To ensure the compatibility of processes in **the education system**, as well as the compatibility of education systems in different countries, they began to form international standards for e-learning (IMS, ADL), as well as systems for nostrification of education documents. The European Union has addopted the law belatedly that obliges, from 2017, to start producing one, unified charger for all types of portable electronics: phones, tablets, laptops, photo and video cameras, GPS navigators. **Science** has not yet comprehended the high importance of knowledge codification (as an element of knowledge compatibility). The need for compatibility of different **cultures, religions**, smoothing out huge differences in citizens' incomes is growing more and more.

Full name of the most important human need	Option of the most important human need	An example of implementing option of the most important human need
		A cheap **laptop** (Fig. A4.31) has been developed by the Massachusetts Institute of Technology (USA) for developing countries (price no more than $ 100). One of its distinctive features is a handle, **by turning it**, you can **charge the battery**, and at the same time **pump up the muscles**.  **Fig. A4.31. Laptop with manual power converter**
6. Self-transformation and storage of energy	**6.1.** Self-transformation and storage of energy using muscle strength	A **mechanism** has been developed that is attached to the knee joint, and when walking, **mechanical energy is converted into electrical energy**, sufficient for the operation of a mobile phone, a flashlight. A similar mechanism is built into shoe sole (Fig. A4.32).  **Fig. A4.32. Mechanism for converting energy when walking**

<table>
<tr><th>Full name of the most important human need</th><th>Option of the most important human need</th><th>An example of implementing option of the most important human need</th></tr>
<tr><td rowspan="2">7. To arrange the order of objects and their structural elements (including signs, symbols, sounds); the shape of objects; types and characteristics of movements; connections between objects and their structural elements, replace the existing order with another one</td><td rowspan="2">7.1. To arrange the order of objects and their structural elements, and then replace the existing order with another one</td><td>One of the fast food restaurants in South Africa has begun to decorate hamburgers with sesame seeds in the form of Braille inscriptions for the blind (Fig. A4.33) – this is how the composition of the product is reported[20].

Fig. A4.33. Sesame seed lettering on burger</td></tr>
<tr><td>In the walls of the tunnels, compactly folded containers made of multilayer synthetic fabric are placed in advance, which, in the event of a fire or flood, inflate in a few minutes, turning into giant (several meters long) insulating bridges (Fig. A4.34).

Fig. A4.34. Subway Tunnel Flood Protection System</td></tr>
</table>

Full name of the most important human need	Option of the most important human need	An example of implementing option of the most important human need
8. To change the properties and parameters of technologies, equipment and products during their operation, transportation, storage	**8.1.** To change the aerodynamic properties of an object at different operation stages	 **Fig. A4.35. Aircraft with variable-geometry wing**
	8.2. To change parameters (geometric dimensions) during transportation and storage of equipment	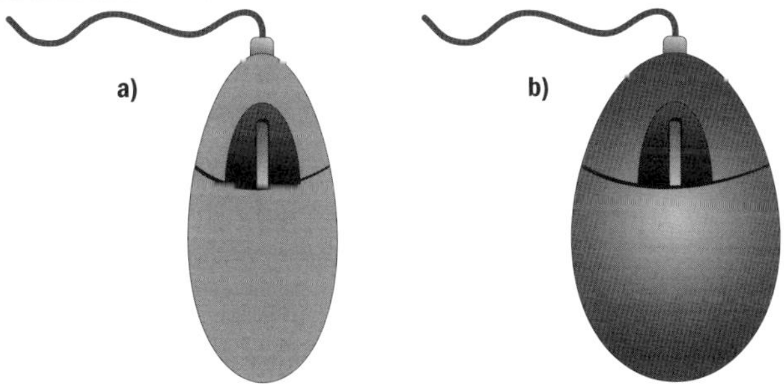 **Fig. A4.36. Inflatable computer mouse: a - during transportation; b — in operating order**
		An antenna made of **an alloy with a "shape memory"** is tightly packed for delivery to orbit and in space it is deployed into a multi-meter structure, "recalling" its original shape (Fig. A4.37).  **Fig. A4.37. Shape memory alloy antenna**

Full name of the most important human need	Option of the most important human need	An example of implementing option of the most important human need
	8.3. To change the parameters (geometric dimensions) during t functioning (operatiing) of the product	Toddlers grow faster than their shoes wear out. Parents can save money by buying **sliding children's boots** from the British company INCHworm (Fig. A4.38), the boots can be lengthened by one full size by pressing the button on the outside of the heel. The joints are made in the form of "accordions" that can stretch.  **Fig. A4.38. Sliding children's boots**
9. To reserve (to increase the reliability of a technical system or organizational structure, viability of an object of living nature) technical systems, the values of their strength, deformation and other characteristics, communications, energy, time, space, information, personnel, organs of living organisms	**9.1.** To reserve technical system elements	There are tights with three "legs" (Fig. A4.39). **The reserve third "leg"** is folded into a special pocket and can be used to replace the torn "leg", which, in turn, is hidden in the pocket. 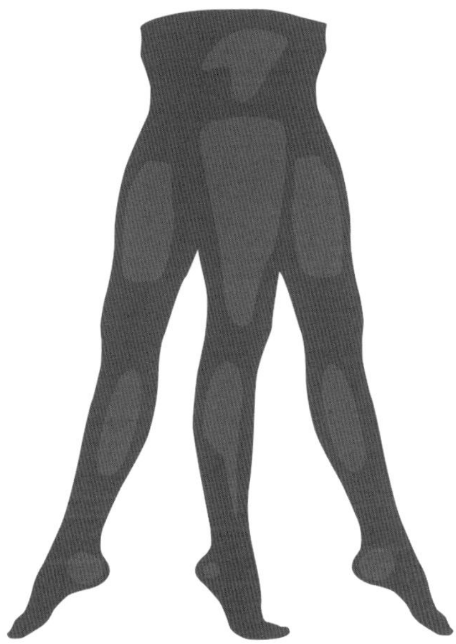 **Fig. A4.39. Tights with the third leg**

Full name of the most important human need	Option of the most important human need	An example of implementing option of the most important human need
9. To reserve (to increase the reliability of a technical system or organizational structure, viability of an object of living nature) technical systems, the values of their strength, deformation and other characteristics, communications, energy, time, space, information, personnel, organs of living organisms	**9.2.** To reserve technical systems	**A parachutist** must jump with **two parachutes** (Fig. A4.40) – the main and the emergency (reserve) one.  **Fig. A4.40. Parachutist with two parachutes**
	9.3. To reserve information	It is widely used **to back up data** from a server to a remote network folder on another server (Fig. A4.41).  **Fig.A4.41. Data backup**
	9.4. To reserve organs of living organisms	**Eyes reservation (duplication)** provides stereoscopic vision, and **ear** reservation (duplication) allows you to identify the sound source (binaural effect)

Full name of the most important human need	Option of the most important human need	An example of implementing option of the most important human need
10. To develop mental, creative, physical abilities, as well as to develop spiritually	**10.1.** To develop physical ability	**Shopping carts** have appeared in British supermarkets (Figure A4.42), transforming shopping into **fitness activities**. The fifth wheel under the bottom of the cart slows down its movement, creating an additional load on the customer, and the degree of braking can be adjusted from 1 to 10. The sensor on the handle of the cart measures the heart rate, the mini-computer calculates the calorie consumption[21] and displays the indicators on the screen. At maximum load, for a 20-minute walk along the shelves, the buyer will spend 120 kilocalories, which is equivalent to 10 minutes of breaststroke swimming.  **Fig. A4.42. Fifth wheel shopping carts**
11. To have pleasant smells, tactile properties, sound (or soundlessness), color (or colorlessness)	**11.1.** To have pleasant smells	Guests of hotels in Edinburgh, Newcastle, Birmingham and Nottingham (UK) can **choose the smell** of bed **linen**: sea breeze, freshly cut grass, apple pie, milk chocolate and even baby powder[22]. **Technologies and means of "transfer" of smells via the Internet** are developing. **Smell-O-Vision** cinemas have appeared that show movies with the effect of smells
12. To protect and rationally use nature	**12.1.** To restore the lawn of the football field	The Graduate School of Design and Communication's student in Trier (Germany) suggests filling the **spikes of football boots** with liquid lawn fertilizer. As the players run across the field, they **can fertilize the soil**, stimulating grass recovery after the match. At the same time, the most trampled areas of the field can be fertilized most of all.

Full name of the most important human need	Option of the most important human need	An example of implementing option of the most important human need
13. To have convenient and efficient recycling technology	**13.1.** To implement biodegradation of packaging	Many cosmetic products are usually packaged in boxes with the picture of flowers. But sooner or later, this beauty goes to garbage containers. The specialists of the cosmetics company Cargo Cosmetics have decided to give the painted flowers a second life. Their development – Plantlove lipstick – is packaged in a biodegradable tube made from corn (Fig. A4.43). And the case is made of colored paper with flower seeds embedded in it. You need to put the case in water, let it swell and plant the resulting "biomass" in a pot, after a few weeks a sprout appears.  Fig. A4.43. Biodegradable lipstick
		Packaging made from expanded polystyrene will be replaced by packaging based on biological materials (Fig. A4.44) – bran from cereals (buckwheat, rice, oats, wheat, barley, corn) and cotton seeds, fastened by the mycelium of fungi. Such packaging **decomposes after a month, turning into fertilizer.** 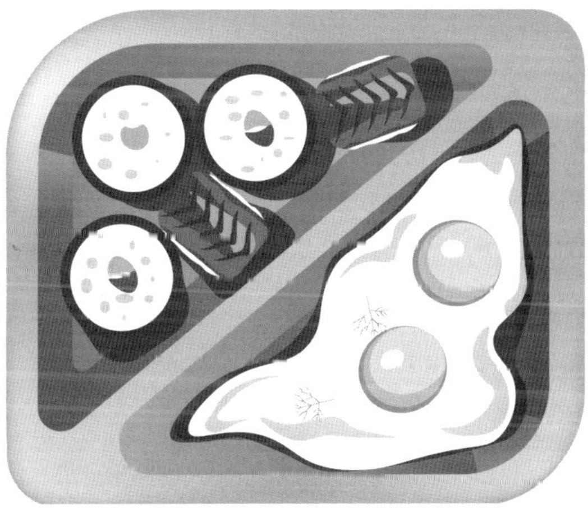 Fig. A4.44. Biomaterial packaging

Full name of the most important human need	Option of the most important human need	An example of implementing option of the most important human need
13. To have convenient and efficient recycling technology	**13.1.** To implement biodegradation of packaging	Some types of **packaging** are **edible** (Fig. A4.45).  **Fig. A4.45. Edible packaging** The famous English designer Helen Storey offers **clothes** made from **biodegradable polymers** that dissolve completely in water (Fig. A4.46). For comparison: polyethylene decomposes in 200 years, hard plastic – in 1000 years.  **Fig. A4.46. Biodegradable Polymer Clothing**

Appendix 5
List of nouns recommended for use in the method of focal objects

Turtle	Glass	Watch	Acid	Cave
Mirror	Antenna	Earth	Map	Telephone
Sea	Thread	Bell	Soap	Weight
Wood	Umbrella	Submarine	Feather	Elevator
Tank	Arrow	Star	Eye	Heart
Fire	Wind	Mountain	Tablet	Point
Glue	Time	Ring	Guitar	Wave
Ball	Bake	Lightning	Towel	Leather
Book	Metal	Telescope	Matryoshka	Rubik's Cube
Bird	Caterpillar	Town	Spinning top	Eraser
Space	Pie	Food	Web	Fireworks
Spring	Gates	Traffic lights	Milk	Bridge
Bag	Volcano	Waterfall	Dust	Computer
Smoke	Corn	Pendulum	Anchor	Key
Sword	Human	Field	Moon	Zebra
House	Flower	Diamond	Powder	Cloud
Cupboard	Wall	Road	Machine	Onion
Snowflake	Ray	Perfume	Train	Cork
Chameleon	Balloon	Fan	Magnet	Dye
Virus	Rainbow	Satellite	Wheel	Coffee

Appendix 6
Characteristics of traditional resource groups

Material resources — raw materials, materials, components, products, fuel, energy (objects of labor), as well as machines, equipment and devices (means of labor) necessary for the production of products[1].

There are secondary material resources — materials and products that can be reused as raw materials, components or energy carriers. The main sources of secondary material resources are production and consumption waste products[2].

Labor resources — a part of the population with physical development, intellectual (mental) abilities and professional competencies necessary for labor activity. The labor force includes both working and unemployed, but actively looking for a job and ready to start it[3].

Financial resources are funds at the disposition of an economic entity. Financial resources are directed to the development of production (production and trade process), the maintenance and development of non-production facilities, consumption, and can also be stored as reserves[4].

Intellectual resources include the protected results of intellectual activity and means of individualization.

Information resources — data collection organized to effectively obtain reliable information. Also, this type of resources means individual documents and arrays of documents in information systems: libraries, archives, funds, databases, etc[5].

Natural (inartificial) resources are called a part of the entire set of natural conditions for the existence of mankind and the most important components of the natural environment that surround it, used in the process of social production for the purpose of satisfying the material and cultural needs of society[6].

Natural resources are classified on different grounds: origin, method of application in material production, exhaustion[7]. In turn, exhaustible resources are divided into renewable and non-renewable[8].

Appendix 7
Examples of existing and potential implementations of non-traditional (hidden) resources

Resource class	Class characteristics	Existing or potentially possible technical solutions
Material resources	Any waste of production and consumption, as well as the waste of human life and other representatives of wildlife	The UN estimates that 1.3 billion tons of usable food is thrown away worldwide every year. This is four times more than is necessary to solve the problem of hunger on Earth. Every American or European annually throws an average of 95–115 kg of quite high-quality food into the trash[1]. The space station operates systems with a closed cycle of all waste
	Substances of natural origin in TS environment that play a harmful or neutral role in the performance of useful TS functions	Raindrops and snow obstruct the driver's view. In the rain sensor (Fig. A7.1), light-emitting diodes emit infrared rays, which, in the absence of drops, are almost completely reflected from the windshield onto the photodiode (Fig. A7.1a). When drops appear on glass, the radiation is completely or partially scattered (Fig. A7.1b). The photodiode reacts to the change in the amount of reflected radiation, which remotely turns on the wipers. 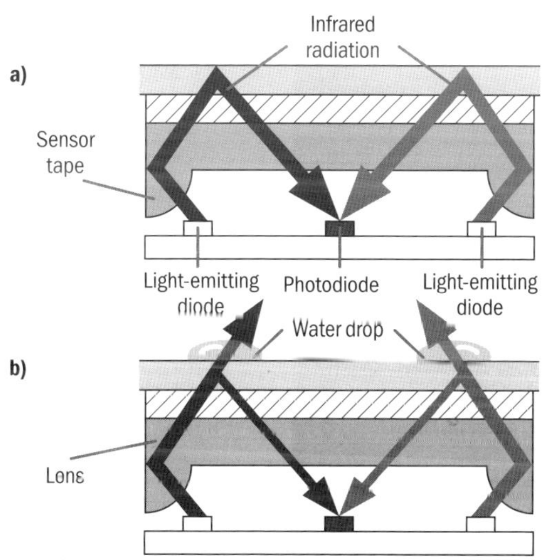 **Fig. A7.1. The principle of operation of the rain sensor: a — no drops; b — drops on glass**

Resource class	Class characteristics	Existing or potentially possible technical solutions
Material resources	Material parts of TS and the environment that can be used as part of other TS	The body of a truck, as a rule, is higher than the cab, therefore, when driving, it creates strong resistance to air-flow. To reduce it, an air fairing is installed on the roof of the cabin (Fig. A7.2), in which a berth can also be placed (Fig. A7.3).  Fig. A7.2. Air fairing design: 1 — cab; 2 - air fairing; 3 — body  Fig. A7.3. Air fairing — berth

Resource class	Class characteristics	Existing or potentially possible technical solutions
Material resources	Substances of natural origin in TS environment that play a harmful or neutral role in the performance of useful TS functions	The role of an aerodynamic fairing can also be played by a compacted incoming air jet (Fig. A7.4). 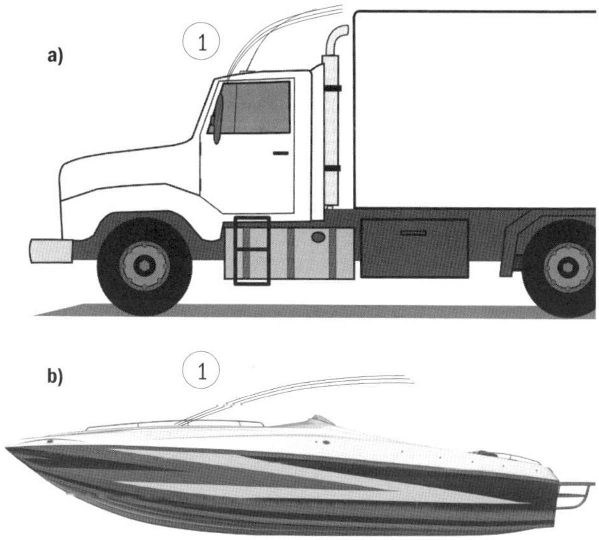 **Fig. A7.4. An air fairing formed by a compacted incoming air jet (1): a — in a truck; b — instead of the windshield on the boat** To create, due to the incoming air flow, the downforce of the car to the road, special aerodynamic devices are used – spoilers (Fig. A7.5).  **Fig. A7.5. Spoiler**

Resource class	Class characteristics	Existing or potentially possible technical solutions
Material resources	Material part of TS and the environment, which can be used as components of another TS	The pipeline elements were used to create a new transport system (for example, in the tundra) (Fig. A7.6). 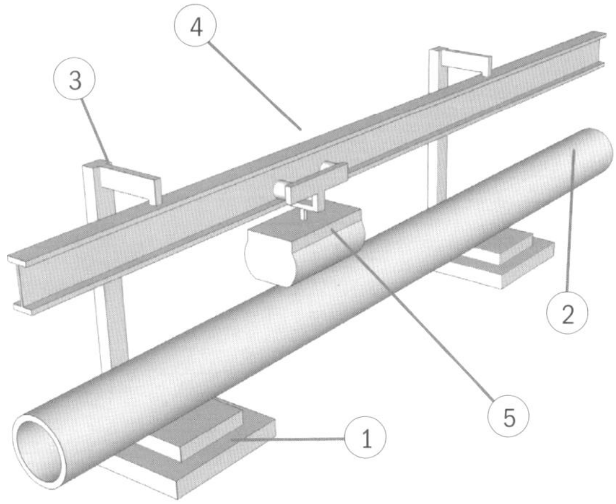 **Fig. A7.6. New transportation system built on the pipeline: 1 — pipeline support; 2 — pipeline; 3 — support of the new transport system; 4 — guide beam; 5 — transported cargo** Houses built on a tree, rocks, under water (Fig. A7.7, A7.8).  **Fig. A7.7. House built on a tree**

Resource class	Class characteristics	Existing or potentially possible technical solutions
Material resources	Material part of TS and the environment, which can be used as components of another TS	 **Fig. A7.8. Houses built: a — on the rocks; b — under water** Lake Ladoga ice road was of great importance in the defense of Leningrad during the Second World War (Fig. A7.9).  **Fig. A7.9. Ice road on Lake Ladoga**

Resource class	Class characteristics	Existing or potentially possible technical solutions
Material resources	Free spaces in TS or its surroundings (including at the micro level), filled with liquid, gas or other filler, or being a vacuum	Non-pressure pipeline is used for transportation of cargo containers (Fig. P7.10). 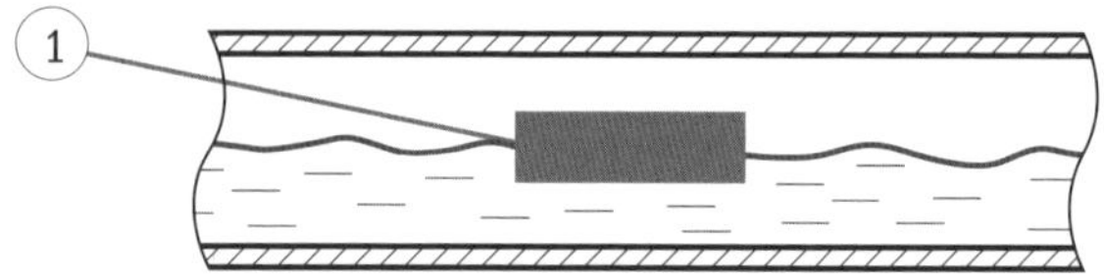 **Fig. A7.10. Transportation of cargo containers in a non-pressure pipeline: 1 — container with positive buoyancy**  **Fig. A7.11. Jumeirah Island in Dubai (2006) — one of the islands of the artificial archipelago. It resembles a giant 2-kilometer-long palm tree with 16 branches and is surrounded by an 11-kilometer crescent. It has taken 85 million cubic meters of sand to create it.** Another example. It is possible to fill the free spaces of the porous medium with another substance, which can ensure the formation of new properties of the porous medium (material)
Spatial resources	Part of TS space that can be used (as a rule, temporarily) to implement new useful functions and properties	In emergency situations, cargo containers with zero buoyancy can be transported via the pressure pipeline (Fig. A7.12). 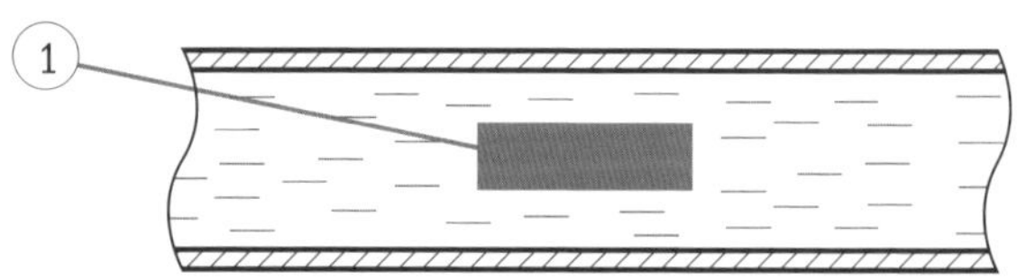 **Fig. A7.12. Transporting a cargo container (1) with zero buoyancy in a pressure pipeline**

Resource class	Class characteristics	Existing or potentially possible technical solutions
Spatial resources	Spatial changes and transfer of TS functional structure elements	For example, if a large-diameter pipeline is shaped elliptically, its strength along the vertical axis can increase sharply, which can increase the distance between the supports in the above-ground pipeline (Fig. A7.13) and the maximum length of the overhead pipeline (Fig. A7.14). I have received a patent for this technical solution. 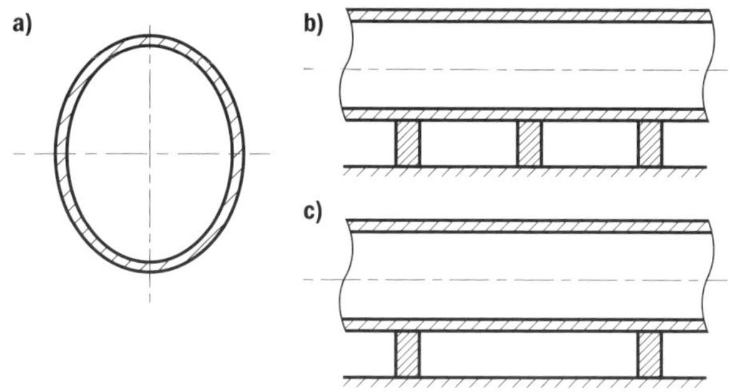 **Fig. A7.13. Elliptical pipeline: a — pipeline cross-section; b — round ground pipeline on supports; c — a decrease in the number of supports for an elliptical pipeline** 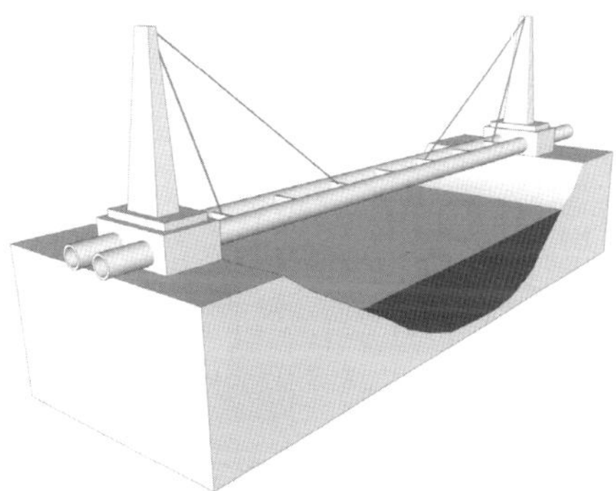 **Fig. A7.14. Suspended pipeline** The pipeline acquires new strength properties due to the arched shape (Fig. A7.15). 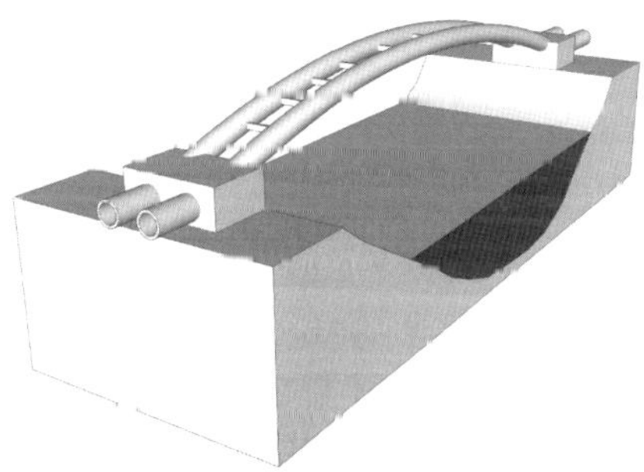 **Fig. A7.15. Arched pipeline**

Resource class	Class characteristics	Existing or potentially possible technical solutions
Spatial resources	Spatial changes and transfer of TS functional structure elements	After access space station orbit, the power supply system deploys the solar panels in order to fully provide the station with energy (Fig. A7.16).  **Fig. A7.16. Space station power supply system** The revolution in the organization of libraries and archives storage was made by a simple but effective technical solution – mobile shelving, which made it possible to use all the aisles (except one) to install additional mobile shelves (Fig. A7.17). 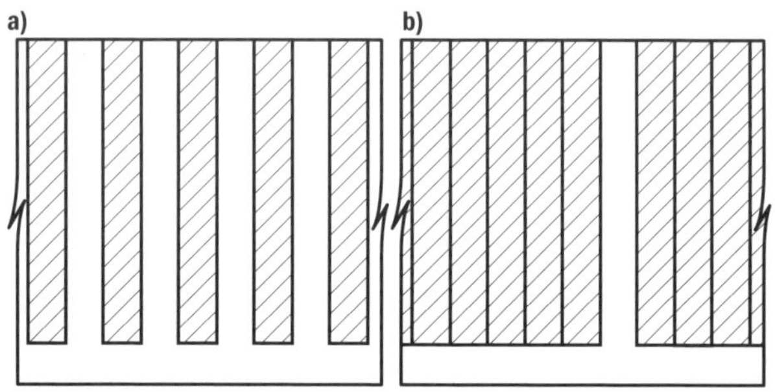 **Fig. A7.17. Stealth placement plans: a — how it was; b — how it has become**
Time resources	Unused time in TS before or after the end of a technological cycle, process, operation, reception, action	If we consider a cultural event (a performance, a film show, a sports competition) as a technology, then it is possible to conduct a sociological express survey of visitors (questionnaires, interviews), advertising campaigns, spectator competitions, a charity auction, etc. before and after the event

Resource class	Class characteristics	Existing or potentially possible technical solutions
Time resources	The execution time of TS of the technological cycle, process, operation, reception, action, which can be used for other technological cycles, processes, operations, techniques, actions in parallel	For example, during the movement of a mail train or carriage, the processing (sorting, registration) of postal items is carried out. Combine harvester (Fig. P7.18) can simultaneously cut the stalks, thresh and clean the grain, unload it into the hopper, and then into the vehicle, collect straw and chaff (chaff, waste), which are then thrown out in the form of bunches or crushed and scattered across the field in the form of green manure (natural fertilizers).  **Fig. A7.18. Combine harvester**
Informational resources	Any information characterizing the state of TS and the environment	For example, extraneous noise in the engine, indicating wear and tear of certain parts, violation of the fuel, oil supply etc. The use of information on the state of the environment, its pollution and information products in Russia is determined by the Federal Law of July 19, 1998 No. 113-FZ "On the Hydrometeorological Service" (with amendments and additions)
	Possible changes in TS properties during operation, storage, utilization, disposal of TS	For example, during the accumulation, storage of crushed brown and bituminous coals, oil shale, sulfide ores, carbonaceous (mudstones), bituminous rocks, their spontaneous combustion may occur due to intense oxidation with oxygen. In this case, more heat is generated than is dissipated into the environment
	TS properties indicating the presence of other TS properties	The color of the hot steel indicates its temperature – from dark brown (visible in the dark), which corresponds to 530-580°C, to bright white, indicating a temperature of 1250-1350°C
	The presence of certain environmental objects or their characteristics, indirectly indicating the presence of other environmental objects or possible changes in the environment	For example, thickets of a perennial herb with pale pink flowers – *Kachim Patrena* (Fig. P7.19) – most likely indicate a copper deposit, and the *European Starflower*, a small plant with snow-white flowers and, as a rule, seven petals (Fig. A7.20) is an indicator of tin deposits. There are many such pointers in nature, and experienced geologists use them, distinguishing some even from an airplane.  **Fig. A7.19. Kachim Patrena**

Resource class	Class characteristics	Existing or potentially possible technical solutions
Informational resources	The presence of certain environmental objects or their characteristics, indirectly indicating the presence of other environmental objects or possible changes in the environment	 **Fig. A7.20. European Starflower** Scientists have identified around 600 earthquake precursors that can be observed without instruments: changes in water levels in wells and boreholes, unusual animal behavior, sparking power lines, the sudden smell of gas in areas where it was previously absent, scattered light flashes and luminous pillars, a bluish glow on indoor surfaces, ground tremors and low-frequency rumbling, etc.
Functional resources	Harmful and neutral functions and properties arising in TS during its operation, storage, utilization, disposal or under the influence of the environment	During the underground mining of minerals, voids are formed, the pressure of the overlying (surrounding) rocks increases sharply, which can lead to massive collapses (and even lateral ejections) of rocks. However, with a certain order of mimning, it is possible to concentrate the increasing rock pressure in certain areas of the field (for example, in areas of an oil-bearing layer with reduced oil recovery) and, by controlling the increased rock pressure, to dramatically increase mining efficiency. This technical solution has been patented by me (in co-authorship) in Russia
	Useful TS functions or TS properties that can be used (usually temporarily) as useful, principal, main and auxiliary properties of another TS	For example, a small-sized generator that emits vibrations of an ultrasonic (inaudible to humans) frequency can effectively scare away stray dogs. Ultrasound (above 20 kHz) operates at a distance of 15 m. Explosives detectors use a method of detecting explosives by their smells. Drug detectors are similarly designed
Energetic resources	All types of energy arising from TS operation, storage, utilization, disposal or under the influence of the environment as harmful or neutral factors	For example, the mechanical vibrations of the seat when the truck is moving are converted into electrical energy used to condition the air in the cab. The kinetic energy of the oncoming air flow when the vehicle is moving is used in direct or transformed form for the operation of some devices and other equipments
	Surplus energy not used during TS operation, storage, utilization, disposal or energy of natural origin available from the environment	For example, when driving a car, especially in a city, you often have to pick up speed and brake. During acceleration, the engine power is spent on increasing the speed, and during braking, the kinetic energy of the accelerated car is lost. In order to partially use this energy, an energy recovery system has been developed, due to which the battery is charged. The future of humanity is the use of non-traditional (renewable) energy sources (sun, wind, sea waves, currents, tides, biomass, geothermal sources, etc.)

Resource class	Class characteristics	Existing or potentially possible technical solutions
Energetic resources	Part of the energy applied in TS that can be used (usually temporarily) to implement new useful functions and properties	For example, the energy of the oncoming air flow, which creates buoyancy, is temporarily used to brake the aircraft during landing (Fig. A7.21).   **Fig. A7.21. Aircraft braking during landing: a — with brake parachutes; b — spoilers (interceptors)** In addition, almost all passenger and transport aircrafts are equipped with engine thrust reversers, which are used for braking during landing (Fig. A7.22).  **Fig. A7.22. Using reverse thrust when landing**
Field resources	All types of physical fields arising during TS operation, storage, utilization, disposal or under the influence of the environment as harmful or neutral factors	For example, electrostatic fields are used to determine the properties of agricultural products, in electrostatic filters, separators, classifiers, insulating coatings and other technical systems

<table>
<tr><th>Resource class</th><th>Class characteristics</th><th>Existing or potentially possible technical solutions</th></tr>
<tr><td rowspan="2">Field resources</td><td>Physical fields in TS environment acting as harmful or neutral factors</td><td>For example, in gravity shelving, the shelves are set at an angle of 6° to the horizontal. Loads (boxes) placed on roller strips, under the action of gravity due to the weight of the load, move towards the inclination of the shelf (Fig. A7.23).

Fig. A7.23. Gravity shelving</td></tr>
<tr><td>Physical fields used in TS and caused by the environment, which can be used (usually temporarily) to implement new functions and properties</td><td>For example, a backpack with a solar battery (Fig. A7.24). The energy stored in it can be used to recharge a phone, tablet, any other electronics with a USB output, or for electrically heated clothing.

Fig. A7.24. Solar Backpack
The electrostatic fields arising for various reasons in TS and in TS environment (see Fig. 195) have a harmful effect on human nervous and cardiovascular systems, as well as TS functioning. To ensure electrostatic safety, various TS have been developed to allow the electrostatic charge to "drain" to earth. However, many TS have been developed (their list is constantly expanding), in which the harmful energy of the electrostatic field is used as a resource for charging batteries or powering low-power devices with high resistance: electronic watches, sensors, etc. In addition, the resulting electrostatic field is temporarily used in some technological processes. For example, when laminating wooden tiles, the film is fixed before it enters the press (it used to slide off during transportation)</td></tr>
</table>

Resource class	Class characteristics	Existing or potentially possible technical solutions
Resources of phenomena and processes	Phenomena and processes occurring in the atmosphere, space, the earth's crust, as well as the consequences of these processes (volcanism, emissions of hot water, steam, mountain falls, tsunamis, avalanches, etc.)	Scientists believe that it is very promising (especially for the military) to control atmospheric electricity, the ozone layer, earthquakes, landslides, avalanches, mudflows in specified areas (geophysical weapons). For peaceful purposes, huge lightning energy can be accumulated. Geothermal energy is developing (Fig. A7.25), the use of ocean energy is expanding (due to sea waves, tides, salinity, currents, differences in water temperatures, etc.). 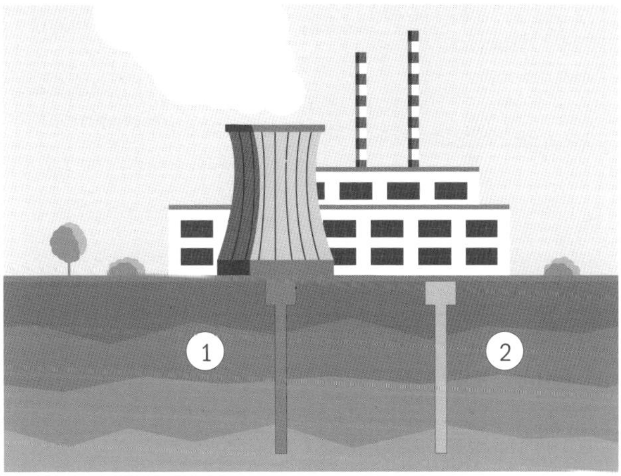 **Fig. A7.25. Geothermal station: 1 — cold water; 2 — hot water**
	Abiotic and biotic cycles of substances on Earth	Thanks to solar energy, the internal energy of the Earth, continuous biotic and abiotic processes of formation, transformation and decomposition of many chemical compounds, as well as the transfer of substances around the planet, take place in nature. The most striking example of the abiotic cycle of substances that can be used as resources is the circulation of water (evaporation of water, transfer of water vapor by air flows, condensation into atmospheric precipitation)[2]. For example, in the Canary Islands, with the help of huge plates, they learned to collect moisture from clouds driven by trade winds (Fig. A7.26).  **Fig. A7.26. Transfer of clouds by trade winds across the mountain range on the island of Gomera (Canary archipelago)**

Resource class	Class characteristics	Existing or potentially possible technical solutions
Resources of phenomena and processes	Abiotic and biotic cycles of substances on Earth	In the biotic cycle, living organisms act as a kind of catalysts that quickly synthesize new ones, transform and destroy existing organic compounds[3]. Many processes in the biotic cycle (including rotting, decomposition, fermentation, etc.) are widely used in industrial biological technologies (Fig. A7.27).

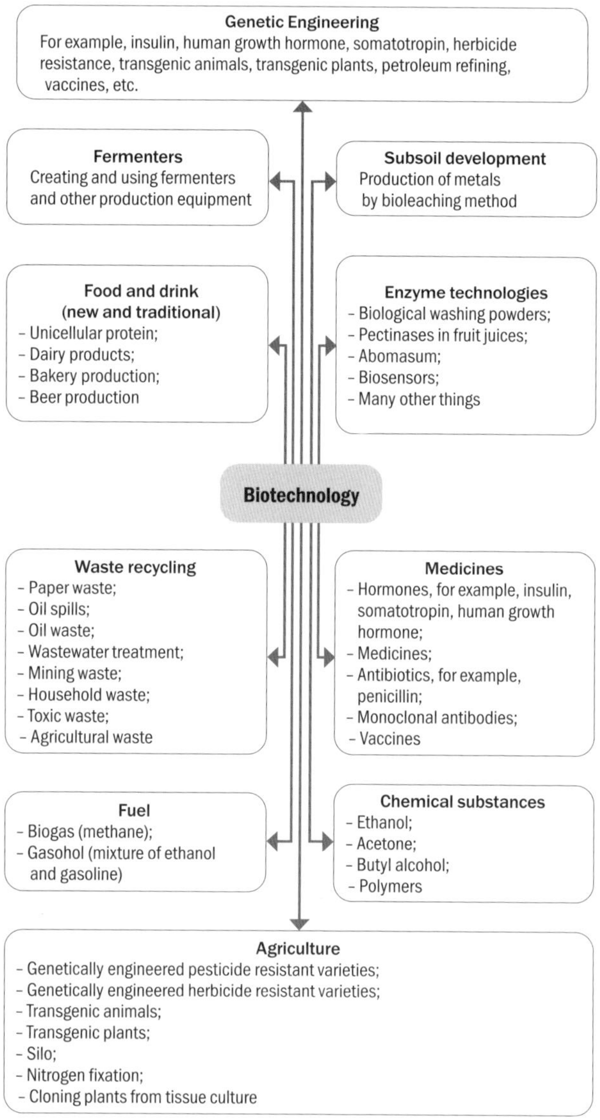

Fig. A7.27. Examples of biotechnology

Resource class	Class characteristics	Existing or potentially possible technical solutions
Resources of phenomena and processes	Phenomena and processes occurring during the operation, storage, utilization, disposal of TS, including those caused by harmful and neutral functions, as well as the interaction of the environment with TS	For example, the harmful function "disintegrate the substance" occurs with operation of electrical switches, when contacts are destroyed due to an electric discharge. The process of electrical erosion has become the basis for an effective electrospark (electrical pulse) method of processing solid conductive materials without cutting tools use Under special conditions, a cavitation process that is harmful to many technologies occurs in liquids, which consists in the formation and collapse of a large number of bubbles (voids, caverns) in the liquid. However, the artificial creation of a cavitation pocket in front of a missile fired from a submarine led to a sharp increase in its speed (more than 100 m/s) due to a decrease in water resistance. Cavitation processes are widely used in medicine (for example, crushing stones and sand in the kidneys) (Fig. A7.28). 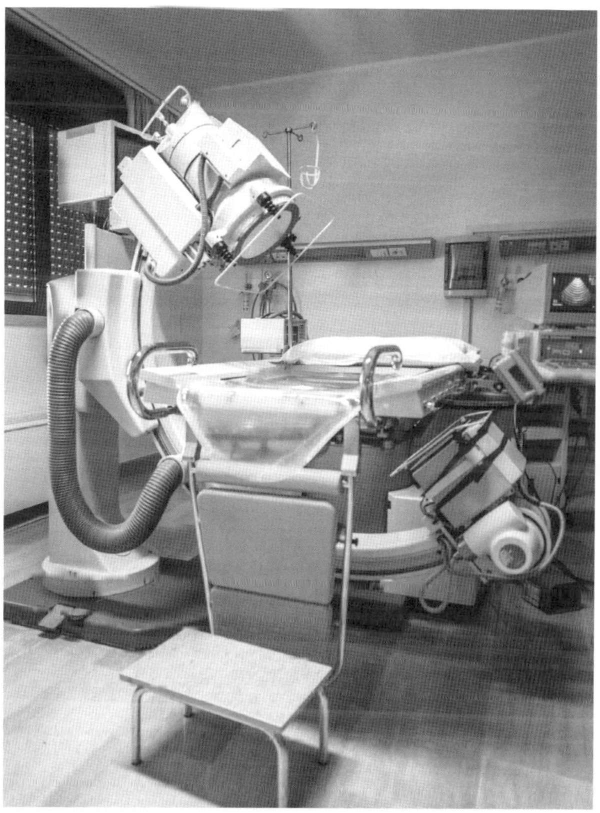 **Fig. A7.28. Plant for crushing stones and sand in kidneys**

Resource class	Class characteristics	Existing or potentially possible technical solutions
System resources	Complexes of interrelated harmful and neutral functions and properties arising during TS operation, storage, utilization, disposal, as well as under the influence of the environment	A complex of neutral functions and properties arising in the process of underground salt mining (dry aerosol of sodium chloride with a dispersion of aerosol particles of 3-5 microns and a high concentration of active substance, increased air ionization, complete absence of allergens and pathogenic microorganisms, low relative humidity, etc.)[4] has a beneficial effect on patients with pulmonary and / or skin diseases, increases the overall defense of the body. Therefore, in many countries, halotherapy for sick people (salt treatment in the former salt mines), as well as preventive measures for healthy people, is carried out with great success (Fig. A7.29).  **Fig. A7.29. Treatment and rehabilitation sanatorium on the basis of the dead salt pit "Wieliczka" (Poland)** Such hospitals on the basis of salt mines, caves and workings have been created in Russia, Ukraine, Belarus, Kyrgyzstan, Armenia and Azerbaijan. In Europe, halotherapy is used in the Czech Republic (Javorzycko and Zlate Gori caves), Slovakia (Bystryanskaya cave), Romania (Tirgu Okna salt mines), Austria (Bad Gastein thermal caves, Solbad salt mine), Hungary (Mir cave), Germany (Ennepetal caves and Klütert cave), Slovenia (Sezana caves)

Resource class	Class characteristics	Existing or potentially possible technical solutions
System resources	Complexes of interrelated harmful and neutral functions and properties arising during TS operation, storage, utilization, disposal, as well as under the influence of the environment	In 1941-1942, the Moscow metro and other underground structures of Metrostroy served as a shelter during air raids for more than 400 thousand people (Fig. P7.30). During air raids, an entire city was unfolding underground: shops, hairdressers, libraries were opened (for example, at the Kurskaya station, see Fig. P7.31). Almost the entire range of useful functions and properties of the metro was used (underground spaces, transport infrastructure, ventilation facilities, etc.).  **Fig. A7.30. Mayakovskaya metro station during an air raid**  **Fig. A7.31. Library at the Kurskaya metro station**

Appendix 8
List of environmental laws[1]

1. The law of biogenic migration of atoms (or V. Vernadsky's law).
2. The law of correlation (formulated by J. Cuvier).
3. The energy maximization law (formulated by G. and Y. Odum and supplemented by M. Raimers).
4. The law of internal dynamic equilibrium.
5. The law of genetic diversity.
6. The law of historical irreversibility.
7. The law of irreversibility of evolution.
8. The law of constancy (formulated by V. Vernadsky).
9. The law of a maximum of biogenic energy (Vernadsky-Bauer law).
10. The law of minimum (formulated by J. Liebig).
11. The law of natural resources scarcity.
12. The law of unidirectional flow of energy.
13. The law of optimality.
14. The law of pyramid of energy (formulated by R. Lindemann).
15. The law of equivalence of living conditions.
16. The law of environmental development.
17. The law for the reduction of energy efficiency in the use of natural resources.
18. The law of the combined action of natural factors (the law of Mitcherlich — Tinemann — Baule).
19. The law of tolerance (Shelford's law).
20. The law of soil depletion (decrease in fertility).
21. The law of physical and chemical unity of living matter (formulated by V. Vernadsky).
22. The law of Competitive Exclusion.
23. The basic law of ecology.
24. The law of emergence.
25. The law of requisite variety.
26. The law of arrangement complication.

27. The biogenetic law (formulated by E. Haeckel).
28. The law of uneven development of system's parts.
29. The law of preservation of life.
30. The principle of order preservation (formulated by J. Prigogine).
31. Le Chatelier — Brown principle.
32. The principle of energy saving (formulated by L. Onsager).
33. The law of maximizing energy and information.
34. The law of system's development at the expense of the environment.
35. Schrödinger's rule "about feeding" the organism with negative entropy.
36. The rule of accelerating evolution.
37. The law of relative independence of adaptation.
38. The principle of minimum population size.
39. The rule of genus representation by one species.
40. The law of depletion of living matter in its island thickenings (G. F. Khilmi).
41. The rule of compulsory filling of ecological niches.
42. The principle of ecosystem formation.
43. The boomerang law.
44. The law of the indispensability of the biosphere.
45. The law of shagreen skin.
46. The principle of remoteness of the event.
47. Hopkins' bioclimatic law.
48. The law of the optimum (in ecology).
49. The classical biological optimal foraging theory, independently developed by Robert MacArthur and Eric Pianca in 1966.
50. Gloger's Rule.
51. Bergman's Rule

Appendix 9
Unusualness criteria for materials properties

1. There is (a characteristic that determines property unusualness).
2. There is no (characteristic that determines property unusualness).
3. Abnormally / extremely / unusually high value (of characteristics, property unusualness).
4. Abnormally / extremely / unusually low value (of characteristics, property unusualness).
5. Abnormally / unusually wide values range (of characteristics, property unusualness).
6. Abnormally / unusually narrow values range (of characteristics, property unusualness).
7. Unique / unusual value or values range (characteristics, which determine property unusualness).
8. The presence of a relationship (between characteristics, which determine property unusualness).
9. Lack of a relationship (between characteristics, which determine property unusualness).
10. Relationship (between characteristics) — direct proportionality (which determines the property unusualness).
11. Relationship (between characteristics) — inverse proportionality (which determines the property unusualness).
12. Strong relationship (between characteristics, which determines property unusualness).
13. Weak relationship (between characteristics, which determines property unusualness).
14. One-way relationship (between characteristics, which determines property unusualness).
15. Bidirectional relationship (between characteristics, which determines property unusualness).

16. Linear relationship (between characteristics, which determines property unusualness).
17. Non-linear relationship (between characteristics, which determines property unusualness).
18. Monotonic dependence (between characteristics, which determines property unusualness).
19. Non-monotonic dependence (between characteristics, which determines property unusualness).
20. Continuous dependence (between characteristics, which determines property unusualness).
21. Dependency with gaps (between characteristics, which determines property unusualness).
22. Dependence of a certain type (between characteristics, which determines property unusualness).
23. Appearance of (characteristics which determines property unusualness).
24. Disappearance of (characteristics which determines property unusualness).
25. Preservation, stability of (characteristics which determines property unusualness).
26. Change of (characteristics which determine property unusualness).
27. Concentration of (characteristics which determines property unusualness).
28. Alignment of (characteristics which determines property unusualness).
29. Distribution of (characteristics which determines property unusualness).
30. Isotropy of (characteristics which determines property unusualness).
31. Anisotropy of (characteristics that determines property unusualness).
32. Accumulation of (characteristics that determines property unusualness).
33. The material absorbs the external object (which determines property unusualness).
34. The material does not absorb the external object (which determines property unusualness).
35. The material emits (generates, reproduce) an external object (which determines property unusualness).
36. The material does not emit an external object (which determines property unusualness).
37. The material is absorbed by the external object (which determines property unusualness).
38. The material is not absorbed by an external object (which determines property unusualness).

39. Material is emitted (generated, generated) by an external object (which determines property unusualness).
40. The material is not emitted by an external object (which determines property unusualness).
41. The material interacts with an external object (which determines property unusualness).
42. The material does not interact with an external object (which determines property unusualness).

Appendix 10
Examples for filling the database "Materials with unusual properties"

This section contains materials with unusual properties, collected on the basis of media reports and open sources from the Internet.

1. Graphene

Unusual property: two-dimensional structure, minimum thickness (the thinnest material in the world), record high thermal conductivity, high strength and electrical conductivity.

Brief description of material and unusual property

At present it is the thinnest material in the world, with only one layer of carbon atoms. In its structure, graphene resembles a honeycomb, that is, the atoms are organized here in a hexagonal crystal lattice, which is "packed" as tightly as possible. You can get it from graphite, which is well known to us from pencil rods[1]. Graphene can be represenred as a single graphite plane separated from a bulk crystal. According to estimates, graphene has high mechanical rigidity and record high thermal conductivity (~1 TPa and ~5•103 $W{\bullet}m^{-1}{\bullet}K^{-1}$, respectively). High mobility of charge carriers (the maximum electron mobility among all known materials) makes it a promising material for use in a variety of applications, in particular, as a future basis for nanoelectronics and a possible replacement for silicon in integrated chips[2].
Possibilities of material use: electronics (flexible electronics, high-frequency transistors, logic transistors), photonics (photodetectors, optical modulators, mode-locked lasers, THz generators and optical polarizers), composite materials, paints and coatings, energy storage and generation applications, sensors and applications for metrology, biological applications, etc[3].
Information about the developer (manufacturer): developers — Nobel Prize Winners A. K. Game, K. S. Novoselov (University of Manchester, UK); main manufacturers — Ningbo Morsh Technology[4], Samsung, Fujitsu[5].

2. Tearproof fabric

Unusual property: super thinness, super strength, super lightness.

Brief description of material and unusual property

A material made of supertough filaments — carbon nanotubes, which are thousands of times thinner than a human hair[6] and at the same time are able to withstand loads much more than steel wire of the same thickness. The fabric is very light, much lighter than metal foil: a square kilometer of fabric weighs no more than 30 kg[7]. It conducts electric current well.
Information about the developer (manufacturer): University of Texas, USA.

3. Glycoporphyrins

Unusual property: prevents the negative effects of chemotherapy.

Brief description of material and unusual property

They are obtained on the basis of chlorophyll synthesis. Compounds accumulate only in cancer cells; upon irradiation, photochemical processes develop only in diseased tissues. They do not cause complications in patients, as after conventional chemotherapy[8].
Possibilities of using the material: replacement of conventional chemotherapy in the treatment of cancer.
Information about the developer (manufacturer): Moscow State Academy of Fine Chemical Technology named after M. V. Lomonosov[9].

4. Covalent organic frameworks

Unusual property: resistance to high temperatures, large internal surface area.

Brief description of material and unusual property

They do not differ from traditional plastics in chemical composition, but at the molecular level they have a regular polymer-crystalline structure. They withstand temperatures up to 500–600 °C, have a large internal surface area —

from 700 to 1600 m2 per gram of substance. This allows storing a large amount of hydrogen the cells[10].
Possibilities of using the material: fuel storage in hydrogen cars.
Information about the developer (manufacturer): USA, University of Michigan, Adrian Kote's group[11].

5. A solution of alpha-cyclodextrin and 4-methylpyridine in water

Unusual property: a compound that hardens when heated and melts again when cooled.
Manifestation of the property: when it's heated to 45–75 °C, the liquid turns into a milky-white solid, when it's cooled, it passes into the initial liquid state. The exact junction temperature is determined by the ratio of the components[12].

Brief description of material and unusual property

Liquids that harden when heated have been known before. This effect is observed in a number of polymerization processes, but this process is irreversible.
The reversibility of the "liquid — solid — liquid" transition for the aqueous solution of alpha-cyclodextrin and 4-methylpyridine is provided by rearranging hydrogen bonds between the molecules of two organic substances and water. Neutron transmission test has showed that molecular motion in the solution slows down during solidification. This is how a new composition appears — a liquid that "freezes" when heated[13].
Possibilities of using the material: in the military industry.
Information about the developer (manufacturer): Joseph Fourier University, France.

Appendix 11
100 Outstanding Achievements of Humanity[1]

1. Battery
2. Alphabet and writing
3. Biotechnology
4. Paper
5. Bicycle
6. Screw (drill, thread, etc.)
7. Gyroscope
8. Potter's wheel, lathe tool
9. Internal combustion engine
10. Money
11. Extraction and processing of minerals
12. Residential house (dwelling)
13. Railroad
14. Agriculture
15. Gear train
16. Games (toys)
17. Measuring instruments
18. Internet
19. Artificial rubber
20. Use of X-rays
21. Geographic maps
22. Cinema
23. Brick
24. Glue
25. Wedge (ax, knife, shovel, plow, cutter, saw)
26. Book, typography, library
27. Wheel
28. Compass
29. Conservation

30. Dyeing
31. Cooking food
32. Laser
33. Medicines
34. Lens (microscope, telescope)
35. Elevator
36. Boat, ship
37. Location
38. Bow
39. Magnetic recording
40. Pendulum
41. Mill
42. Hammer
43. Bridge
44. Musical instrument
45. Soap
46. Pump
47. Thread
48. Music notes
49. Firearms
50. Clothes
51. Space exploration
52. Steam engine (steam locomotive, steamer)
53. Sail
54. Oven
55. Plastic
56. Dam
57. Pneumatic design
58. Cart, car
59. Bearing
60. Semiconductor devices
61. Receiving and transmission of electrical energy
62. Getting fire
63. Gunpowder
64. Piston
65. Flow line production
66. Energy converters
67. Prophylactic vaccinations (immunization)
68. Domestication of wild animals

69. Prostheses
70. Spring
71. Radio
72. Jet Propulsion
73. Lever
74. Airplane
75. Welding
76. Breeding
77. Networks
78. Synthesis of organic substances
79. Automatic control systems
80. Number systems
81. System of measures and weights
82. Spacesuit
83. Sports activities, competitions
84. Grease
85. Glass
86. Means of communication (mail, telephone, telegraph, radio, etc.)
87. Television
88. Fabric, weaving machine
89. Brake
90. Pipeline
91. Controlled nuclear reaction
92. Photography
93. Bread
94. Cement
95. Clock
96. Hinge
97. Sewing needle, sewing machine
98. Electric lighting lamp
99. Electric motor
100. Electronic computer

Appendix 12
Creativity criteria of pedagogical technologies, textbooks and manuals[1]

1. Formation of ideas about the levels of innovative tasks and solutions, criteria of novelty.
2. Presence of a holistic view of innovation stages and phases of — from generating ideas to introducing a new product or service to the market.
3. Presence of recommendations for setting innovative tasks in the form of the ideal final result, as well as examples of constructive and technological solutions close to ideal.
4. Presence of methods and means for the development of imaginative thinking, visualization of the studied thesis.
5. Reflection of studied material and practical problems theoretical theses relationship in which these theses can be used.
6. Formation of learning goal as creative result achievement.
7. Presentation of the educational problems system, reflecting the real contradictions between science and practice.
8. Presence of methods and means that provide a subject-subjective relationship between the teacher and the student.
9. Presence of methods, means and tasks to overcome psychological barriers to thinking.
10. Presence of special methods and means that increase motivation to learn.
11. Presentation of heuristic strategies, tactics, methods, techniques and tasks involving their use.
12. Presence of methods, techniques, tasks that develop associative thinking.
13. Presentation of the theoretical foundations of grouping objects and examples of creative classifications.
14. Presence of methods, means and tasks for the identification and use of TS analogies and analogues in living, inanimate nature, society in technical systems (TS).

15. Using a functional approach in the study of objects and the tasks of their improvement.
16. Presence of methods, means and tasks of revealing and formation of hidden productive knowledge by students.
17. Presence of methods and means of convergence (closing-in, transfer) of knowledge related to different subject areas, as well as the formation of interdisciplinary connections.
18. Presence of control and assessment methods and means of development students' creative abilities level.
19. Presence of methods and means for the development of abstract thinking in students.
20. Presence of methods, examples and tasks for the formation of metaphorical thinking in students.
21. Presence of methods and means of choosing and / or generating new human needs while improving or creating new generations of technologies, equipment, products.
22. Presence of methods and means of selection and / or generation of new functions and consumer properties of technologies, equipment, products.
23. Presence of methods and means for modeling the subject and social content of the future professional activity of a student (elements of sign-contextual learning).
24. Presence of business game elements (quasi-professional activity).
25. Presence of educational creative tasks, puzzles.
26. Presence of methods and means for reducing and / or leveling the rigidity and conformity of students.
27. Presence of methods and means of evaluating acquired creative solutions.
28. Presence of methods and means of psychophysiological creative activity activation.
29. Presentation of work examples ("secrets of creativity") of great creators in technology, science, education, art.
30. Presentation of information on the laws and patterns of TS development in educational materials, as well as instructions and examples of their use in innovation.
31. Presence of methods and means for the implementation of the heuristics basal principle.
32. Presence of methods and means for the implementation of the main goal of education, which provides competencies that ensure the formation and implementation of important designs in the sectors of the economy, as well as in everyday life.

33. Presence of beauty and harmony examples in living and inanimate nature, art, architecture, technology, science for the formation the most important needs of students for beauty, harmony, as well as protection and rational use of natural resources.
34. Presence of information and methods of identifying non-traditional ready-made resources hidden in technologies, equipment, products and the environment, about methods of modification, transformation and technologies of using resources.
35. Presence of information on the criteria for technologies, equipment, products progressive development on the methodology of their formation and use in creative activity.
36. Presence of information about the causes of various types of contradictions and ways to resolve them.
37. Presence of information about psychological effects and illusions of visual perception.
38. Formation of students' awareness of the high social significance of creative activity, the need for creativity, awareness of themselves as a subject of creativity.
39. Presence of information about the prejudices and delusions of mankind in science, creativity.
40. Presence of methods and means of forming the sense of optimism, the ability to prevent frustrations, to defend against the "collapse of the last hope" in students.

Appendix 13
Definitions of terms

Definition of term	Source	Section
Abstraction is a mental distraction, isolation from certain sides, properties or connections of objects and phenomena in order to highlight their essential features.	Explanatory dictionary of the Russian language by S. I. Ozhegov, N. Yu. Shvedov	1.4.1
Additivity (*from Lat.* Additio – addition) is a property of a quantity, which consists in the fact that the sum of the values of quantities corresponding to parts of an object is equal to the value of a quantity corresponding to the whole object. For example, the additivity of volume means that the volume of a whole body is equal to the sum of the volumes of its parts.	Encyclopedic Dictionary	6.2.2.2
Analogue is something that represents a correspondence to another object, phenomenon or concept.	Russian dictionary of synonyms	1.1
Antonyms (*from Greek* ἀντί – against and ὄνομα – name) are words with opposite meanings that serve to denote contrasting phenomena. For example, quiet – loud, appear – disappear, many – few, up – down.	Great Soviet Encyclopedia	1.3
Aspiration – persistent attraction, fixation on something, decisive proclivity for something.	Great Encyclopedia of Psychology	1.1
Attraction is striving to satisfy any need of a living organism. In lower organisms, attraction is expressed in instinctive reactions.	Great Encyclopedia of Psychology	1.1
Categorization is a mental operation aimed at forming a category as concepts that ultimately generalize and classify the results of human cognitive activity; the creation of categories, the process of classifying objects based on apparent similarities.	Dictionary of Philosophy	5.1.2
Classification is a general scientific method of systematizing knowledge, aimed at organizing a certain set (variety) of studied objects of various areas of reality, knowledge and activity, into a system of subordinate groups (classes), over which these objects are distributed based on their similarity in certain essential properties.	Humanitarian encyclopedia	5.1.4
Cognitive psychology is one of the leading areas of modern foreign psychology, which studies the structure and course of human cognitive processes.	Financial Terms Dictionary, 2012	1.2.3

Definition of term	Source	Section
Competitiveness of a product is a comparative characteristic of consumer and cost parameters of a given product in relation to a competing product.	Great Soviet Encyclopedia	1.2.3
Concept (*from Lat.* Conceptio – concept, idea, system) is a certain way of understanding, interpreting an object, phenomenon, process, the main point of view on the subject, etc., a guiding idea. The term "concept" is also used to denote a leading concept, a constructive principle in scientific, artistic, technical and other activities.	Great Soviet Encyclopedia	1.2
Conditioned reflexes are individually acquired complex adaptive reactions of the organism of animals and humans that arise under certain conditions (hence the name) on the basis of the formation of a temporary connection between the conditioned (signal) stimulus and the unconditionally reflex act that reinforces this stimulus.	Great Soviet Encyclopedia	5.1.5
Conformity is temporization, thoughtless adherence to general opinions, popular tendencies.	Dictionary of the Russian language by S. I. Ozhegov	6.2.4
The ***consumer value (cost)*** of a product is a set of properties associated directly with both the product itself and related services. These properties, one way or another, appear before the buyer and determine his intention to buy the product and become a regular customer of this production.	Business terms dictionary	1.4.2
Desire (in psychology) is an experience characterized by a more or less conscious idea of the aspiration to perform an act (action). The development of desire is determined both by the object of desire and by the stability, duration, and strength of desire itself. Depending on this, desires are distinguished: feasible, unreal, contradictory, reckless, targeted, etc.	Great Encyclopedia of Psychology	1.1
Didactics is a scientific area of pedagogy that explores the principles, values, patterns of functioning and development of the process of education and training.	Pedagogical Dictionary	7.1
Differentiation is dismemberment, division, stratification of the whole into different parts, forms, steps. Distinguishing a certain part of something from the total number (objects or values) according to some criteria.	Dictionary of Logic	2.3.1
Diversification (Late Latin diversificatio – change, variety *from Lat.* diversus – different and facio – I do) – an increase in the number of industries and the range of goods (services) produced by certain enterprises in new areas for them.	Financial Terms Dictionary, 2012	1.2.3
Dominant (*from Lat.* dominans – prevailing) is a temporarily prevailing reflex system that ensures the functioning of the nerve centers at the moment and thereby gives the behavior a certain direction.	Psychological Dictionary	2.3.5

Definition of term	Source	Section
Ecological and ergonomic TS properties are indicators characterizing the impact of a technical system (TS) on the environment, as well as meeting the hygienic, anthropometric, psychological, psychophysiological needs of a man interacting with TS during the creation, operation, storage, transportation, utilization or disposal of TS.	Author	1.4.2
The economic properties of TS are indicators that characterize the costs of creation, operation, storage, transportation, utilization or disposal of a technical system (TS), as well as its exchange value.	Author	1.4.2
Effect (*from Lat.* Effectus – performance, action, from efficio – act, execute, produce) – a result, consequence of any reasons, actions.	Great Soviet Encyclopedia	1.4.4
Endogenous (*from Greek* ἔνδον – inside, γένος – genus, origin) – of internal origin; in medicine – originating from the causes that lie in the internal environment of the body.	Encyclopedic Dictionary	1.2.1
Exogenous (*from Greek* ἔξω – out, outside, γένος – genus, origin) – of external origin; in medicine – originating from causes that lie outside the body.	Encyclopedic Dictionary	1.2.1
Frustration (*from Lat.* Frustratio – deception, disappointment, destruction of plans) – 1) a mental state, expressed in the characteristic features of experiences and behavior, caused by objectively insurmountable (or subjectively so understood) difficulties arising on the way to achieving a goal or task; 2) a state of collapse and depression caused by the experience of failure.	Great Encyclopedia of Psychology	7.7
Functional properties of TS are indicators characterizing functions performance of a technical system (TS) during its operation, storage, transportation, utilization or disposal.	Author	1.4.2
Functional structure of TS – spatial arrangement of interacting elements of a technical system (TS); a set of interconnected functional elements or subsystems that ensure the implementation of TS functions.	Author	1.4.3
Genesis is origin, uprising; in a broader sense – the inception and the subsequent process of development, which has led to a certain state, type, phenomenon.	Great Soviet Encyclopedia	5.1.4
Graph theory is a branch of finite mathematics, a feature of which is a geometric approach to the study of objects. The main concept of the theory is a graph. A graph is defined by a set of vertices (points) and a set of edges (links) connecting some (and perhaps all) pairs of vertices. In this case, pairs of vertices can be connected by several edges.	Great Soviet Encyclopedia	1.4.3

Definition of term	Source	Section
Heuristics (*from Greek* Εὑρίσκω – looking for, discovering) – 1) a scientific area that studies the specifics of creative activity; 2) the principle, law or pattern of technical system (TS) development, strategy, tactics, method, technique based on productive knowledge and methods of their use.	Author	2.3.2
Insight – a sudden and not derived from past experience understanding of essential relationships and the structure of the situation as a whole, through which a meaningful solution to the problem is achieved; sudden enlightenment, understanding the essence of a problem situation in the course of a strong inner experience.	Psychological Dictionary	6.3.8
Intellectual experiment is a special theoretical procedure, which consists in obtaining a new one or checking idealized objects and manipulating them in artificially conditioned situations.	The latest philosophical dictionary	6.3.5
Interest (*from Lat.* interest – has meaning, important) social – the real cause of social actions, events, accomplishments, behind the immediate impulses – motives, thoughts, ideas – of the individuals, social groups, classes participating in these actions.	Great Encyclopedia of Psychology	1.1
Inversion (*from Lat.* Inversio – overturning, rearrangement) – in psychology, the process and result of a violation of the normal order and sequence of elements, their rearrangement or replacement up to the opposite.	The latest philosophical dictionary	1.4.3
Man-made – associated with the technical and technological human activities.	Dictionary of Ecology	1.2.2
Motivation is a dynamic process of physiological and psychological control of human behavior, which determines its orientation, organization, activity and stability.	Great Encyclopedia of Psychology	1.1
Motive (*from Lat.* moveo – I move, I set in motion) is an incentive for a man's behavior and actions: 1) arising under the influence of his needs and interests; 2) representing an image of the good desired by a man.	Great Encyclopedia of Psychology	1.1
Need – the need for something or lack of something necessary to maintain the vital activity of the organism, human personality, social group, society as a whole; internal stimulus of activity.	Great Encyclopedia of Psychology	1.1
Operation principle (existence principal) of TS – cause-and-effect transformations of flows of energy, matter and information in the elements of the flow of a technical system (TS) (effects and phenomena), interconnected on the basis of the identical qualitative and accurate or close (in the range) values of quantitative proporties (parameters) characteristics of flow elements.	Author	1.4.3

Definition of term	Source	Section
Optimization of property parameters, or parametric optimization, is the process of passing parameter values at which the maximum efficiency of the procedure is achieved or the best characteristics of a device or system are provided from the point of view of the established criterion. Optimization parameter is a quantitative characteristic of a goal.	Russian-English Dictionary of Normative and Technical Terminology	1.4.6
Phenomenon (physical) – an action, process and property of the world around us, which do not depend on human action. Types of physical phenomena: mechanical; electrical; magnetic; optical; thermal; sound; atomic.	Explanatory dictionary by D. V. Dmitriev	1.4.4
Productive knowledge is reproductive knowledge, revealed hidden in the technical sphere, living and inanimate nature, as well as systematized and selected according to a certain criterion, with an increased ability to generate new knowledge.	Author	2.3.2
Property is an attribute of an object, a side of manifestation of any quality; quality, a sign that make a distinctive feature of someone, something.	Dictionary of the Russian language by S. I. Ozhegov	1.4.1
Reflection (*from Lat.* reflexus) is a form of mental activity, manifested in the desire for a constant analysis of own thoughts, actions, experiences, emotions.	The Great Explanatory Dictionary of the Modern Russian Language by D. N. Ushakov	4.2
Reproductive knowledge is the result of the cognizing reality process based on the study of descriptions of creation, operation, storage, transportation, utilization and disposal of a technical system (TS) processes.	Author	2.3.2
The ***specific*** is something multifaceted, considered in all a variety of properties and relationships.	A Brief Dictionary of Philosophical Terms	1.4.1
Stereotype is something that is repeated unchanged; a pattern of action, behavior, etc., applied without hesitation, reflection, even unconsciously.	Psychological Dictionary	2.3.4
Synonym is a word that differs from another in sound or spelling, but coincides or is close to it in meaning.	New dictionary of the Russian language by T. F. Efremova	6.2.11
Tactics – a set of techniques and means for someone to achieve their intended goal; part of the strategy.	The Great Explanatory Dictionary of the Modern Russian Language by D. N. Ushakov	3.2.3
Tactile – touchable	New dictionary of the Russian language by T. F. Efremova	6.2.2.1
Technical system (TS) is an artificial material object or human action with material objects (including natural ones) to satisfy a need.	Author	1.1
Technically realizable need is a need that can be satisfied by creating and / or using a technical system.	Author	1.1

Definition of term	Source	Section
Technological TS properties – indicators characterizing the efficiency of the processes of creation, operation, storage, transportation, utilization or disposal of a technical system (TS).	Author	1.4.2
Technology (*from Greek* τέχνη – art, mastery, skill, λόγος – word, doctrine) – a set of techniques and methods for obtaining, processing or recycling raw materials, materials, semi-finished products or products carried out in various industries; a scientific discipline that develops and improves such techniques and methods. Technology (or technological processes) is also called the operations of extraction, processing, recycling, transportation, stockpiling, storage.	Great Soviet Encyclopedia	1.1
Thesaurus (*from Lat.* thesaurus – a storehouse, a treasury) is a set of semantic-expressing units of a certain language with a system of semantic relations set on it.	Great Soviet Encyclopedia	5.1.5
TS function is an action (interaction or reaction) of a material object or phenomenon aimed at changing or preserving the properties of another material object or phenomenon.	Author	1.4.1
Utilization – the use of resources that are not directly used for their intended purpose, secondary resources, production and consumption waste.	Modern economic dictionary by B. A. Reiseberg, L. Sh. Lozovsky, E. B. Starodubtseva	6.2.2.1
Verification (*from Lat.* verus – true and facio – I do) – examination, empirical confirmation of the theoretical positions of science by comparing them with observed objects, sensory data, experiment.	Encyclopedic Dictionary	6.2.2.2

Appendix 14
Recommendations for using a bookmark for personal growth training "Find an idea!"

The aim of the training is to open up new opportunities and perspectives, get out of a problem situation, and resolve contradictions. During the training, strong internal changes often occur, manifested in the form of a kind of breakthrough, insight (enlightenment), which in general leads to the development of creative thinking, a sharp increase in self-esteem, positivity and efficiency of activity. Therefore, such trainings are called transformational or personal growth trainings.

The training device is made in the form of a bookmark. There is a through window in the upper part. There are heuristics under the window, which in general form represent productive knowledge that effectively generate new knowledge and include advice or a methodology for using this knowledge to develop new technologies, equipment, products (technical systems — TS).

Heuristics can be, for example, descriptions of groups of the most important human needs (16 items) or characteristics of classes of non-traditional (hidden) resources (9 items). Descriptions of heuristics are provided on both sides of the bookmark.

Use an attachment bookmark as follows. At arm's length, you should look through the window at any element of any TS (for example, on a door — it is a handle, keyhole or door hinge) and conduct a thought experiment, trying to find a technical solution in order to implement any heuristic in TS or TS element in generally visible through the through window. After some time (usually several days or weeks) after acquiring the skills to use heuristics for TS or TS elements in general, visible through the window at arm's length, the view should be gradually increased by bending the arm in which the bookmark-device is located and bringing it closer to your eye.

At the same time, the view expands, other TS elements, entire TS, and even an organized set of TS (for example, a part of a car, an entire car, an automobile parking) will gradually fall into the field of view. Accordingly, technical solutions suggested by heuristics can be more complex and varied. It is recommended to immediately record all emerging ideas (write down briefly, sketch).

Testing the bookmark-device by children and adults, consisting in the search for improvement ideas using heuristics of an object (or its element) that comes into the field of view (through the window), has showed the active development of creative thinking, manifested in increasing effectiveness (compared to traditional methods scientific and technical creativity) of such ideas generation. The skills of heuristics practical use acquired in the course of the training are consolidated and tone up the creative process even without the bookmark-device.

List of notes

Preface

1 Kreativnaya pedagogika. Metodologiya, teoriya, praktika [Creative pedagogy. Methodology, theory, practice] / ed. V. V. Popova, Yu. G. Kruglov. 3rd ed., rev. and add. M.: BINOM. Knowledge Laboratory, 2011. p. 18.

2 *Molyako V. A.* Psychology of creative activity. [Psihologiya tvorcheskoj deyatel'nosti.] M.: Znamya, 1978.

3 *Popov V. V.* Sozdanie i razvitie tekhnicheskih sistem na osnove potrebnostej cheloveka. [Creation and development of technical systems based on human needs.] (2nd ed.), Rev. and add. Moscow: Russian State University of Oil and Gas named after I. M. Gubkin, 2012. p. 39.

4 *Popov V. V.* (2016). Nauchno-tekhnicheskoe tvorchestvo, osnovannoe na ispol'zovanii evristik [Scientific and technical creativity based on the use of heuristics]. Yunye tekhniki i izobretateli. https://xn——itbbmalqd7b5a5d8a.xn — plai/nauchno-tehnicheskoe-tvorchestvo-osnovannoena-ispolzovanii-evristik/ (accessed May 29, 2025).

5 *Dryden G., Vos J.* Revolyuciya v obuchenii [The New Learning Revolution] / tr. from English E. Sigaeva, V. Chupin, H. Shagieva. M.: Publishing house "PARVIN", 2003.

Chapter 1

1 Tvorchestvo: ot biologicheskih osnovanij k social'nym i kul'turnym fenomenam [Creativity: from biological foundations to social and cultural phenomena] / Ed. D. V. Ushakov. M.: Publishing house "Institute of Psychology RAS", 2011.

2 Asotsial'noe povedenie [Antisocial behavior]. (n.d.). I-Goddess.RU — v mire zhenskoi krasoty. Archived by Wayback Machine. https://web.archive.org/web/20180617131747/http://www.i-goddess.ru/v-garmoniis-soboj/asocialnoe-povedenie (accessed May 29, 2025).

3 For more details see: Popov V. V. Sozdanie i razvitie tekhnicheskih sistem na osnove potrebnostej cheloveka. [Creation and development of technical systems based on human needs] (2nd ed.), rev. and add. Moscow: Russian State University of Oil and Gas named after I. M. Gubkin, 2012. p. 8.

4 *Popov V. V.* Sozdanie i razvitie tekhnicheskih sistem na osnove potrebnostej cheloveka… [Creation and development of technical systems based on human needs] pp. 8–10.

5 *Illarionova E. A., Syrovatsky I. P.* Ekologiya kak nauka. Sovremennoe sostoyanie ekologii. Ekosistema. Biosfera, gidrosfera, atmosfera, litosfera: [The current state of ecology. Ecosystem. Biosphere, hydrosphere, atmosphere, lithosphere] A manual on the basics of ecology and nature conservation for students of the Faculty of Pharmacy. Irkutsk: GOU VPO ISMU Roszdrav, 2008. pp. 53–55.

6 *Popov V. V.* Sozdanie i razvitie tekhnicheskih sistem na osnove potrebnostej cheloveka… [Creation and development of technical systems based on human needs] pp. 10–16.

7 *Nekrasova N. A., Nekrasov S. I.* Filosofiya tekhniki. [Philosophy of technology] Textbook. M.: MIIT, 2010. pp. 18–19.

8 Tekhnicheskoe tvorchestvo: Teoriya, metodologiya, praktika. Enciklopedicheskij slovar'-spravochnik [Technical creativity: Theory, methodology, practice. Encyclopedic Dictionary] / ed. A. I. Polovinkina, V. V. Popov. M.: NPO "Inform-sistema", 1995. p. 208; *Nekrasova N. A., Nekrasov S. I.* Filosofiya tekhniki. [Philosophy of technology] Textbook. M.: MIIT, 2010. pp. 18–19

9 *Popov V. V.* Sozdanie i razvitie tekhnicheskih sistem na osnove potrebnostej cheloveka… [Creation and development of technical systems based on human needs] p. 18–19.

10 *Popov V. V.* Sozdanie i razvitie tekhnicheskih sistem na osnove potrebnostej cheloveka… [Creation and development of technical systems based on human needs] p. 20.

11 On the personality theory developed by Henry Alexander Murray more detailed see: *H. Heckhausen*, Motivaciya i deyatel'nost'. [Motivation and Action.] M.: Higher school, 1986.p. 126.

12 *Popov V. V.* Sozdanie i razvitie tekhnicheskih sistem na osnove potrebnostej cheloveka… [Creation and development of technical systems based on human needs] pp. 20–21.

13 *Popov V. V.* Sozdanie i razvitie tekhnicheskih sistem na osnove potrebnostej

cheloveka… [Creation and development of technical systems based on human needs] pp. 21–22.

14 *Hill P.* The Science of Engineering Design, Holt, Rinehart & Winston, 1968. [*Hill P.* Design Science and Art. Design Methods, Scientific Rationale of Solutions] / tr. from English E. G. Kovalenko, ed. and with a preface. V. F. Venda. Moscow: Mir, 1973

15 *Popov V. V.* Sozdanie i razvitie tekhnicheskih sistem na osnove potrebnostej cheloveka… [Creation and development of technical systems based on human needs] p. 23.

16 ib.

17 *Popov V. V.* Sozdanie i razvitie tekhnicheskih sistem na osnove potrebnostej cheloveka… [Creation and development of technical systems based on human needs] pp. 23–25.

18 ib.

19 Problema 2000 goda [Year 2000 problem]. (2025). Wikipedia. https://ru.wikipedia.org/?curid=206810&oldid=142744073 (accessed May 29, 2025).

20 *Vernadsky V. I.* Biosfera i noosfera [Biosphere and noosphere] M.: Nauka, 1989. p. 145.

21 *Meadows D. H., Meadows D. L., Randers J.* Za predelami rosta [Beyond The Limits To Growth]: a manual / tr. from English M.: Progress; Pangea, 1994.

22 *Moiseev N. N.* Ekologiya chelovechestva glazami matematika. [Ecology of mankind through the eyes of a mathematician] M.: Molodaya gvardiya, 1988. pp. 73–81.

23 *Popov V. V.* Sozdanie i razvitie tekhnicheskih sistem na osnove potrebnostej cheloveka… [Creation and development of technical systems based on human needs] pp. 26–28.

24 *Obukhovsky K.* Psihologiya vlechenij cheloveka. [Psychology of human drives] M.: Progress, 1972.p. 75.

25 *Ershov P. M.* Iskusstvo tolkovaniya v 2 ch. [The art of interpretation in 2 hours] Dubna, 1997. Part 2: Rezhissura kak hudozhestvennaya kritika. [Art direction as art criticism] p. 57.

26 *Maslow A. H.* Motivation and personality. N.Y.: Harper, 1954.

27 *Kamaev V. A.* Formirovanie potrebnostej v novyh tekhnicheskih sistemah. Ispol'zovanie abduktivnogo podhoda // Mezhdunarodnaya nauchnaya shkola dlya molodezhi "Shkola nauchno-tekhnicheskogo tvorchestva i konceptual'nogo proektirovaniya". Regional'naya nauchno-prakticheskaya konferenciya "Issledovaniya molodyh uchenyh — vklad v innovacionnoe

razvitie Rossii". [Formation of requirements for new technical systems. Using the abductive approach // International scientific school for youth "School of scientific and technical creativity and conceptual design". Regional scientificpractical conference "Research of young scientists — a contribution to the innovative development of Russia."] Collection of lectures /comp. R. M. Zaripov. Astrakhan: Publishing House "Astrakhan University", 2011. p. 70.

28 *Polovinkin A. I.* Zakony stroeniya i razvitiya tekhniki. [The laws of technology structure and development] Volgograd: VPI, 1995.p. 131.

29 *Popov, V. V.* (n.d.). Sozdanie i razvitie tekhnicheskikh sistem na osnove potrebnostei cheloveka [Creation and development of technical systems based on human needs]. Yunye tekhniki i izobretateli. https://xn—— itbbmalqd7b5a5d8a.xn — p1ai/wp-content/uploads/2015/11/Popov-V.V.-Sozdanie i-razvitie-tehnicheskih-sistem-na-osnove-potrebnostejcheloveka.pdf (accessed May 29, 2025).

30 *Popov, V. V.* (n.d.). Sozdanie i razvitie tekhnicheskikh sistem na osnove potrebnostei cheloveka [Creation and development of technical systems based on human needs]. Yunye tekhniki i izobretateli. https://xn—— itbbmalqd7b5a5d8a.xn — p1ai/wp-content/uploads/2015/11/Popov-V.V.-Sozdanie-i-razvitie-tehnicheskih-sistem-na-osnove-potrebnostejcheloveka.pdf (accessed May 29, 2025).

31 Ponyatie potrebitel'skoi tsennosti s pozitsii teorii marketinga [The concept of consumer value from the perspective of marketing theory]. (n.d.). Studopedia.Org. Archived by Wayback Machine. https://web.archive.org/web/ 20170801114753/https://studopedia.org/8–104765.html (accessed May 29, 2025).

32 *Novozhenov V. A.* Koncepcii sovremennogo estestvoznaniya [Concepts of modern natural science]: Manual. Barnaul: Altay State University, 2001.

33 *Popov, V. V., et al.* (2008). O sisteme [About the system]. Mezhotraslevaya Internet-sistema poiska i sinteza fizicheskikh printsipov deistviya preobrazovatelei energii. http://fpd.informsystema.ru/home/ (accessed May 29, 2025).

34 Himicheskie transportnye reakcii // Bol'shoj enciklopedicheskij slovar': A–YA [Chemical transport reactions // Big encyclopedic dictionary: A — Z /] ed. A. M. Prokhorov. (2nd ed.), rev. and add. M. — SPb. BSE, 2000.p. 1355.

35 Spirali Arhimeda [Archimedes' spirals]. (n.d.). Velikie fiziki. Archived by Wayback Machine. https://web.archive.org/web/20170801114753/ https://studopedia.org/8–104765.html (accessed May 29, 2025).

36 *Popov V. V., et al.* (2008). O sisteme [About the system]. Mezhotraslevaya Internet-sistema poiska i sinteza fizicheskikh printsipov deistviya preobrazovatelei energii. http://fpd.informsystema.ru/home/ (accessed May 29, 2025).

37 Bol'shoj enciklopedicheskij politekhnicheskij slovar' [Big encyclopedic polytechnic dictionary] [Electronic resource]. M.: Publishing house "Multigrade", 2004. 1CD-ROM.

38 Tekhnicheskoe tvorchestvo: Teoriya, metodologiya, praktika. Enciklopedicheskij slovar'-spravochnik [Technical creativity: Theory, methodology, practice. Encyclopedic Dictionary] / ed. A. I. Polovinkina, V. V. Popov. M.: NPO "Inform-system", 1995.

39 Velosiped [Bicycle]. (n.d.). Wikipedia. https://ru.wikipedia.org/wiki/Велосипед (accessed May 29, 2025).

40 Obshchaya kharakteristika zapravochnykh supov [General characteristics of filling soups]. (n.d.). Cooke.ru. Archived by Wayback Machine. https://web.archive.org/web/20180322143306/http://www.cooke.ru/modules/Article/files/ulasevich/7_3.htm (accessed May 29, 2025).

41 Shtamm bakterii LACTOBACTERIUM ACIDOPHILUM, ispol'zuemyi dlya proizvodstva kislomolochnykh produktov [Strain of LACTOBACTERIUM ACIDOPHILUM bacteria used for the production of fermented dairy products]. (n.d.). Baza patentov Kazakhstana. Archived by Wayback Machine. https://web.archive.org/web/20200428223656/http://kzpatents.com/5-545-shtamm-bakterijj-lactobacterium-acidophilum-ispolzuemyjjdlya-proizvodstva-kislomolochnyh-produktov.html (accessed May 29, 2025).

42 Nauchno-tekhnicheskoe prognozirovanie dlya promyshlennosti i pravitel'stvennyh uchrezhdenij [Scientific -technical forecasting for industry and government agencies] / ed. prof. James Bright., tr. from English M.: Progress, 1972; *Kamenev A. F.* Tekhnicheskie sistemy: zakonomernosti razvitiya. [Technical systems: patterns of development] M.: Mechanical engineering. Leningrad branch, 1985.

Chapter 2

1 *Hill P.* The Science of Engineering Design. Holt, Rinehart & Winston, 1968 [The Science and Art of Design: Design Methods, Scientifically Based Decisions] (*Hill P.* Nauka i iskusstvo proektirovaniya: Metody proektirovaniya, nauchnoe obosnovanie reshenij. ())/ tr. from English E. G. Kovalenko, ed. and with a preface. V. F. Venda. M.: Mir, 1973)

2 The world's 10 biggest ideas. (2005). New Scientist. https://www.newscientist.com/article/mg18725171–300-the-worlds-10-biggest-ideas/ (accessed May 29, 2025).

3 *Popov V. V.* (2016). Nauchno-tekhnicheskoe tvorchestvo, osnovannoe na ispol'zovanii evristik [Scientific and technical creativity based on the use of heuristics]. Yunye tekhniki i izobretateli. https://xn itbbmalqd-7b5a5d8a.xn — p1ai/nauchno-tehnicheskoe-tvorchestvo-osnovannoe-a-ispolzovanii-evristik/ (accessed May 29, 2025).

4 *Molyako V. A.* Psihologiya tvorcheskoj deyatel'nosti. Kiev: Znamya, 1978. p. 23. [Psychology of creative activity].

5 *Popov V. V.* (2016). Nauchno-tekhnicheskoe tvorchestvo, osnovannoe na ispol'zovanii evristik [Scientific and technical creativity based on the use of heuristics]. Yunye tekhniki i izobretateli. https://xn itbbmalqd-7b5a5d8a.xn — p1ai/nauchno-tehnicheskoe-tvorchestvo-osnovannoe-na-ispolzovanii-evristik/ (accessed May 29, 2025).

6 *Davydov V. V., Zaporozhets A. V., Lomov B. F. et al.* Psihologicheskij slovar'. M.: Pedagogika M.: Pedagogy, 1983. pp. 152–153 [Psychological dictionary].

7 *Bashmakov A. I., Zhedyaevsky D. N., Popov V. V. et al.* Kreativnaya pedagogika i kompetencii tvorcheskogo specialista. M.: RGU nefti i gaza imeni I. M. Gubkina Moscow: Russian State University of Oil and Gas named after I. M. Gubkin, 2009. p. 47 [Creative pedagogy and the competence of a creative specialist].

8 Tvorchestvo: ot biologicheskih osnovanij k social'nym i kul'turnym fenomenam [Creativity: from biological foundations to social and cultural phenomena] / Ed. D. V. Ushakov. M.: Publishing house "Institute of Psychology RAS", 2011. p. 333

9 *Enikeev M. I.* Osnovy obshchej i yuridicheskoj psihologii. M.: Yurist [Fundamentals of General and Legal Psychology]. M.: Jurist, 1996. pp. 161–162.

10 *Andreichenko G. V., Pavlova I. N.* Concepts of modern natural science. Handbook for students. Stavropol: SSU, 2005. p. 171.

11 Tvorchestvo: ot biologicheskih osnovanij k social'nym i kul'turnym fenomenam [Creativity: from biological foundations to social and cultural phenomena] / Ed. D. V. Ushakov. M.: Publishing house "Institute of Psychology RAS", 2011. p. 351

12 Zakon Yerksa — Dodsona [Yerkes-Dodson law]. (2024). Wikipedia. https://ru.wikipedia.org/?curid=544368&oldid=140854549 (accessed May 29, 2025).

13 *Ukhtomsky A. A.* Dominant. M. — SPb.: Peter, 2002. p. 55. [Dominanta].

14 Motivatsiya [Motivation]. (2025). Wikipedia. https://ru.wikipedia.org/?curid=3732&oldid=144697593 (accessed May 29, 2025).

15 *Roset I. M.* Psihologiya fantazii: Eksperimental'no-teoreticheskoe issledovanie vnutrennih zakonomernostej produktivnoj umstvennoj deyatel'nosti. [Psychology of fantasy: An experimental and theoretical study of the internal patterns of productive mental activity.] (2nd ed.), Rev. and add. Minsk: Universitetskoe, 1991. p. 234.

16 *Gordeeva T. O.* Motivacionnye faktory, vliyayushchie na dostizheniya v uchebnoj deyatel'nosti, ili motivaciya obucheniya: pyat' uslovij uspekha // Psihologiya v vuze [Motivational factors influencing achievements in educational activity, or motivation for learning: five conditions for success // Psychology at the university]. 2005. No. 4. p. 4.

17 Motivatsiya [Motivation]. (2025). Wikipedia. https://ru.wikipedia.org/?curid=3732&oldid=144697593 (accessed May 29, 2025).

18 Motivatsiya [Motivation]. (2025). Wikipedia. https://ru.wikipedia.org/?curid=3732&oldid=144697593 (accessed May 29, 2025).

19 Motivatsiya [Motivation]. (2025). Wikipedia. https://ru.wikipedia.org/?curid=3732&oldid=144697593 (accessed May 29, 2025).

20 Motivatsiya [Motivation]. (2025). Wikipedia. https://ru.wikipedia.org/?curid=3732&oldid=144697593 (accessed May 29, 2025).

21 *Vedeneeva I. G.* (2010). Rol' prepodavatelya v formirovanii motivatsii dostizheniya uspekha studenta [The role of the teacher in shaping students' achievement motivation]. Festival' pedagogicheskikh idey "Otkrytyi urok". https://urok.1sept.ru/articles/573011 (accessed May 29, 2025).

22 Motivatsiya [Motivation]. (2025). Wikipedia. https://ru.wikipedia.org/?curid=3732&oldid=144697593 (accessed May 29, 2025).

23 Motivatsiya [Motivation]. (2025). Wikipedia. https://ru.wikipedia.org/?curid=3732&oldid=144697593 (accessed May 29, 2025).

Chapter 3

1 Sovremennyj tolkovyj slovar'. M.: BSE, 2003 [Modern explanatory dictionary].

2 *Archer B.* The three Rs. // Design Studies. 1979. Vol. 1. No 1. p. 16

3 GOST 2.103–2013. Edinaya sistema konstruktorskoj dokumentacii. Stadii razrabotki. M.: Standartinform, 2015. p. 5. GOST 2.103–2013.

[GOST 2.103–2013. Unified system for design documentation. Development stages].

4 *Polovinkin A. I.* Osnovy inzhenernogo tvorchestva. 2-e izd, pererab. i dop. M.: Mashinostroenie, 1988. p. 201 [Fundamentals of engineering creativity].

5 *Popov V. V.* Metodologiya konceptual'nogo proektirovaniya podzemnoj razrabotki rudnyh mestorozhdenij. Dis. … d-ra tekhn. nauk v vide nauchnogo doklada. M.: Korporaciya "Universitetskie seti znanij", 1995. p. 10 [Conceptual design methodology of ore deposits underground mining. Dis. … of D.Sc. in the form of a scientific report].

6 *Popov V. V.* Metodologiya konceptual'nogo proektirovaniya podzemnoj razrabotki rudnyh mestorozhdenij. Dis. … d-ra tekhn. nauk. p. 8 [Conceptual design methodology of ore deposits underground mining].

7 ib.

8 *Drucker P. F.* Management Challenges for the 21st Century: Textbook / Tr. from English M.: Publishing house "Williams", 2002. p. 40. [Zadachi menedzhmenta v XXI veke: Uchebnoe posobie / Per. s angl].

Chapter 4

1 Bol'shaya sovetskaya enciklopediya v 30 t. M.: Sovetskaya enciklopediya, 1975. T. 21: Proba–Remensy. p. 132 [Great Soviet Encyclopedia in 30 volumes].

2 *Volkov V. V.* Filosofiya: Osnovnye problemy, ponyatiya, terminy. Uchebnoe posobie. SPb.: Sankt-Peterburgskij institut psihologii i akmeologii, 2010. p. 17. [Philosophy: Basic problems, concepts, terms. Textbook].

3 *Alekseev P. V., Panin A. V.* Filosofiya: Uchebnik. 3-e izd., pererab. i dop. M.: TK Velbi, Izd-vo Prospekt, 2003. p. 423 [Philosophy: Textbook. 3rd ed., Rev. and add].

4 *Alekseev P. V., Panin A. V.* Filosofiya… pp. 423–424 [Philosophy].

5 Elektroliticheskaya dissociaciya: metodicheskie ukazaniya k laboratornoj rabote po kursu "Obshchaya i neorganicheskaya himiya". Komsomol'sk-na-Amure: FGBOU VPO "KnAGTU", 2015. p. 3. [Electrolytic dissociation: guidelines for laboratory work on the course "General and inorganic chemistry"].

6 Filosofskaya Enciklopediya. v 5 tt. / Pod red. F. V. Konstantinova i dr. M.: Sovetskaya enciklopediya, 1962. T. 2: Diz"yunkciya Komicheskoe. p. 20. [Philosophical Encyclopedia. In 5 vols].

7 *Altshuller G. S.* Kak nauchit'sya izobretat'. Tambov: Kn. izd-vo, 1961 [How to learn inventing]; *The same author:*. Osnovy izobretatel'stva. Voronezh: Centr.-CHernozem. kn. izd-vo, 1964 [Fundamentals of Invention]; *The same author:*. Algoritm izobreteniya. 2-e izd., ispr. i dop. M.: Mosk. rabochij, 1973 [Algorithm of the invention. (2nd ed.), Rev. and add.]; *The same author:*. Tvorchestvo kak tochnaya nauka: Teoriya resheniya izobretatel'skih zadach. M.: Sov. radio, 1979 [Creativity as an Exact Science: Theory of Inventive Problem Solving.]; *Altshuller G. S., Selyuckij A. B.* Kryl'ya dlya Ikara: Kak reshat' izobretatel'skie zadachi. Petrozavodsk: Kareliya, 1980 [Wings for Icarus: How to Solve Inventive Problems]; *Altshuller G. S., Zusman A. V., Filatov V. I.* Poisk novyh idej: Ot ozareniya k tekhnologii (Teoriya i praktika resheniya izobretatel'skih zadach). Kishinev: Kartya Moldavenyaske, 1989 [Search for new ideas: From insight to technology (Theory and practice of inventive problem solving)]; *Altshuller G. S.* Kak najti ideyu: Vvedenie v teoriyu resheniya izobretatel'skih zadach / Pod red. A. K. Dyunina. 2-e izd. ispr. i dop. Novosibirsk: Nauka, 1991 [How to find an idea: Introduction to the theory of inventive problem solving]; *Altshuller G. S., Vertkin I. M.* Kak stat' geniem: Zhiznennaya strategiya lichnosti. Minsk: Belarus', 1994 [How to become a genius: The life strategy of a personality]; *Altov G.* I tut poyavilsya izobretatel'… M.: Det. lit., 2000 [And then the inventor appeared…].

8 In more detail about heuristic techniques in general, as well as about the mechanism for resolving typical contradictions using typical heuristic techniques, which was developed by *G. S. Altshuller* see in section 6.3.2.

Chapter 5

1 Bol'shoj enciklopedicheskij slovar' / Pod red. A. M. Prohorova. 2-e izd., pererab. i dop. M.; SPb., 2000. p. 545. [Big Encyclopedic Dictionary]

2 *Atkinson T., Claxton G.* The Intuitive Practitioner: On the Value of Not Always Knowing what One is Doing. Buckingham, 2000.

3 *Il'in E. P.* Psihologiya tvorchestva, kreativnosti i odarennosti. SPb.: Piter, 2009. p. 77–78. [Psychology of creativity and giftedness].

4 i.b.p.81.

5 Enciklopediya epistemologii i filosofii nauki / Sost. i obshch. red.

I. T. Kasavin. M.: Kanon+ ROOI "Reabilitaciya", 2009. p. 281 [Encyclopedia of Epistemology and Philosophy of Science].

6 Bol'shaya sovetskaya enciklopediya. M.: Sovetskaya enciklopediya, 1973. T. 12: Kvarner–Kongur. p. 492 [Great Soviet Encyclopedia].

7 Novejshij filosofskij slovar' / Pod red. A. A. Gricanova. Minsk: Knizhnyj Dom, 2003. p. 124 [The latest philosophical dictionary].

8 Filosofskij enciklopedicheskij slovar' / Pod red. L. F. Il'ichyova, P. N. Fedoseeva, S. M. Kovalyova, V. G. Panova. M.: Sovetskaya enciklopediya, 1983. p. 661. [Philosophical Encyclopedic Dictionary]

9 *Bush G. YA.* Analogiya i tekhnicheskoe tvorchestvo. Riga: Avots, 1981. p. 131 [Analogy and technical creativity].

10 *Vojshvillo E. K., Degtyarev M. G.* Logika. M.: Vlados-Press, 2001. p. 413. [Logic].

11 *Bush G. YA.* Analogiya i tekhnicheskoe tvorchestvo. Riga: Avots, 1981. pp. 126–127 [Analogy and technical creativity].

12 Bol'shaya sovetskaya enciklopediya. M.: Sovetskaya enciklopediya, 1970. T. 2: Angola–Barzas. p. 331. [Great Soviet Encyclopedia].

13 Filosofskij enciklopedicheskij slovar' / Pod red. L. F. Il'ichyova, P. N. Fedoseeva, S. M. Kovalyova, V. G. Panova. M.: Sovetskaya enciklopediya, 1983. p. 701. [Philosophical Encyclopedic Dictionary]

14 Bol'shaya sovetskaya enciklopediya. M.: Sovetskaya enciklopediya, 1971. T. 5: Veshin–Gazli. p. 351 [Great Soviet Encyclopedia].

15 Filosofskij enciklopedicheskij slovar' / Pod red. L. F. Il'ichyova, P. N. Fedoseeva, S. M. Kovalyova, V. G. Panova. M.: Sovetskaya enciklopediya, 1983. p. 91 [Philosophical Encyclopedic Dictionary].

16 *Vygotskij L. S.* Razvitie vysshih psihicheskih funkcij. M.: Nauka, 1960. pp. 93–94 [Development of higher mental functions].

17 *Filatov F. R.* Psihologiya dlya studentov vuzov. Rostov-na-Donu: Feniks, 2004. p. 66 [Psychology for university students].

18 *Filatov.* Psihologiya dlya studentov vuzov... p. 68 [Psychology for university students...].

19 *Filatov.* Psihologiya dlya studentov vuzov... p. 66 [Psychology for university students...].

20 *Filatov.* Psihologiya dlya studentov vuzov... pp. 66–67 [Psychology for university students...].

21 *Filatov.* Psihologiya dlya studentov vuzov... p. 67 [Psychology for university students...].

22 ib.

23 *Stepanov S. S.* Populyarnaya psihologicheskaya enciklopediya. M.: Eksmo, 2005. p. 12 [Popular psychological encyclopedia].

24 Kumulyativnyi effekt [Cumulative effect]. (2024). Wikipedia. https://ru.wikipedia.org/?curid=292938&oldid=141637577 (accessed May 29, 2025).

25 *Maklakov A. G.* Obshchaya psihologiya. SPb.: Piter, 2001. p. 290 [General psychology].

26 Bol'shoj enciklopedicheskij slovar': A–YA / Pod red. A. M. Prohorova. 2-e izd., pererab. i dop. M. — SPb.: BSE, 2000. p. 1226.= [Big Encyclopedic Dictionary: A — Z].

27 *Altshuller G. S.* Algoritm izobreteniya. 2-e izd., ispr. i dop. M.: Mosk. rabochij, 1973. p. 239 [Invention algorithm. (2nd ed.)].

28 Bol'shaya sovetskaya enciklopediya. M.: Sovetskaya enciklopediya, 1972. T. 10: Iva–Italiki. p. 34 [Great Soviet Encyclopedia].

29 *Gareev T. F.* Korporativnye strategii v stile Agile. Prakticheskoe rukovodstvo. Razrabotka, vnedrenie i monitoring realizacii innovacionnyh korporativnyh strategij. M.: Izdatel'skie resheniya, 2016. p. 45. [Corporate strategies in the style of Agile. A practical guide. Development, implantation and monitoring of the implementation of innovative corporate strategies].

30 *Altshuller G. S.* Algoritm izobreteniya. 2-e izd., ispr. i dop. M.: Mosk. rabochij, 1973. p. 84 [Invention algorithm. (2nd ed.)]; *The same author.* Tvorchestvo kak tochnaya nauka: Teoriya resheniya izobretatel'skih zadach. M.: Sov. radio, 1979. p. 95 [Creativity as an Exact Science: Theory of Inventive Problem Solving]; *Altshuller G. S., Zlotin B. L., Zusman A. V., Filatov V. I.* Poisk novyh idej: ot ozareniya k tekhnologii (Teoriya i praktika resheniya izobretatel'skih zadach). Kishinev: Kartya Moldavenyaske, 1989. p. 50 [Search for new ideas: from insight to technology (Theory and practice of inventive problem solving)]; *Altshuller G. S.* Kak najti ideyu: Vvedenie v teoriyu resheniya izobretatel'skih zadach / Otv. red. A. K. Dyunin. SO AN SSSR. 2-e izd. ispr. i dop. Novosibirsk: Nauka, 1991. p. 19 [How to find an idea: Introduction to the theory of inventive problem solving].

31 *Popov V. V.* (2023). Metodicheskie rekomendatsii po formirovaniyu aktual'nykh tvorcheskikh zadach: dlya obuchayushcheisya molodezhi i nachinayushchikh predprinimatelei [Methodological recommendations for forming relevant creative tasks: For students and novice entrepre-

neurs] (2nd ed.). Rusevroprint.

32 Radical fantasy leads to abrupt changes in the parameters of features or the emergence of completely new features of objects.

33 *Popov V. V.* (2023). Metodicheskie rekomendatsii po formirovaniyu aktual'nykh tvorcheskikh zadach: dlya obuchayushcheisya molodezhi i nachinayushchikh predprinimatelei [Methodological recommendations for forming relevant creative tasks: For students and novice entrepreneurs] (2nd ed.). Rusevroprint.

34 ib.

35 ib.

36 ib.

37 ib.

38 Bol'shaya sovetskaya enciklopediya. M.: Sovetskaya enciklopediya, 1973. T. 13: Konda–Kun. p. 594 [Great Soviet Encyclopedia].

39 *Hill P.* The Science of Engineering Design. Holt, Rinehart & Winston, 1968 [The Science and Art of Design: Design Methods, Scientifically Based Decisions] (*Hill P.* Nauka i iskusstvo proektirovaniya: Metody proektirovaniya, nauchnoe obosnovanie reshenij.)/ tr. from English E. G. Kovalenko, ed. and with a preface. V. F. Venda. M.: Mir, 1973).

Chapter 6

1 *Popov V. V.* Metodologiya konceptual'nogo proektirovaniya podzemnoj razrabotki rudnyh mestorozhdenij. Dis. … d-ra tekhn. nauk v vide nauchnogo doklada. M., 1995. p. 7–9 [Conceptual design methodology of underground ore deposits mining. Dis. … D.Sc. in the form of a scientific report]; *Popov V. V.* Razvitie tekhnicheskih sistem na osnove potrebnostej cheloveka. M.: Russian State University of Oil and Gas named after I. M. Gubkin, 2008. p. 43 [Development of technical systems based on human needs].

2 *Popov.* Sozdanie i razvitie tekhnicheskih sistem na osnove potrebnostej cheloveka… pp. 51–52 [Creation and development of technical systems based on human needs…]

3 *Bush G. YA.* Problemnye zadachi i regulyativy poiska ih resheniya: Uchebnoe posobie. M.: VNIIPI, 1989. p. 46 [Problematic tasks and regulators of the search for their solution: Textbook].

4 *Levitas A.* (2005, June 9). Rassylka "Bol'she deneg ot Vashego bizne-

sa!". Vypusk № 38 [Mailing list "More money from your business!" Issue No. 38]. Archived by Wayback Machine. https://web.archive.org/web/20240421170641/http://www.levitas.ru/BD/bd38.htm (accessed May 29, 2025).

5 *Tring M., Lejtuejt E.* Kak izobretat'? M.: Mir, 1980. p. 97. [How to Invent?]

6 *Popov.* Sozdanie i razvitie tekhnicheskih sistem na osnove potrebnostej cheloveka... p. 38 [Creation and development of technical systems based on human needs...].

7 *Druker P. F.* Zadachi menedzhmenta v XXI veke: Uchebnoe posobie / Per. s angl. M.: Izdatel'skij dom "Vil'yams", 2002. p. 37 [Management Challenges for the 21st Century: Textbook].

8 *Popov.* Razvitie tekhnicheskih sistem na osnove potrebnostej cheloveka... pp. 38–39 [Development of technical systems based on human needs...].

9 *Popov.* Razvitie tekhnicheskih sistem na osnove potrebnosti cheloveka... pp. 34–36 [Development of technical systems based on human needs...]; *The same author.* Sozdanie i razvitie tekhnicheskih sistem na osnove potrebnostej cheloveka... p. 39 [Creation and development of technical systems based on human needs...].

10 *Popov.* Sozdanie i razvitie tekhnicheskih sistem na osnove potrebnostej cheloveka... p. 39 [Creation and development of technical systems based on human needs...].

11 *Popov V. V., Chikichev D. N.* Razvitie tvorcheskih potencialov shkol'nikov i razrabotka novyh tekhnicheskih reshenij na osnove ispol'zovaniya ustojchivyh potrebnostej cheloveka. M.: OOO "Izdatel'stvo "NEFT' i GAZ", 2016. pp. 11–12 [Development of the creative potential of schoolchildren and the development of new technical solutions based on the use of sustainable human needs].

12 *Antonov O. K.* Krasivyj samolet letaet horosho... // Russkij yazyk za rubezhom. 1976. No 6. pp. 93–97 [A beautiful plane flies well... // Russian language abroad].

13 *Popov V. V., Chikichev D. N.* Razvitie tvorcheskih potencialov shkol'nikov... p. 13 [Development of the creative potential of schoolchildren...].

14 Quoted from: *Popov V. V.* Razvitie tekhnicheskih sistem na osnove potrebnosti cheloveka. M.: Russian State University of Oil and Gas named after I. M. Gubkin, 2008. p. 36 [Development of technical systems based on human needs]; *Popov, Chikichev.* Razvitie tvorcheskih potencialov shkol'nikov... p. 14 [Development of creative potential of schoolchildren...].

15 *Bashmakov A. I., & Popov V. V.* (n.d.). Kriterii progressivnogo razvitiya tekhnicheskikh sistem [Criteria for the progressive development of technical systems]. Elektronnyi entsiklopedicheskii slovar'-spravochnik "Kreativnaya pedagogika". Archived by Wayback Machine. https://web.archive.org/web/20181005023945/http://pedagogics.ru:80/content/kriteriiprogressivnogo-razvitiya-tekhnicheskikh-sistem (accessed May 29, 2025).

16 *Brahman T. R.* Mnogokriterial'nost' i vybor al'ternativy v tekhnike. M.: Radio i svyaz', 1984. p. 7 [Multicriteria and the choice of alternatives in technology].

17 *Dzhons Dzh.K.* Metody proektirovaniya. M.: Radio i svyaz': Mir, 1986. p. 303.

18 *Bashmakov A. I., & Popov V. V.* (n.d.). Kriterii progressivnogo razvitiya tekhnicheskikh sistem [Criteria for the progressive development of technical systems]. Elektronnyi entsiklopedicheskii slovar'-spravochnik "Kreativnaya pedagogika". Archived by Wayback Machine. https://web. archive.org/web/20181005023945/http://pedagogics.ru:80/content/kriteriiprogressivnogo-razvitiya-tekhnicheskikh-sistem (accessed May 29, 2025).

19 This method is described in more detail in the book: *Zlotin B. L., Zusman A. V.* Reshenie issledovatel'skih zadach. Kishinev: MNTC Progress, Kartya Moldavenyaske, 1991. p. 84 [Solution of research problems].

20 Quoted from: *Brahman T. R.* Mnogokriterial'nost' i vybor al'ternativy v tekhnike. M., 1984. p. 99 [Multicriteria and the choice of alternatives in technology].

21 *Veblen T.* Teoriya prazdnogo klassa / Per. S. Sorokinoj. M.: Progress, 1984. [Theory of the leisure class].

22 *Popov.* Metodologiya konceptual'nogo proektirovaniya podzemnoj razrabotki rudnyh mestorozhdenij. Dis. … d-ra tekhn. nauk… pp. 14–16 [Methodology for conceptual design of underground ore deposits mining. Dis. … D.Sc. …].

23 Filosofskaya enciklopediya: v 5 t. / Pod red. F. V. Konstantinova. M.: Sovetskaya enciklopediya, 1960–1970. [Philosophical Encyclopedia: in 5 volumes].

24 *Altshuller G. S.* Algoritm izobreteniya. 2-e izd., ispr. i dop. M.: Mosk. rabochij, 1973 [Invention algorithm. (2nd ed.)]; The same author. Tvorchestvo kak tochnaya nauka: Teoriya resheniya izobretatel'skih zadach. M.: Sov. radio, 1979 [Creativity as an Exact Science: Theory of Inventive Problem

Solving].

25 *Orlov M. A.* Osnovy klassicheskoj TRIZ. Vvodnyj kurs vysokoeffektivnogo innovacionnogo myshleniya. 3-e izd., ispr. i dop. M.: SOLON-PRESS, 2008 [Foundations of classical TRIZ. An introductory course on highly effective innovative thinking. (3rd ed.)]; *Revenkov A. V., Rezchikova E. V.* Teoriya i praktika resheniya tekhnicheskih zadach: Uchebnoe posobie. M.: FORUM, 2008 [Theory and practice of solving technical problems: Textbook]; *Shpakovskij N. A.* Derev'ya evolyucii. Analiz tekhnicheskoj informacii i generaciya novyh idej: Uchebnometodicheskoe posobie. M.: TRIZ-profi, 2006 [Trees of Evolution. Analysis of technical information and generation of new ideas: Study guide].

26 *Tovmasyan S. S.* Filosofskie problemy truda i tekhniki. M.: Mysl', 1972. [Philosophical problems of labor and technology].

27 *Popov.* Metodologiya konceptual'nogo proektirovaniya podzemnoj razrabotki rudnyh mestorozhdenij. Dis. … d-ra tekhn. nauk… pp. 20–21 [Methodology for conceptual design of underground ore deposits mining. Dis. … D.Sc. …].

28 *Bashmakov A. I., Zhedyaevskiy D. N., Popov V. V., Chikichev D. N.* Metodologiya ispol'zovaniya zakonov i zakonomernostej razvitiya tekhnicheskih sistem kak sredstvo dlya intensifikacii i povysheniya effektivnosti konceptual'nogo proektirovaniya tekhnologij, tekhniki i izdelij. M.: Izdatel'stvo "NEFT' i GAZ", 2013. p. 18 [Methodology of using the laws and patterns of technical systems development as a means for intensifying and increasing the conceptual design efficiency of technologies, equipment and products].

29 *Bashmakov A. I., Zhedyaevskiy D. N., Popov V. V., Chikichev D. N.* Metodologiya ispol'zovaniya zakonov i zakonomernostej razvitiya tekhnicheskih sistem. pp. 19–20 [Methodology of using the laws and patterns of technical systems development].

30 *Feofanov YU. A.* Funkcional'nyj analiz tekhnicheskih ob"ektov. Uchebnoe posobie dlya studentov vuzov. M., 2005. p. 22 [Functional analysis of technical objects. Textbook for university students].

31 *Bashmakov A. I., Zhedyaevskiy D. N., Popov V. V., Chikichev D. N.* Metodologiya ispol'zovaniya zakonov i zakonomernostej razvitiya tekhnicheskih sistem. pp.19–20 [Methodology of using the laws and patterns of technical systems development].

32 *Bashmakov A. I., Zhedyaevskiy D. N., Popov V. V., Chikichev D. N.*

Metodologiya ispol'zovaniya zakonov i zakonomernostej razvitiya tekhnicheskih sistem. pp. 19–20 [Methodology of using the laws and patterns of technical systems development].

33 Pat. 2280562 Rossijskaya Federaciya, MPK v 60 v 19/00, MPK v 60 k 17/14. Kolesnyj dvizhitel' [Wheel propulsor] / Denisenko G. I.; zayavitel' i patentoobladatel' Denisenko G. I. No 2004131015/11; zayavl. 13.10.04; opubl. 20.03.06, Byul. No 21.

34 *Bashmakov A. I., Zhedyaevskiy D. N., Popov V. V., Chikichev D. N.* Metodologiya ispol'zovaniya zakonov i zakonomernostej razvitiya tekhnicheskih sistem. pp. 43–44 [Methodology of using the laws and patterns of technical systems development].

35 *Bashmakov A. I., Zhedyaevskiy D. N., Popov V. V., Chikichev D. N.* Metodologiya ispol'zovaniya zakonov i zakonomernostej razvitiya tekhnicheskih sistem p. 48. [Methodology of using the laws and patterns of technical systems development].

36 *Bashmakov A. I., Zhedyaevskiy D. N., Popov V. V., Chikichev D. N.* Metodologiya ispol'zovaniya zakonov i zakonomernostej razvitiya tekhnicheskih sistem. p. 52. [Methodology of using the laws and patterns of technical systems development].

37 *Bashmakov A. I., Zhedyaevskiy D. N., Popov V. V., Chikichev D. N.* Metodologiya ispol'zovaniya zakonov i zakonomernostej razvitiya tekhnicheskih sistem. p. 54. [Methodology of using the laws and patterns of technical systems development].

38 *Bashmakov A. I., Zhedyaevskiy D. N., Popov V. V., CHikichev D. N.* Metodologiya ispol'zovaniya zakonov i zakonomernostej razvitiya tekhnicheskih sistem. pp. 19–20. [Methodology of using the laws and patterns of technical systems development].

39 Drawing by V. V. Popov

40 *Borshch-Kompaniec V. I.* Kompleksnye issledovaniya proyavleniya gornogo davleniya pri razrabotke Dzhezkazganskogo mednorudnogo bassejna kamerno-stolbovymi sistemami: Avtoref. dis. … d-ra tekhn. nauk. M., 1967 [Complex studies of the manifestation of rock pressure in the development of the Dzhezkazgan copper ore basin with chamber-andpillar systems: Author's abstract. dis. … D.Sc.].

41 Drawing by V. V. Popov

42 *Bashmakov A. I., Zhedyaevskiy D. N., Popov V. V., Chikichev D. N.* Vyyavlenie zakonomernostej razvitiya tekhnicheskih sistem i formirovanie ih opisanij

// Vestnik komp'yuternyh i informacionnyh tekhnologij. 2014. No 7. p. 25. [Identification of technical systems development patterns and the formation of their descriptions // Bulletin of computer and information technologies].

43 *Bashmakov A. I., Zhedyaevskiy D. N., Popov V. V., Chikichev D. N.* Metodologiya ispol'zovaniya zakonov i zakonomernostej razvitiya tekhnicheskih sistem. pp. 43–44 [Methodology of using the laws and patterns of technical systems development].

44 *Bashmakov A. I., Zhedyaevskiy D. N., Popov V. V., Chikichev D. N.* Metodologiya ispol'zovaniya zakonov i zakonomernostej razvitiya tekhnicheskih sistem. p. 45. [Methodology of using the laws and patterns of technical systems development].

45 *Bashmakov A. I., Zhedyaevskiy D. N., Popov V. V., Chikichev D. N.* Metodologiya ispol'zovaniya zakonov i zakonomernostej razvitiya tekhnicheskih sistem. pp. 45–46 [Methodology of using the laws and patterns of technical systems development].

46 *Bashmakov A. I., Zhedyaevskiy D. N., Popov V. V., Chikichev D. N.* Metodologiya ispol'zovaniya zakonov i zakonomernostej razvitiya tekhnicheskih sistem. p. 45 [Methodology of using the laws and patterns of technical systems development]

47 *Bashmakov A. I., Zhedyaevskiy D. N., Popov V. V., Chikichev D. N.* Metodologiya ispol'zovaniya zakonov i zakonomernostej razvitiya tekhnicheskih sistem. p. 25 [Methodology of using the laws and patterns of technical systems development].

48 *Bashmakov A. I., Zhedyaevskiy D. N., Popov V. V., Chikichev D. N.* Metodologiya ispol'zovaniya zakonov i zakonomernostej razvitiya tekhnicheskih sistem. p. 48. [Methodology of using the laws and patterns of technical systems development].

49 *Bashmakov A. I., Zhedyaevskiy D. N., Popov V. V., Chikichev D. N.* Metodologiya ispol'zovaniya zakonov i zakonomernostej razvitiya tekhnicheskih sistem. pp. 50–52. [Methodology of using the laws and patterns of technical systems development].

50 *Bashmakov A. I., Zhedyaevskiy D. N., Popov V. V., Chikichev D. N.* Metodologiya ispol'zovaniya zakonov i zakonomernostej razvitiya tekhnicheskih sistem. p. 52 [Methodology of using the laws and patterns of technical systems development].

51 *Bashmakov A. I., Zhedyaevskiy D. N., Popov V. V., Chikichev D. N.* Metodologiya ispol'zovaniya zakonov i zakonomernostej razvitiya tekh-

nicheskih sistem. p. 54 [Methodology of using the laws and patterns of technical systems development].

52 *Bashmakov A. I., Zhedyaevskiy D. N., Popov V. V., Chikichev D. N.* Metodologiya ispol'zovaniya zakonov i zakonomernostej razvitiya tekhnicheskih sistem. p. 54. [Methodology of using the laws and patterns of technical systems development].

53 *Bashmakov A. I., Zhedyaevskiy D. N., Popov V. V., Chikichev D. N.* Metodologiya ispol'zovaniya zakonov i zakonomernostej razvitiya tekhnicheskih sistem. p. 56 [Methodology of using the laws and patterns of technical systems development].

54 *Bashmakov A. I., Zhedyaevskiy D. N., Popov V. V., Chikichev D. N.* Metodologiya ispol'zovaniya zakonov i zakonomernostej razvitiya tekhnicheskih sistem. pp. 52–54 [Methodology of using the laws and patterns of technical systems development].

55 *Rejmers N. F.* Ohrana prirody i okruzhayushchej cheloveka sredy: Slovar'spravochnik. M.: Prosveshchenie, 1992 [Protection of nature and human environment: Dictionary-reference book].

56 *Koberg D., Bagnall J.* The Universal Traveler: A Soft-Systems Guide to Creativity, Problem-Solving, and the Process of Reaching Goals. Ann Arbor, 1974.

57 *Popov V. V.* (n.d.). Kurs dopolnitel'nogo obrazovaniya detei "Naidi ideyu!". Prezentatsiya k zanyatiyu No 8 [Children's supplementary education course "Find an idea!". Presentation for lesson No. 8]. Myslitel'noe karate. https://xn-80aanacnqecgpm3aerj3lya.xn — p1ai/learning/program_01/ presentations (accessed May 29, 2025).

58 *Gagliano M., Grimonprez M., Depczynski M., Renton M.* Turn in: plant roots use sound to locate water // Oecologia. May 2017. Vol. 184. No 1. pp. 151–160.

59 Umnye rasteniya [Smart plants]. (n.d.). Vash Sad. Archived by Wayback Machine. https://web.archive.org/web/20200218050251/http://www.vashsad.ua/plants/interesting_plants/show/7116/ (accessed May 29, 2025).

60 *Tamarchenko N. D.* Teoreticheskaya poetika: ponyatiya i opredeleniya. Hrestomatiya dlya studentov filologicheskih fakul'tetov. M.: RGGU, 1999. p. 81 [Theoretical poetics: concepts and definitions. Reader for students of philological faculties].

61 *Popov V. V., & Chikichev D. N.* (2016). Metod fokal'nykh ob"ektov i ego ispol'zovanie dlya preodoleniya psikhologicheskikh bar'erov myshleniya [The method of focal objects and its use for overcoming psychological barriers of thinking]. OOO "Izdatel'stvo "NEFT' i GAZ".

62 *Popov V. V.* (2016). Nauchno-tekhnicheskoe tvorchestvo, osnovannoe na ispol'zovanii evristik [Scientific and technical creativity based on the use of heuristics]. Yunye tekhniki i izobretateli. https://xn——itbbmalqd-7b5a5d8a.xn — p1ai/nauchno-tehnicheskoe-tvorchestvoosnovannoe-na-ispolzovanii-evristik/ (accessed May 29, 2025).

63 *Polovinkin A. I.* Osnovy inzhenernogo tvorchestva. M.: Mashinostroenie, 1988 [Fundamentals of engineering creativity].

64 *Altshuller G. S.* Tvorchestvo kak tochnaya nauka: Teoriya resheniya izobretatel'skih zadach. M.: Sov. radio, 1979. p. 54. [Creativity as an Exact Science: Theory of Inventive Problem Solving].

65 *Altshuller G. S.* Algoritm izobreteniya. 2-e izd., ispr. i dop. M.: Mosk. rabochij, 1973. p. 182 [Invention algorithm. (2nd ed.)].

66 Bol'shaya sovetskaya enciklopediya v 30 t. M.: Sovetskaya enciklopediya, 1975. V.22. p. 49 [Great Soviet Encyclopedia in 30 volumes].

67 *Rejmers N. F.* Prirodopol'zovanie: Slovar'-spravochnik. M.: Mysl', 1990. p. 451 [Nature management: Dictionary-reference book].

68 *Altshuller G. S.* Algoritm izobreteniya… [Algoritm izobreteniya…]; *The same author.* Tvorchestvo kak tochnaya nauka… [Creativity as an Exact Science].

69 *Popov V. V.* Metodologiya ispol'zovaniya resursov, imeyushchihsya v tekhnicheskih sistemah i okruzhayushchej ih srede kak mezhotraslevoe sredstvo dlya intensifikacii i povysheniya effektivnosti innovacionnoj deyatel'nosti // Vestnik komp'yuternyh i informacionnyh tekhnologij. 2011. No 8. pp. 27–33 [Methodology of using the resources available in technical systems and their environment as an intersectoral means for intensifying and increasing the efficiency of innovation activity // Bulletin of computer and information technologies].

70 Federal Law No. 89-FZ of June 24, 1998 "Ob othodah proizvodstva i potrebleniya". Stat'ya 12. Trebovaniya k ob"ektam razmeshcheniya othodov. ["On Production and Consumption Waste". Article 12. Requirements for waste disposal facilities].

71 *Altshuller G. S.* Tvorchestvo kak tochnaya nauka. p. 71 [Creativity as an Exact Science].

72 Bol'shaya rossijskaya enciklopediya: v 35 t. M.: Izd-vo Bol'shaya rossijskaya enciklopediya, 2002. Vol. 15, p. 127. [The Great Russian Encyclopedia: in 35 volumes].

73 ib.

74 *Bashmakov A. I., Zhedyaevskiy D. N., Popov V. V., CHechenov H. D., Filatova M. N., Sychevoj M. D., Vorob'eva E. V., Kashapova D. A.* Metody i sredstva sistematizacii svedenij o materialah s neobychnymi svojstvami i ih ispol'zovaniya dlya obucheniya professional'nomu tvorchestvu: Uchebnoe i proizvodstvenno-prakticheskoe posobie. M.: Russian State University of Oil and Gas named after I. M. Gubkin, 2009, pp. 9–10 [Methods and means of systematizing information about materials with unusual properties and their use for teaching professional creativity: Educational and production-practical manual].

75 *Gasanov A. I., Gomhan B. M., Efimochkin A. P. i dr.* Rozhdenie izobreteniya: Strategiya i taktika resheniya izobretatel'skih zadach. M.: Intepraks, 1995. p. 103 [The birth of an invention: Strategy and tactics for solving inventive problems]; Zajniev G. A. Evolyuciya i izobretatel'. M.: Academia, 2007. p. 240 [Evolution and the Inventor]; The same author. Inkubator pervichnyh izobretenij. M.: Academia, 2007. pp. 25, 242 [Incubator for primary inventions].

76 Grazhdanskij kodeks Rossijskoj Federacii (chast' chetvertaya) ot 18.12.2006 No 230-FZ. St. 1225. Ohranyaemye rezul'taty intellektual'noj deyatel'nosti i sredstva individualizacii. [Civil Code of the Russian Federation (part four) dated 18.12.2006 No. 230-FZ. Art. 1225. Protected results of intellectual activity and means of individualization].

77 Grazhdanskij kodeks Rossijskoj Federacii (chast' chetvertaya) ot 18.12.2006 No 230-FZ. St. 1350. Usloviya patentosposobnosti izobreteniya.. [Civil Code of the Russian Federation (part four) dated 18.12.2006 No. 230-FZ. Art. 1350. Conditions of patentability of an invention].

78 *Ryzhov K. V.* 100 velikih izobretenij. M.: Veche, 1999. p. 5. [100 great inventions.

79 *Popov V. V.* (2016). Nauchno-tekhnicheskoe tvorchestvo, osnovannoe na ispol'zovanii evristik [Scientific and technical creativity based on the use of heuristics]. Yunye tekhniki i izobretateli. https://xn——itbbmalqd7b5a5d8a.xn — p1ai/nauchno-tehnicheskoe-tvorchestvo-osnovannoena-ispolzovanii-evristik/ (accessed May 29, 2025).

80 Mirovoi opyt vnedreniya proektov v sfere Industrial'nogo (Promyshlennogo) Interneta veshchei [Global experience in implementing Industrial Internet of Things projects]. (n.d.). IT Channel News. https://www.novosti-itkanala.ru/news/detail.php? ID=115959 (accessed May 29, 2025).

Chapter 7

1 Kreativnaya pedagogika. Metodologiya, teoriya, praktika / Pod red. V. V. Popova, YU.G. Kruglova. M.: BINOM. Knowledge Laboratory, 2012, p. 14 [Creative pedagogy. Methodology, theory, practice].

2 *Popov V. V., & Kruglov Yu. G.* (2014). Kreativnaya pedagogika. Metodologiya, teoriya, praktika [Creative pedagogy. Methodology, theory, practice]. Laboratoriya znanii.

3 *Popov V. V., CHernegov YU. A., Vzyatyshev V. F., Polovinkin A. I.* O "tret'ej kul'ture" ili "Dizajne s bol'shoj bukvy" // Sovremennaya vysshaya shkola. 1991. No. 1. pp. 13–14 [On the "third culture" or "Design with a capital letter" // Modern Higher School].

4 Kreativnaya pedagogika. Metodologiya, teoriya, praktika / Pod red. V. V. Popova, YU.G. Kruglova. M.: BINOM. Knowledge Laboratory, 2012. [Creative pedagogy. Methodology, theory, practice / Ed. V. V. Popov, Yu. G. Kruglov].

5 Kreativnaya pedagogika. Metodologiya, teoriya, praktika... p. 197 [Creative pedagogy. Methodology, theory, practice...].

6 *Alejnikov G. S.* O kreativnoj pedagogike // Vestnik vysshej shkoly. 1989. No. 12. pp. 24–34 [About creative pedagogy // Vestnik vysshei shkoly].

7 Kreativnaya pedagogika. Metodologiya, teoriya, praktika... p. 198 [Creative pedagogy. Methodology, theory, practice...].

8 ib.

9 ib.

10 Kreativnaya pedagogika. Metodologiya, teoriya, praktika... p. 199 [Creative pedagogy. Methodology, theory, practice...].

11 Kreativnaya pedagogika. Metodologiya, teoriya, praktika... p. 201 [Creative pedagogy. Methodology, theory, practice...].

12 Kreativnaya pedagogika. Metodologiya, teoriya, praktika... p. 22 [Creative pedagogy. Methodology, theory, practice...].

13 Kreativnaya pedagogika. Metodologiya, teoriya, praktika... p. 194 [Creative pedagogy. Methodology, theory, practice...].

Chapter 8

1 Genrikh Saulovich Altshuller. 15.10.1926–24.09.1998 [Genrikh Saulovich Altshuller. 10/15/1926–09/24/1998]. (2003). Sait ofitsial'nogo fonda

G. S. Altshullera. https://www.Altshuller.ru/biography/ (accessed May 29, 2025).

2 *Altshuller G. S.* Creativity as an Exact Science: The Theory of the Solution of Inventive Problems / Translated by Anthony Williams. Amsterdam: Gordon and Breach Publishers, 1984.

3 *Altshuller G. S., Vertkin I. M.* Kak stat' geniem: Zhiznennaya strategiya tvorcheskoj lichnosti. Minsk: Belarus, 1994.p. 146 [How to become a genius: Life strategy of a creative personality].

4 *Altshuller G. S.* Algoritm izobreteniya. M.: M.: Mosk. worker, 1973, p. 5 [Invention algorithm]; *The same author.* Tvorchestvo kak tochnaya nauka: Teoriya resheniya izobretatel'skih zadach. M.: Sov. radio, 1979. p. 15 [Creativity as an Exact Science: Theory of Inventive Problem Solving].

5 Algoritm izobreteniya… p. 84 [Algorithm of invention…]; *The same author.* Tvorchestvo kak tochnaya nauka… p. 95 [Creativity as an Exact Science…]; *Altshuller G. S., Zlotin B. L., Zusman A. V., Filatov V. I.* Poisk novyh idej: ot ozareniya k tekhnologii (Teoriya i praktika resheniya izobretatel'skih zadach). Kishinev: Kartya Moldavenyaske, 1989. p. 50 [Search for new ideas: from insight to technology (Theory and practice of inventive problem solving)]; *Altshuller G. S.* Kak najti ideyu: Vvedenie v teoriyu resheniya izobretatel'skih zadach / Otv. red. A. K. Dyunin. SO AN SSSR. 2-e izd. ispr. i dop. Novosibirsk: Nauka, 1991. p. 19 [How to find an idea: Introduction to the theory of inventive problem solving].

6 *Altshuller.* Algoritm izobreteniya… [Algorithm of invention…]; *The same author.* Tvorchestvo kak tochnaya nauka… [Creativity as an exact science…]; *Altshuller, Zlotin, Zusman, Filatov.* Poisk novyh idej: ot ozareniya k tekhnologii… [Search for new ideas: from insight to technology…]; *Altshuller.* Kak najti ideyu… [How to find an idea…].

7 *Altshuller.* Algoritm izobreteniya… [Algorithm of invention…]; *The same author.* Tvorchestvo kak tochnaya nauka… [Creativity as an exact science…]; *Altshuller G. S., Selyuckij A. B.* Kryl'ya dlya Ikara: Kak reshat' izobretatel'skie zadachi. Petrozavodsk: Kareliya, 1980 [Wings for Icarus: How to Solve Inventive Problems]; *Altshuller, Zlotin, Zusman, Filatov.* Poisk novyh idej: ot ozareniya k tekhnologii… [Search for new ideas: from insight to technology…].

8 *Altshuller.* Algoritm izobreteniya… [Algorithm of invention…]; *The same author.* Tvorchestvo kak tochnaya nauka… [Creativity as an exact science…].

9 *Altshuller.* Algoritm izobreteniya... p. 5 [Algorithm of invention...]; *The same author.* Najti ideyu. Vvedenie v TRIZ — teoriyu resheniya izobretatel'skih zadach. 2-e izd. M.: Al'bina Pablisher, 2008. p. 53 [Find an idea. Introduction to TRIZ — Theory of Inventive Problem Solving. (2nd ed.)].

10 *Altshuller, Vertkin.* Kak stat' geniem... p. 5 [How to become a genius...].

11 *Altshuller, Vertkin.* Kak stat' geniem... p. 63 [How to become a genius...].

12 *Altshuller.* Tvorchestvo kak tochnaya nauka... [Creativity as an exact science...].

13 *Altshuller, Zlotin, Zusman, Filatov.* Poisk novyh idej: ot ozareniya k tekhnologii... [Search for new ideas: from insight to technology...].

14 ib.

15 *Altshuller, Vertkin.* Kak stat' geniem... [How to become a genius...].

16 *Altshuller G. S.* Osnovy izobretatel'stva. Voronezh: Centr.-CHernozem. kn. izd-vo, 1964 [Fundamentals of invention].

17 *Altshuller.* Algoritm izobreteniya... [Algorithm of invention...].

18 *Altshuller.* Osnovy izobretatel'stva... [Fundamentals of Invention...]; *The same author.* Algoritm izobreteniya... [Algorithm of invention...]; *The same author.* Tvorchestvo kak tochnaya nauka... [Creativity as an exact science...].

19 *Altshuller, Zlotin, Zusman, Filatov.* Poisk novyh idej: ot ozareniya k tekhnologii... [Search for new ideas: from insight to technology...].

20 *Altshuller.* Tvorchestvo kak tochnaya nauka... [Creativity as an exact science...].

21 *O. Genri.* Supruzhestvo kak tochnaya nauka: Izbr. rasskazy. SPb.: Azbukaklassika, 2004. [The Exact Science of Matrimony: Fav. stories].

Appendix 2

1 *Popov.* Metodologiya konceptual'nogo proektirovaniya podzemnoj razrabotki rudnyh mestorozhdenij. Dis. ... d-ra tekhn. nauk... p. 34; Kreativnaya pedagogika. Metodologiya, teoriya, praktika... p. 270 and ff.) [Conceptual design methodology for underground mining of ore deposits. Thesis. ... Dr. Tech. Sc. ...]; [Creative pedagogy. Methodology, theory, practice...]

Appendix 3

1 *Polovinkin A. I.* Metody inzhenernogo tvorchestva. Uchebnoe posobie. Volgograd, 1984 [Methods of engineering creativity. Study manual]. Prilozhenie 4: Mezhotraslevoj fond evristicheskih priemov preobrazovaniya ob"ekta. pp. 339–348. [Appendix 4: Interindustry fund of heuristic methods of object transformation].

2 *Alekseev V. P., Ozerkin D. V.* Ozerkin D. V. Sistemnyj analiz i metody nauchnotekhnicheskogo tvorchestva [System analysis and methods of scientific and technical creativity]. Tomsk, 2012. CH. 2: Uchebnoe posobie po izucheniyu discipliny dlya studentov distancionnogo obucheniya special'nostej [Part 2: Manual on the discipline study for students of distance learning specialties] 20.08.00, 20.14.00, 20.15.00, 01.31.00. p. 271; *Polovinkin A. I.* Osnovy inzhenernogo tvorchestva. Uchebnoe posobie dlya studentov VTUZov. M.: Mashinostroenie, 1988. p. 180 [Fundamentals of engineering creativity. Study manual for students of higher technical educational institutions].

3 Mezhotraslevoj fond evristicheskih priemov preobrazovaniya ob"ekta [Interindustry fund of heuristic methods of object transformation]; *Polovinkin.* Osnovy inzhenernogo tvorchestva… p. 302 [Fundamentals of engineering creativity…].

Appendix 4

1 *Popov V. V.* (2011). Mobil'noe sidene. Dlya doma i ofisa [Mobile seat. For home and office]. Mobil'noe sidene. http://m-seat.ru/assigning_seats/home_and_office/ (accessed May 29, 2025).

2 *Lyubeshkina E.* Triumf upakovki // Nauka i zhizn'. 2006. No. 10. p. 104 [Packaging Triumph // Science and Life].

3 BINTI (Byuro inostrannoj nauchno-tekhnicheskoj informacii) // Nauka i zhizn'. 2008. No. 3. p. 15. [BINTI (Bureau of Foreign Scientific and Technical Information) // Science and Life].

4 Sozdana futbolka, kotoraya umeet preduprezhdat' o chrezmernom zagryaznenii vozdukha [A T-shirt has been created that can warn about excessive air pollution]. (n.d.). Uznai vse! https://uznayvse.ru/nauka-i-tchnologii/sozdana-futbolka-kotoraya-umeet-preduprezhdat-o-chrezmernomzagryaznenii-vozduha-27229.html (accessed May 29, 2025).

5 BINTI (Byuro inostrannoj nauchno-tekhnicheskoj informacii) // Nauka i zhizn'. 2008. No. 3. p. 9. [BINTI (Bureau of Foreign Scientific and Technical Information) // Science and Life].

6 ib.

7 BINTI (Byuro inostrannoj nauchno-tekhnicheskoj informacii) // Nauka i zhizn'. 2008. No. 3. p. 7 [BINTI (Bureau of Foreign Scientific and Technical Information) // Science and Life].

8 *Popov.* Sozdanie i razvitie tekhnicheskih sistem na osnove potrebnostej cheloveka… p. 44 [Creation and development of technical systems based on human needs…].

9 ib.

10 "Firefly" — tyubik s taimerom i zhidkokristallicheskim ekranom ot Dr. Fresh ["Firefly" — a tube with a timer and LCD screen from Dr. Fresh]. (2007). upakovano.ru. http://www.upakovano.ru/news/5893 (accessed May 29, 2025).

11 *Popov.* Sozdanie i razvitie tekhnicheskih sistem na osnove potrebnostej cheloveka… p. 45 [Creation and development of technical systems based on human needs…].

12 Chasy Citizen podruzhili s mobil'nikom [Citizen watches paired with a mobile phone]. (2010). LIFE.RU. https://life.ru/p/21918 (accessed May 29, 2025).

13 ib.

14 Opredelit' zrelost' fruktov obeshchaet novaya elektronnaya etiketka [A new electronic label promises to determine the ripeness of fruits]. (2007). Unipack.ru. https://news.unipack.ru/17405/ (accessed May 29, 2025).

15 Kunstkamera // Nauka i zhizn'. 2008. No. 3. p. 29 [Kunstkamera // Science and Life].

16 Uchenye izobreli odezhdu, kotoruyu ne nuzhno stirat' [Scientists have invented clothes that don't need to be washed]. (2008). Setevoe izdanie "VZGLYAD.RU". https://vz.ru/news/2008/2/12/144320.html (accessed May 29, 2025).

17 *Popov.* Sozdanie i razvitie tekhnicheskih sistem na osnove potrebnostej cheloveka… p. 46 [Creation and development of technical systems based on human needs…].

18 *Popov.* Sozdanie i razvitie tekhnicheskih sistem na osnove potrebnostej cheloveka… pp. 46–47 [Creation and development of technical systems based on human needs…].

19 *Popov.* Sozdanie i razvitie tekhnicheskih sistem na osnove potrebnostej cheloveka… p. 46 [Creation and development of technical systems based on human needs…].

20 Kunstkamera // Nauka i zhizn'. 2012. No. 7. p. 66 [Kunstkamera // Science and Life].

21 Kunstkamera // Nauka i zhizn'. 2005. No. 6. p. 114 [Kunstkamera // Science and Life].

22 *Popov V. V.* (n.d.). Sozdanie i razvitie tekhnicheskikh sistem na osnove potrebnostei cheloveka [Creation and development of technical systems based on human needs]. Yunye tekhniki i izobretateli. https://xn—— itbbmalqd7b5a5d8a.xn — p1ai/wp-content/uploads/2015/11/Popov— V.V.-Sozdanie-i-razvitie-tehnicheskih-sistem-na-osnove-potrebnostejcheloveka.pdf (accessed May 29, 2025).

Appendix 6

1 Bol'shoj enciklopedicheskij politekhnicheskij slovar' [Elektronnyj resurs] M.: Mul'titrejd, 2004. [Great Encyclopedic Polytechnic Dictionary] [Electronic resource].

2 Resursy vtorichnye material'nye [Secondary material resources]. (n.d.). EconomyLit.online. Archived by Wayback Machine. https://web.archive. org/web/20190212200834/https://economylit.online/obschie-rabotyi_719/ resursyi-vtorichnyie-materialnyie-32803.html (accessed May 29, 2025).

3 Trudovye resursy [Labor resources]. (2024). Wikipedia. https://ru.wikipedia.org/?curid=1590371&oldid=140934432 (accessed May 29, 2025).

4 *Balabanov I. T., Goncharuk O. V., Savinskaya N. A.* Den'gi i finansovye instituty: Uchebnoe posobie. SPb.: Piter, 2000. p. 16 [Money and financial institutions: study manual].

5 Slovar' finansovykh terminov [Dictionary of financial terms]. (n.d.). "FINAM". https://www.finam.ru/publications/section/dictionary/ (accessed May 29, 2025).

6 Prirodnye resursy [Natural resources]. (n.d.). Bol'shaya Sovetskaya Entsiklopediya (BSE). https://bse.sci-lib.com/article092891.html (accessed May 29, 2025).

7 Resursnaya ekologiya [Resource ecology]. (n.d.). KiberPedia. https://

cyberpedia.su/11x621a.html (accessed May 29, 2025).

8 Ischerpaemye prirodnye resursy [Exhaustible natural resources]. (n.d.). FB.ru. https://fb.ru/article/242763/ischerpaemyie-prirodnyieresursyi-primeryi-ischerpaemyie-prirodnyie-resursyi-vozobnovimyie-inevozobnovimyie (accessed May 29, 2025).

Appendix 7

1 Kunstkamera // Nauka i zhizn'. 2013. No. 8. p. 103 (Kunstkamera // Science and Life].

2 *Zaveruha N. M., Serebryakov V. V., Skiba YU. A.*Osnovy ekologii. Navch. Posibnik. Kiiv: Karavela, 2006. pp. 90–91 [Fundamentals of Ecology. Study manual].

3 *Tejlor D., Grin N., Staut U.* Biologiya: v 3 t. / Pod red. R.Sopera. M.: Mir, 2004. Vol. 2. p. 40 [Biology: in 3 volumes]; Zaveruha N. M., Serebryakov V. V., Skiba YU. A. Osnovy ekologii… p. 92 [Fundamentals of Ecology…].

4 *Chervinskaya A. V.* Suhoj aerozol' hlorida natriya — glavnyj dejstvuyushchij faktor speleoi galoterapii // Kurortnye vedomosti. 2004. No. 4–5. pp. 21–22 [Dry aerosol of sodium chloride is the main active factor of speleoand halotherapy // Kurortnye vedomosti].

Appendix 8

1 *Rejmers N. F.* Ekologiya: Teorii, zakony, pravila principy i gipotezy. M., 1994. pp. 47, 65, 67, 71, 72, 75, 78, 82, 87, 90, 96, 98, 103, 107, 110, 116, 117, 123, 127, 131, 140, 151. [Ecology: Theories, laws, rules, principles and hypotheses].

Appendix 10

1 Grafen i grafenovaya revolyutsiya [Graphene and the graphene revolution]. (n.d.). Fotokomok.ru. https://fotokomok.ru/grafen-i-grafenovayarevolyuci-ya/ (accessed May 29, 2025).

2 Grafen [Graphene]. (2025). Wikipedia. https://ru.wikipedia.org/?curid=325880&oldid=145073572 (accessed May 29, 2025).

3 Grafenovyi bum: itogi [The graphene boom: Results]. (2012). Nanonewsnet.ru. Archived by Wayback Machine. https://web.archive.org/web/20200217173935/http://www.nanonewsnet.ru/articles/2012/grafenovy-ibum-itogi (accessed May 29, 2025).

4 Grafen i ego primenenie. Otkrytie grafena. Nanotekhnologii v sovremennom mire [Graphene and its applications. The discovery of graphene. Nanotechnologies in the modern world]. (n.d.). FB.ru. https://fb.ru/article/222039/grafen-i-ego-primenenie-otkryitie-grafena-nanotehnologiiv-sovremennom-mire (accessed May 29, 2025).

5 Grafen otsenili avansom [Graphene was evaluated in advance]. (n.d.). OPF NGU. https://opf.nsu.ru/ru/node/36 (accessed May 29, 2025).

6 O nanotrubkakh i o kosmicheskom lifte [About nanotubes and the space elevator]. (2007). Nanonewsnet.ru. Archived by Wayback Machine. https:// web.archive.org/web/20241008220359/https://www.nanonews-net.ru/ articles/2007/o-nanotrubkakh-o-kosmicheskom-lifte (accessed May 29, 2025).

7 Iz nanotrubok sdelali tkan' dlya "lestnitsy v nebo" [Nanotubes were used to make fabric for a "ladder to the sky"]. (2005). Lenta.Ru. https://lenta.ru/news/2005/08/19/nano/ (accessed May 29, 2025).

8 Polozhitel'nyi i otritsatel'nyi tropizm onkologicheskikh kletok [Positive and negative tropism of cancer cells]. (n.d.). Stati po onkologii Garbuzova G. A. Archived by Wayback Machine. https://web.archive.org/web/20190717141239/http://www.fotostim.bewell.ru/kn017.htm (accessed May 29, 2025).

9 Glikoporfiriny — porfiriny, soderzhashchie fragmenty sakharov [Glycoporphyrins — porphyrins containing sugar fragments]. (n.d.). Materialy s neobychnymi svoistvami. http://anomal.informsystema.ru/effects/catalog/est/byId/description/431/index.html (accessed May 29, 2025).

10 Kovalentnye organicheskie karkasy — perspektivnye materialy dlya khraneniya vodoroda [Covalent organic frameworks — promising materials for hydrogen storage]. (2005). Elementy. https://elementy.ru/novosti_nauki/164988 (accessed May 29, 2025).

11 Kovalentnye organicheskie karkasy — perspektivnye materialy dlya khraneniya vodoroda [Covalent organic frameworks — promising materials for hydrogen storage]. (2005). Elementy. https://elementy.ru/novosti_nauki/164988 (accessed May 29, 2025).

12 Poluchen sostav, tverdeyushchii pri nagrevanii i tayushchii pri okhlazhdenii [A composition has been obtained that hardens when heated and melts when cooled]. (2004). Setevoe izdanie "Utro.ru". https://utro.ru/

news/2004/09/27/355182.shtml (accessed May 29, 2025).

13 ib.

Appendix 11

1 This list is based on the catalogue formed by A. I. Polovinkin.

Appendix 12

1 First published: Kreativnaya pedagogika. Metodologiya, teoriya, praktika... pp. 194–196 [Creative Pedagogy. Methodology, theory, practice...].

Bibliography

1. *Aleynikov G. S.* (1989). O kreativnoy pedagogike [On creative pedagogy]. Vestnik vysshey shkoly, 12, 24–34.
2. *Altov G.* (2000). I tut poyavilsya izobretatel... [And then the inventor appeared...]. Detskaya literatura.
3. *Altshuller G. S.* (1961). Kak nauchitsya izobretat [How to learn to invent]. Knizhnoe izdatelstvo.
4. *Altshuller G. S.* (1964). Osnovy izobretatelstva [Fundamentals of invention]. Tscntralno Chernozemnoye knizhnoye izdatelstvo.
5. *Altshuller G. S.* (1973). Algoritm izobreteniya [Algorithm of invention] (2nd ed.). Moskovskiy rabochiy.
6. *Altshuller G. S.* (1979). Tvorchestvo kak tochnaya nauka: Teoriya resheniya izobretatelskikh zadach [Creativity as an exact science: Theory of inventive problem solving]. Sov. radio.
7. *Altshuller G. S.* (1991). Kak nayti ideyu: Vvedenie v teoriyu resheniya izobretatelskikh zadach [How to find an idea: Introduction to the theory of inventive problem solving] (2nd ed.). Nauka.
8. *Altshuller G. S.*, & *Selyutsky A. B.* (1980). Krylya dya Ikara: Kak reshat izobretatelskie zadachi [Wings for Icarus: How to solve inventive problems]. Kareliya.
9. *Altshuller G. S.*, & *Vertkin I. M.* (1994). Kak stat geniem: Zhiznennaya strategiya lichnosti [How to become a genius: Life strategy of personality]. Belarus.
10. *Altshuller G. S.*, *Zlotin B. L.*, *Zusman A. V.*, & *Filatov V. I.* (1989). Poisk novykh idey: ot ozareniya k tekhnologii (Teoriya i praktika resheniya izobretatelskikh zadach) [Search for new ideas: from insight to technology (Theory and practice of solving inventive problems)]. Kartya Moldavenyaske.
11. *Archer B.* (1979). The three Rs. Design Studies, 1(1), 16.
12. *Bashmakov A. I.*, *Zhedyaevsky D. N.*, *Popov V. V.*, *et al.* (2009). Kreativnaya pedagogika i kompetentsii tvorcheskogo spetsialista [Creative pedagogy

and competencies of creative specialists]. Gubkin Russian State University of Oil and Gas.
13. *Bashmakov A. I., Zhedyaevsky D. N., Popov V. V., et al.* (2009). Metody i sredstva sistematizatsii svedeniy o materialakh s neobychnymi svoystvami i ikh ispolzovaniya dlya obucheniya professionalnomu tvorchestvu [Methods and means of systematizing information about materials with unusual properties and their use for teaching professional creativity]. Gubkin Russian State University of Oil and Gas.
14. Bolshaya rossiyskaya entsiklopediya [Great Russian Encyclopedia] (Vol. 15). (2002). Bolshaya rossiyskaya entsiklopediya.
15. Bolshaya sovetskaya entsiklopediya [Great Soviet Encyclopedia] (3rd ed.). (1975). Sovetskaya entsiklopediya.
16. Bolshoy entsiklopedicheskiy politekhnicheskiy slovar [Great Polytechnical Encyclopedic Dictionary] [Electronic resource]. (2004). Multitrade.
17. Bolshoy entsiklopedicheskiy slovar [Great Encyclopedic Dictionary] (2nd ed.). (2000). Prokhorov A. M. (Ed.).
18. *Bush G. Ya.* (1981). Analogiya i tekhnicheskoe tvorchestvo [Analogy and technical creativity]. Avots.
19. *Bush G. Ya.* (1989). Problemnye zadachi i regulyativy poiska ikh resheniya [Problem tasks and search regulations for their solution]. VNIIPI.
20. *Davydov V. V., Zaporozhets A. V., Lomov B. F., et al.* (1983). Psikhologicheskiy slovar [Psychological dictionary]. Pedagogika.
21. *Drucker P. F.* (2002). Zadachi menedzhmenta v XXI veke [Management challenges for the 21st century]. Williams.
22. Entsiklopediya epistemologii i filosofii nauki [Encyclopedia of epistemology and philosophy of science]. (2009). Kasavin I. T. (Ed.). Kanon+ ROOI "Reabilitatsiya".
23. *Ershov P. M.* (1990). Potrebnosti cheloveka [Human needs]. Mysl.
24. Filosofskaya entsiklopediya [Philosophical Encyclopedia] (Vols. 1–5). (1960–1970). Konstantinov F. V. (Ed.). Sovetskaya entsiklopediya.
25. *Gagliano M., Grimonprez M., Depczynski M., & Renton M.* (2017). Turn in: plant roots use sound to locate water. Oecologia, 184(1), 151–160. https://doi.org/10.1007/s00442-017-3862-z
26. *Gasanov A. I., Gomkhan B. M., Efimochkin A. P., et al.* (1995). Rozhdenie izobreteniya: Strategiya i taktika resheniya izobretatelskikh zadach [Birth of an invention: Strategy and tactics of solving inventive problems]. Interpraks.
27. *Heckhausen H.* (2003). Motivatsiya i deyatelnost [Motivation and activity] (2nd ed.). Piter; Smysl.

28. *Hill P.* (1968). The Science of Engineering Design. Holt, Rinehart & Winston.
29. *Hill P.* (1973). Nauka i iskusstvo proektirovaniya: Metody proektirovaniya, nauchnoe obosnovanie resheniy [The science and art of design: Design methods, scientific justification of solutions]. (E. G. Kovalenko, Trans.; V. F. Venda, Ed.). Mir.
30. *Jones J. K.* (1986). Metody proektirovaniya [Design methods]. Mir.
31. *Kamenev A. F.* (1985). Tekhnicheskie sistemy: Zakonomernosti razvitiya [Technical systems: Development patterns]. Mashinostroenie.
32. *Koberg D.*, & *Bagnall J.* (1974). The Universal Traveler: A Soft-Systems Guide to Creativity, Problem-Solving, and the Process of Reaching Goals. Ann Arbor.
33. Kreativnaya pedagogika: Metodologiya, teoriya, praktika [Creative pedagogy: Methodology, theory, practice] (3rd ed.). (2012). Popov V. V., & Kruglov Yu. G. (Eds.). BINOM.
34. *Maslow A. H.* (1954). Motivation and personality. Harper.
35. *Meadows D. H.*, *Meadows D. L.*, *Randers J.* (1994). Za predelami rosta [Beyond growth]. Progress; Pangeya.
36. *Molyako V. A.* (1978). Psikhologiya tvorcheskoy deyatelnosti [Psychology of creative activity]. Znamya.
37. Nauchno-tekhnicheskoe prognozirovanie dlya promyshlennosti i pravitelstvennykh uchrezhdeniy [Scientific and technical forecasting for industry and government institutions]. (1972). Bright J. (Ed.). Progress.
38. *Obukhovskiy K.* (1972). Psikhologiya vlecheniy cheloveka [Psychology of human drives]. Progress.
39. *Orlov M. A.* (2008). Osnovy klassicheskoy TRIZ. Vvodnyy kurs vysokoeffektivnogo innovatsionnogo myshleniya [Fundamentals of classical TRIZ. Introductory course in highly effective innovative thinking] (3rd ed.). SOLON-PRESS.
40. *Polovinkin A. I.* (1988). Osnovy inzhenernogo tvorchestva [Fundamentals of engineering creativity]. Mashinostroenie.
41. *Polovinkin A. I.* (1995). Zakony stroeniya i razvitiya tekhniki [Laws of structure and development of technology]. VPI.
42. *Popov V. V.* (2012). Sozdanie i razvitie tekhnicheskikh sistem na osnove potrebnostey cheloveka [Creation and development of technical systems based on human needs] (2nd ed.). Gubkin Russian State University of Oil and Gas.
43. *Reimers N. F.* (1992). Okhrana prirody i okruzhayushchey cheloveka sredy [Protection of nature and human environment]. Prosveshchenie.
44. *Revenkov A. V.*, & *Rezchikova E. V.* (2008). Teoriya i praktika resheniya tekhnicheskikh zadach [Theory and practice of solving technical problems]. FORUM.

45. *Ryzhov K. V.* (1999). 100 velikikh izobreteniy [100 great inventions]. Veche.
46. *Shpakovsky N. A.* (2006). Derevya evolyutsii. Analiz tekhnicheskoy informatsii i generatsiya novykh idey [Trees of evolution. Analysis of technical information and generation of new ideas]. TRIZ-profi.
47. Sovremennyy tolkovyi slovar [Modern explanatory dictionary]. (2003). BSE.
48. Tekhnicheskoe tvorchestvo: teoriya, metodologiya, praktika. Entsiklopedicheskiy slovar-spravochnik [Technical creativity: theory, methodology, practice. Encyclopedic dictionary-reference]. (1995). Polovinkin A. I., & Popov V. V. (Eds.). SPA "Inform-systema".
49. *Tovmasyan S. S.* (1972). Filosofskie problemy truda i tekhniki [Philosophical problems of labor and technology]. Mysl.
50. *Tring M.*, & *Leituayt E.* (1980). Kak izobretat? [How to invent?]. Mir.
51. Tvorchestvo: ot biologicheskikh osnovaniy k sotsialnym i kulturnym fenomenam [Creativity: from biological foundations to social and cultural phenomena]. (2011). Ushakov D. V. (Ed.). Institut psikhologii RAN.
52. *Vernadsky V. I.* (1989). Biosfera i noosfera [Biosphere and noosphere]. Nauka.
53. *Vygotsky L. S.* (2007). Myshlenie i rech [Thinking and speech]. Labirint.
54. *Zayniev G. A.* (2007). Evolyutsiya i izobretatel [Evolution and the inventor]. Academia.
55. *Zayniev G. A.* (2007). Inkubator pervichnykh izobreteniy [Incubator of primary inventions]. Academia.
56. *Zlotin B. L.*, & *Zusman A. V.* (1991). Reshenie issledovatelskikh zadach [Solution of research problems]. Kartya Moldavenyaske.

List of figures

List of tables

Nonfiction Science Book

Valery Popov

MENTAL KARATE

Methodology of scientific and technical creativity
and conceptual design
4rd edition

Published by Pushkin House
(04631) 2F, 10, Toegye-ro 20na-gil, Jung-gu, Seoul, Republic of Korea
tel +82-2-2237-9387
fax +82-2-2238-9388
www.pushkinhouse.co.kr

Published by Pushkin House, 2025